Solid Mechanics for Engineers

Liangchi Zhang

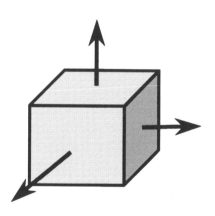

palgrave

© Liangchi Zhang 2001

All rights reserved. No reproduction, copy or transmission of this publication may be made without written permission.

No paragraph of this publication may be reproduced, copied or transmitted save with written permission or in accordance with the provisions of the Copyright, Designs and Patents Act 1988, or under the terms of any licence permitting limited copying issued by the Copyright Licensing Agency, 90 Tottenham Court Road, London W1P 0LP.

Any person who does any unauthorised act in relation to this publication may be liable to criminal prosecution and civil claims for damages.

The author has asserted his right to be identified as the author of this work in accordance with the Copyright, Designs and Patents Act 1988.

First published 2001 by
PALGRAVE
Houndmills, Basingstoke, Hampshire RG21 6XS and
175 Fifth Avenue, New York, N.Y. 10010
Companies and representatives throughout the world

PALGRAVE is the new global academic imprint of
St. Martin's Press LLC Scholarly and Reference Division and
Palgrave Publishers Ltd (formerly Macmillan Press Ltd).

ISBN 0–333–92098–8 paperback

This book is printed on paper suitable for recycling and made from fully managed and sustained forest sources.

A catalogue record for this book is available from the British Library.

10 9 8 7 6 5 4 3 2 1
10 09 08 07 06 05 04 03 02 01

Printed and bound in Malaysia

Solid Mechanics for Engineers

To my wife, Xiafen,
for her understanding, patience and encouragement

Contents

Preface		xiii
Chapter 1	**THE APPROACH TO STUDY**	**1**
1.1	Introduction	2
1.2	The Way of Solution	5
	1.2.1 Amount of interference	6
	1.2.2 Reduction of rolling force	7
1.3	The Approach to Study	9
	References	10
	Questions	11
	Problems	11
Chapter 2	**BASIC ASSUMPTIONS**	**13**
2.1	Scale of Analysis	14
2.2	Basic Assumptions in Elementary Solid Mechanics	16
	2.2.1 Continuity	16
	2.2.2 Homogeneity and isotropy	17
	2.2.3 Small deformation	18
	2.2.4 Absence of initial stresses	18
	References	19
	Important Concepts	19
	Questions	19
	Problems	20
Chapter 3	**STRESS**	**22**
3.1	Forces	23
	3.1.1 Type of forces	23
	3.1.2 Internal forces	23
	3.1.3 External forces	24
3.2	Stress	26
	3.2.1 Mathematical definition	26
	3.2.2 Physical definition	27

	3.3	Commonly Used Notations for Stresses	27
	3.4	Sign of Stresses	31
		3.4.1 The positive direction of normal stresses	31
		3.4.2 The positive direction of shear stresses	32
	3.5	Symmetry of Stress Matrix	33
	3.6	Stress Transformation	35
		3.6.1 Stress in any direction: two-dimensional stress states	36
		3.6.2 Stresses with coordinate axis rotation	38
		3.6.3 Principal stresses and maximum shear stress	39
		3.6.4 Principal stresses as eigenvalues of the stress matrix	43
		3.6.5 Stress in any direction: three-dimensional stress states	45
		3.6.6 Principal stresses in three dimensions	48
		3.6.7 Maximum shear stress in three dimensions	51
	3.7	Equations of Motion and Equilibrium	54
	References		59
	Important Concepts		59
	Questions		60
	Problems		60
Chapter 4		**DISPLACEMENT AND STRAIN**	**65**
	4.1	Displacement	66
	4.2	Strain	67
	4.3	Strain-Displacement Relations under Small Deformation	68
	4.4	Principal Strains and Their Directions	73
	4.5	Compatibility of Strains	75
	References		79
	Important Concepts		80
	Questions		80
	Problems		81
Chapter 5		**STRESS-STRAIN RELATIONSHIP**	**86**
	5.1	Mechanism of Stress-Strain Behaviour	87
	5.2	Generalised Hooke's Law	89
	5.3	Physical Indications of Elastic Constants	98
		5.3.1 Young's modulus E	99
		5.3.2 Poisson's ratio v	100
		5.3.3 Shear modulus G	101
		5.3.4 Bulk modulus K	101

5.4	Effect of Temperature Change		106
5.5	Deformation beyond Elasticity		109
References			109
Important Concepts			110
Questions			110
Problems			110

Chapter 6 MODELLING AND SOLUTION 114

- 6.1 Difficulties 115
- 6.2 Mechanics Modelling 115
  - 6.2.1 Displacement and stress boundary conditions 115
  - 6.2.2 Mixed boundary conditions 120
  - 6.2.3 Understanding of special deformation characteristics 122
    - 6.2.3.1 Symmetry of deformation 122
    - 6.2.3.2 Plane-stress deformation 128
    - 6.2.3.3 Plane-strain deformation 130
    - 6.2.3.4 Similarity and difference in plane-stress and plane-strain 131
  - 6.2.4 Use of the principle of superposition 133
- 6.3 Solution Approaches and Skills 135
  - 6.3.1 Approaches 135
  - 6.3.2 Solution skills 137
    - 6.3.2.1 A cylinder under inner and outer pressure: displacement approach 138
    - 6.3.2.2 A rotating disk under outer Pressure: mixed approach 145
    - 6.3.2.3 A uniform bar under gravity: stress approach 153
- 6.4 Saint-Venant's Principle 158
- References 162
- Important Concepts 163
- Questions 163
- Problems 164

Chapter 7 APPLICATIONS 169

- 7.1 Torsion of a Circular Shaft 170
  - 7.1.1 Modelling and solution 170
  - 7.1.2 Analysis of the solution 173
- 7.2 Beam Bending 175

	7.2.1	Modelling and solution	175
	7.2.2	Discussion	180
7.3	Thermal Stress Analysis		181
	7.3.1	A general method for thermal deformation analysis	182
	7.3.2	Thermal fit of a hollow disk onto a shaft	184
7.4	Stress and Deformation Due to Contact		188
	7.4.1	A half-space under a normal concentrated load	192
	7.4.2	A half-space under a tangential concentrated load	195
	7.4.3	A half-space under a local uniform pressure	197
	7.4.4	Elastic bodies in contact	200
References			202
Important Concepts			203
Questions			203
Problems			204

Chapter 8 STRESS FUNCTION METHOD   209

8.1	Introduction		210
8.2	Airy Stress Function		210
8.3	A Rectangular Plate under Pure Bending		217
8.4	Plate Bending under a Concentrated Shear Force		218
8.5	Stress Concentration around a Circular Hole		223
	8.5.1	Modelling and solution	223
	8.5.2	Analysis of the solution	225
	8.5.3	Alternative methods of solution	227
8.6	Pure Bending of a Curved Beam		230
8.7	Effect of Body Forces		232
8.8	Another Scalar Method		232
References			233
Important Concepts			233
Questions			233
Problems			234

Chapter 9 PLASTICITY AND FAILURE   237

9.1	Introduction		238
9.2	Octahedral Shear Stress		239
9.3	Distortion Energy		240
9.4	Onset of Plasticity and Deformation after Initial Yielding		241
	9.4.1	Plasticity under simple tension	241

		9.4.2	Plasticity under complex stress states	248
			9.4.2.1 Initial yielding: yield criterion	248
			9.4.2.2 Tresca criterion	249
			9.4.2.3 von Mises criterion	250
		9.4.3	Experimental verification	253
			9.4.3.1 Lode's test	253
			9.4.3.2 Test by Taylor and Quinney	254
		9.4.4	Some remarks	256
	9.5	Failure Theories		268
		9.5.1	Ductile materials	269
			9.5.1.1 The maximum shear stress theory	269
			9.5.1.2 The maximum distortion energy theory	269
		9.5.2	Brittle materials	270
			9.5.2.1 The maximum normal stress theory	270
			9.5.2.2 The maximum normal strain theory	271
		9.5.3	The Mohr theory	271
	References			273
	Important Concepts			275
	Questions			275
	Problems			275
Chapter 10	**AN INTRODUCTION TO THE FINITE ELEMENT METHOD**			**280**
	10.1	Introduction		281
	10.2	Fundamentals		281
		10.2.1	Introductory examples	281
		10.2.2	Types of elements	283
		10.2.3	Control volume	285
	10.3	Formulation in the Finite Element Method		286
		10.3.1	Problem description	286
		10.3.2	Finite element solution	288
	10.4	Usual Procedures of the Finite Element Solution		296
	10.5	Some Considerations in Finite Element Modelling		298
	10.6	A Case Study		301
		10.6.1	Background of the problem	301
		10.6.2	Modelling	302
		10.6.3	Solution	306
		10.6.4	Results and analysis	308

		10.6.5 Summary	309
		References	309
		Important Concepts	310
		Questions	311
		Problems	311
Appendix A		**MOHR'S CIRCLE**	**315**
Appendix B		**FORMULAE FOR SPECIAL CONTACT PROBLEMS**	**319**
	B.1	Two Balls in Contact	319
	B.2	A Sphere in Contact with a Flat Half-Space	320
	B.3	A Sphere in Contact with a Large Concave Spherical Surface	321
	B.4	The Contact of Two Cylinders with Parallel Axes	321
	B.5	A Cylinder in Contact with a Flat Half-Space	322
	B.6	The Contact of a Cylinder with a Large Concave Cylindrical Surface	323
	B.7	The Contact of Two Cylinders with Perpendicular Axes	323
Appendix C		**THE THEOREM OF MINIMUM POTENTIAL ENERGY**	**325**
Appendix D		**UNITS AND CONVERSION FACTORS**	**328**
Index			**331**

Preface

This book aims to introduce the basics of solid mechanics for engineers from a practical point of view. There are several reasons why it should be written. First, solid mechanics is and will remain a central branch of engineering science. The success of the application of solid mechanics in a variety of engineering disciplines has led to significant scientific, technological and economic progress. However, there are several ways to approach the subject, depending on whether one is particularly interested in its mathematical beauty or in the physical understanding and practical application. For engineers who aim at solutions to complex engineering problems, it is extremely helpful to understand the subject both physically and mathematically. It is thus important to have a single text to present the physical mechanisms of material properties, the theory of solid mechanics, solution methodology and direct engineering applications as an organic whole.

Secondly, although the completeness of solutions and mathematical methods based on given mechanics models has been well addressed by many books introducing solid mechanics, students and engineers still find it difficult to use solid mechanics in engineering applications, or have less confidence in modelling a problem correctly. For example, when studying the subject, engineering students often raise questions such as 'Why should we establish this set of equations?' 'Why should we solve the given model?' 'What is the use of the solution in engineering practice?' or 'I know all the equations and solution methods but I just have no idea to use them to solve practical problems encountered.' All these are largely due to a low emphasis on the appropriate mechanics modelling of real engineering problems. Therefore, a text that introduces the fundamental skills of mechanics modelling is practically useful.

With the rapid development of powerful computers, the application of numerical methods nowadays, such as the finite element method, seems to have become versatile. In the eyes of many students and engineers the use of numerical software commercially available seems to require little dependence on knowledge about solid mechanics. They feel this because numerical solution methods are usually introduced separately in engineering courses and the correctness, reliability and application of the numerical solutions obtained are not integrated well with the foundations of solid mechanics. In fact, one cannot imagine obtaining a reliable solution to a problem with an inappropriate description and a poor physical understanding of the basic mechanics involved. It is therefore essential to have a single book that not only emphasises the strategies and skills of practical modelling and analytical solution but also demonstrates their central role in solving a problem correctly and efficiently by using a numerical method.

This book aims to overcome the above problems. The text is developed from my lecture notes for a course offered to senior year students at the University of Sydney who have had some exposure to the engineering mechanics of materials or strength of materials. However, I have kept practical engineering readers in mind and tried to minimise the dependence on previous knowledge of solid mechanics and mathematics. On the other hand, to facilitate the study of the subject, the materials have been organised in an approach based on practical problems as illustrated in the following figure, in which engineering examples are often discussed before introducing a new concept or a new method of solution. These examples are then modelled into mechanics problems with particular emphasis on modelling skills and the necessity of theories. It is expected that in this way the reader may understand more directly why the theories are needed and hence become interested in achieving a more in-depth understanding.

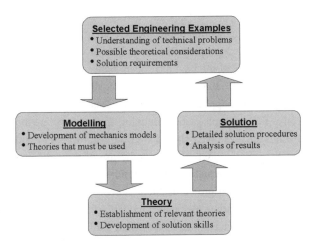

The subject matter is organised into ten chapters. Chapter 1 begins with an introduction to some practical problems in industry so that the reader can realise the importance and role of solid mechanics. It then outlines a method of study to help the reader to gain a general understanding of the material presented in the latter chapters.

The framework of elementary solid mechanics relies on some basic assumptions. Chapter 2 is therefore particularly arranged to explain the necessity of the basic assumptions from the point of view of the scale effect, microstructural effect and formulation requirement. In this way, the reader can become clearer about the scope of the subject and understand how to avoid mistakes in application and how to extend the use of the theory studied.

Stress and strain are the most important concepts in elementary solid mechanics. Chapters 3 and 4 are therefore devoted to establishing a basis. Importantly, specific attention in this text is paid to the physical understanding of these concepts and their relations to material deformation in practical design, manufacturing and assessment of structures and components.

A key issue in a solid mechanics analysis is to describe the constitutive behaviour of a material properly. A simple mathematical derivation will hinder the reader's understanding. Thus in Chapter 5 of the book, the elasticity, plasticity and thermal effect of a material are introduced physically with minimum mathematical formulation. Meanings of material constants are emphasised and treated in terms of their roles in engineering application.

With the understanding gained and the theories developed in the previous chapters, Chapters 6 to 8 concentrate on the development of mechanics models from selected engineering problems and their solutions. The chapters provide essential skills via a practical approach and in this way improve the reader's confidence in solving more complex problems. Throughout the discussions, the reader is always reminded that a correct modelling is central, as otherwise any further effort will become meaningless. The importance of Saint-Venant's principle to modelling and the description of boundary conditions is particularly addressed in detail.

While the previous chapters focus on elastic deformation, Chapter 9 introduces the application of plastic deformation in engineering, discusses some basic methods of plasticity analysis, demonstrates the difference in the solutions to elastic and elastic-plastic problems, and introduces the common theories for failure prediction of structures and elements. With the physical concept of plasticity established in Chapter 5, this chapter explains the different behaviour of a component in elastic and plastic deformation regimes and outlines the use of plasticity via various examples. The criteria and understanding achieved in the first part of the chapter on plastic yielding then naturally bring about the establishment of the failure theories.

In the above solution processes, the reader will have experienced mathematical difficulties. Chapter 10 is then designed to introduce one of the most popular numerical methods of solution, the finite element method. However, it is stressed that any numerical method is only a tool to overcome mathematical difficulties. The understanding of mechanics is still the key to achieve a correct and meaningful solution to a problem. To this end, in addition to some simple examples, which are used to show the principle and formulation of the finite element method, a case study is introduced to illustrate the process of mechanics analysis of a complex problem in modern engineering practice, demonstrate the inherent dependence of the numerical method upon the fundamentals of solid mechanics, and help the reader to achieve an overall view about how to apply the principles and solution skills to an engineering analysis.

The establishment of concepts and the understanding of theories and solution skills need a series of exercises and in-depth thinking. Thus a large number of questions and problems are selected and arranged at the end of each chapter for the reader to practice. Important concepts are also listed for specific attention.

Nevertheless, due to its elementary nature, this book ignores many important topics of solid mechanics, such as variational methods, finite deformation and dynamics. Fortunately, all these can be found in many other textbooks and monographs in the field. The selection of the materials for this book is based on my understanding, which may be subjective, that the materials are appropriate enough for a junior engineer in application but at the same time lay a sound foundation for those who desire further studies in the discipline. On the other hand, for the reader who aims at the most basic knowledge of the subject, the materials presented in Chapters 7 and 8 can be regarded as supplementary ones or appendices because the omission of these chapters does not hinder a qualitative understanding of the later texts. I believe that the reader who takes time to study the fundamentals of solid mechanics and its basic solution methodology presented in this introductory text will find ample reward.

I am indebted to Professor Ding Haojiang, who introduced me to the field of elasticity and computational mechanics, and Professor Yu Tongxi, who guided me to engineering plasticity. Over the years, the lecture notes that this text is based on have been improved by the suggestions of my colleagues and students. Professor Arcady Dyskin read the manuscript and offered some valuable comments. Finally, I wish to thank my family for their support, especially my wife Xiafen and children Margaret and Major.

Liangchi Zhang
January 1999
Sydney

Acknowledgements

The author and publishers are grateful to the following for permission to reproduce copyright material:

Nobuyuki Moronuki for an illustration from *Annals of CIRP*, Vol. 46, no. 1, 1997; McGraw-Hill for an illustration from S. P. Timoshenko, *History of Strength of Materials*, 1953.

Every effort has been made to contact all the copyright-holders but if any have been inadvertently omitted the publishers will be pleased to make the necessary arrangement at the earliest opportunity.

Chapter 1

THE APPROACH TO STUDY

The simple theory that we learned from the engineering mechanics of materials is insufficient to solve most engineering problems. This chapter introduces some practical problems to demonstrate the skills needed. It then outlines the overall structure of this book that uses a practical approach to learning.

The application of engineering components that need solid mechanics analysis is inexhaustible. The above photo shows a sequential combustion gas turbine to be used in a new power plant to be open in late 2000 in the south coast of England (Langston, 1999). The design, manufacture and reliability of most components of the turbine require sophisticated analysis of solid mechanics.

1.1 INTRODUCTION

The analysis of stress and deformation is a central part in the advanced design, manufacturing, reliability assessment and maintenance of engineering structures. Thus *Solid Mechanics*, a subject that deals with the deformation and motion of a solid under external loads and constraints, has been playing a key role in technological development over centuries.

Figure 1.1 The Great Stone Bridge spanning the Zhaozhou River is the world's first segmental arch bridge built in stone by the architect Li Chun in the year 610 and renovated in the twentieth century. The bridge has a span of 123 feet without piers! Precise indeed are the cross-bondings and joints between the stones, masonry blocks delicately interlocking like mill wheels. The semi-circular arch spandrels to either side let through additional flood and allow the structure to be lighter in weight.

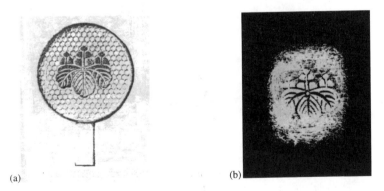

Figure 1.2 This cast bronze magic mirror was made in imitation. Not seen is the polished face, which reflects normally. (a) the pattern cast in relief on the ornamented back of the mirror, (b) the image reflected in bright sunshine upon a dark wall by the polished face. The plant design and some of the honeycomb pattern around it 'pass through' the solid mirror and become visible in the wall reflection. 'Magic mirrors' were invented in China by the fifth century AD (Temple, 1986).

Early applications of solid mechanics can go back to the ancient Chinese civilizations more than a thousand years ago, such as the famous Zhaozhou Stone Bridge built in 610, as shown in **Fig. 1.1**, and the Magic Mirror invented by the fifth century AD[1.1] (Needham, 1965), as shown in **Fig. 1.2**.

The Great Stone Bridge over the Zhaozhou River was the result of the work of the Chinese engineer Li Chun. It is difficult for us today to appreciate how impressive a sight the bridge must have been to pre-modern eyes. Its construction is unusual, and no one knows on what principle Li Chun made it. However, it is an optimal structure even from the point of view of the modern optimisation theory of solid mechanics when considering its structural stability and load carrying capacity.

What is a magic mirror then? The following is the description by Temple (1986). 'On its back it has cast bronze designs – pictures, or written characters, or both. The reflecting side is convex and is of bright, shiny polished bronze that serves as a mirror. In many conditions of lighting, when held in the hand, it appears to be a perfect normal mirror. However, when the mirror is held in bright sunshine, its reflecting surface can be "seen through", making it possible to inspect from a reflection cast onto a dark wall the written characters or patterns on the back. Somehow, mysteriously, the solid bronze becomes transparent, leading to the Chinese name for the objects, "light-penetration mirrors".'

Clearly, solid bronze cannot be transparent. When magic mirrors came to the attention of the West in 1832, dozens of prominent scientists attempted to discover their secret. A satisfactory theory was finally established after a century. The secret was the residual stresses produced when making the mirror and thus was a great application of solid mechanics and materials science by artisans' experience, although solid mechanics was not established in the fifth century. The basic mirror shape, with the design on the back, was cast flat, and the convexity of the surface produced afterwards by elaborate scraping and scratching. The surface was then polished to become shiny. The stresses set up by the process caused the thinner parts of the surface to bulge outwards and become more convex than the thicker portions. Finally, a mercury amalgam was laid over the surface, which created further stresses and preferential buckling. The result was that imperfections of the mirror surface matched the patterns on the back, although they were too minute to be seen by the eye. But when the mirror reflected bright sunlight against a wall, with the resultant magnification of the whole image, the effect was to reproduce the patterns as if they were passing through the solid bronze by way of light beams.

[1.1] Some references state that the invention of the magic mirror dates from 200 B.C., *e.g.*, Noyan and Cohen (1987).

Coming back to the current age, we can easily see the important contributions of solid mechanics to modern technologies, such as in the development of satellites, spaceships, cars, robots, computers, buildings and bridges, medical implants, sport instruments and micromachines.

For example, the micro-rib for a micromachine, as shown in **Fig. 1.3**, was made by micro-grinding by engineers in Japan. Unfortunately, the reliability of the products still relies on grinding conditions and on the skills of machine operators. The ribs may collapse shortly after grinding due to machining induced residual stresses. Investigations into the buckling mechanism of ribs using solid mechanics have been a major concern of engineers in this emerging field.

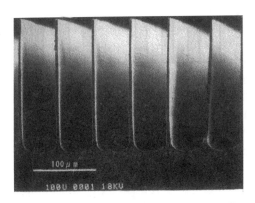

Figure 1.3 A tungsten carbide micro-rib made by micro-grinding (Courtesy of Tetsuya Suto, MEL, Japan).

Figure 1.4 The telerobot for heart valve repair.

An important milestone in medical history was achieved in May 1998 in Paris, France and Leipzig, Germany, when surgeons successfully performed a delicate operation to repair a valve in a patient's heart using the telerobot shown in **Fig. 1.4** (Salisbury, 1998). During much of the complex procedure, the surgeons' hands never entered or touched the patient's body. In fact, they were not even at the operating table. The procedure required one surgical wound eight centimetres long and two wounds only eight millimetres long. A tiny mechanical joint, as shown in **Fig. 1.5** (Salisbury, 1998), is a key component of the telerobot, which gives the surgeons the ability to reach around, beyond, and behind delicate body structures, and is connected to the rest of the robot by sophisticated, mechanical cable transmissions. We can easily understand that putting seven degrees of freedom in a tiny package was a challenge for mechanical design, mechanics analysis and materials selection. A current focus in the further development of the telerobot, which needs a more advanced application of solid mechanics, is to make use of force-feedback to produce the contact forces of surgery and let surgeons feel contact interaction and the work space limit of the robot.

All the above ancient and modern examples show that solid mechanics is indeed needed whenever the design, manufacturing, or application of a solid component involves stresses, deformation or motion.

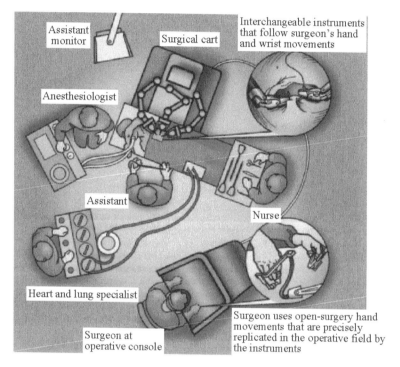

Figure 1.5 A schematic diagram showing the relationship among various components in minimally invasive surgery.

1.2 THE WAY OF SOLUTION

Now let us go through some simple but practical cases in detail in order to understand the major issues involved in solving engineering problems with solid mechanics and see the essential procedures needed to achieve the solutions.

1.2.1 Amount of Interference

Interference fit is a popular mechanical assembly method used in manufacturing. A typical example is the fitting of a cylindrical pin with a collar, as shown in **Fig. 1.6**. A major design consideration in the determination of the amount of interference, $\delta = (R_m - R_i)$, is that the interface stress, σ_{int}, between the pin and collar must be great enough to fasten the two components but in the meantime should not be so large as to create any plastic yielding or cracking.

To investigate stress and deformation, we need to convert the two parts of the assembly into corresponding mechanics models.

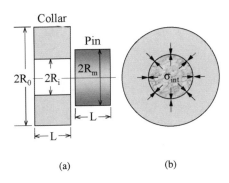

(a)   (b)

Figure 1.6 An example of interference fit. (a) before assembly, (b) after assembly.

Since the assembly is axisymmetrical and the materials of both the pin and collar are metals whose microstructural grains are much smaller than the dimension of the components, it is reasonable to assume that the materials are homogeneous and isotropic and thus the interface stress, σ_{int}, is uniform. Hence, the deformation of the collar can be modelled as a hollow circular cylinder subjected to an inner pressure, σ_{int}. The inner radius of the cylinder, R_i, will certainly increase to, say, R_f after deformation. On the other hand, the pin can be depicted by a solid cylinder under an outer pressure σ_{int}, but its radius R_m will decrease due to deformation and will also become R_f because after the assembly, as shown in **Fig. 1.6b**, the pin and collar must be in contact at a common radius. Although R_f is unknown, it must satisfy the deformation compatibility, *i.e.*,

$$(R_f - R_i) + (R_m - R_f) = \delta, \qquad (1\text{-}1)$$

where δ is the total interference. Clearly, the determination of δ depends finally on the following:
(1) the description of stress and deformation in the collar and pin, and
(2) the deformation upon stress in both the collar and pin materials, including the prediction of the onset of plastic deformation.

These are the tasks of solid mechanics that we are going to study here.

1.2.2 Reduction of Rolling Force

A production engineer for rolling metal plates in a manufacturing company is requested by the company's engineering manager to advise whether the company could contract for the supply of some specially rolled plates to a client.

For a number of reasons, such as the requirement of the microstructure, mechanical properties and the surface finish of the rolled plates, the client had specified the material for the plates and also the following rolling conditions:

(1) The plates must be cold-rolled.
(2) The reduction ratio of the plate thickness, ζ, by each stand of rolling mills must not be less that 30%, where the reduction ratio is defined as ζ = (inlet thickness – outlet thickness) / (inlet thickness).
(3) The width of the plates must not be less than 2 m.
(4) The inlet thickness of plates at the final stand should be $0.88^{+0.1}_{-0.1}$ mm.
(5) The lubricant provided by the client must be applied to the roll-plate interface during rolling.

As can be realized immediately, to obtain an answer one must find out the relationship between the reduction ratio, ζ, and the rolling force. Thus the plate deformation in the rolling gap must first be converted into a mechanics model.

From **Fig. 1.7**, it is clear that the plate must undergo normal stress at the roll-plate interface. When the rolls are rotating, they pull the plate into the rolling zone through interface friction. Thus the rolls must also apply frictional forces onto the plate surfaces.

On the other hand, let us assume that the plate material is incompressible, *i.e.*, the material's volume does not vary during deformation. Since the plate thickness is reducing in the rolling zone from the inlet to the outlet, to keep the volume rate of metal flow constant, the velocity of the plate must increase as it moves through the rolling zone, which is very similar to a fluid flow through a converging channel. This indicates that at the entry of the rolling zone, the plate moves slower than the roll surface so that backward slip occurs. At the exit of the zone, however, the plate moves faster and a forward slip takes place. Hence, there must exist a region between the entry and exit on which the velocity of the plate surface is the same as the surface velocity of the rolls such that no slip happens there. For convenience, let us assume that the region is very small and can be considered as a single cross-section, called the *neutral section*. Therefore the frictional force on the plate surface changes its direction at the neutral section. The deformation of the plate can therefore be modelled as that shown in **Fig. 1.8**.

It is indeed a complex case that involves the contact deformation of both rolls of a rolling mill that are necessary in its elastic regime and a plate that is subjected to severe plastic deformation. Although we cannot solve the problem before studying solid mechanics, we can see the following key steps needed to approach its solution, based on the mechanics model obtained:

(1) the stress and deformation in the rolls and plate, and

(2) the dependence of deformation (deflection of roll surfaces and thickness reduction ratio ζ) upon stresses (pressure and friction) in the rolls and plate. Both elastic and plastic deformation are involved.

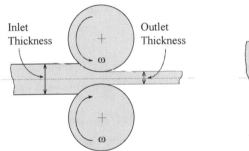

Figure 1.7 Cold rolling of a plate

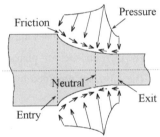

Figure 1.8 Mechanics model of the plate during rolling.

The question to answer is even more difficult. According to the mechanical properties of the material, friction properties when using the lubricant supplied and the geometrical parameters of the rolling mills of the company, the engineer found that the rolling force for this type of plate rolling would definitely exceed the maximum capacity of the existing mills which were all standard four-high ones. The manager could not justify the purchase of any new rolling mill but in the meantime he still requested the production engineer to work out a way of production using the mills available.

Based on the formulation generated from the above two steps, therefore, the engineer must analyse the result again to try to find an alternative way of production.

1.3 THE APPROACH TO STUDY

Engineering problems that need the solution of solid mechanics are inexhaustible. However, the above examples have shown clearly that to solve a general problem, we must know

(1) how to convert a real engineering problem into a mechanics model,

(2) how to obtain the solution to the model in terms of stress and deformation, and

(3) how to analyse the results to understand the mechanism of deformation involved in the problem and find an optimal way of solving the original problem.

These are the basic steps in mechanics analysis. This book is organised to achieve these goals following the practical approach as explained in the Preface. **Fig. 1.9** illustrates the overall structure of the book. We will first discuss the basic assumptions on which the elementary theory of solid mechanics is established. This clarifies the regime of applicability of the theory. After an investigation of stress and deformation analysis and a discussion on their relationships, we will emphasise the method and principle of mechanics modelling. This is so important because, as we can realise through the above examples, *a solution to a mechanics model is meaningless if the model does not capture correctly the characteristics of the original engineering problem*. The method of solution, of course, becomes central when a correct mechanics model is ready. We will discuss various methods, including the semi-inverse method, the stress function method and the finite element method. However, from the point of view of solving an engineering problem, the solution of its mechanics model is only the first step. A further important step is to analyse how to obtain guidelines for the optimisation of the design, manufacturing or maintenance plan of the problem. The plate rolling introduced above is a representative example in engineering practice. As we can see later in Chapter 9, whether the rolling engineer can finally offer an alternative way or not depends entirely on his/her depth of understanding of the solution. Thus whenever possible, the book will discuss how to interpret the solutions physically and their significance to the original problems. This is indeed the merit of the practical approach that we are going to follow.

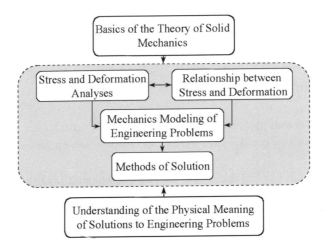

Figure 1.9 The approach to study.

References

Datsko, J (1997), *Materials Selection for Design and Manufacturing: Theory and Practice*, Marcel Dekker, Inc., New York.

Langston, LS (1999), The return of gaslight, *Mechanical Engineering (Supplement Magazine)*, July, 35.

Needham, J (1965), *Science and Civilization in China*, Cambridge University Press, Cambridge.

Noyan, IC and Cohen, JB (1987), *Residual stress: Measurement by Diffraction and Interpretation*, Springer-Verlag, New York, pp.1.

Salisbury, Jr, JK (1998), The heart of microsurgery, *Mechanical Engineering*, **120** (12) 46-51.

Suto, T, Waida, T and Okano, K (1997), High efficiency grinding of difficult-to-machine materials, in: *Advances in Abrasive Technology*, edited by L C Zhang and N Yasunaga, World Scientific, Singapore, pp.86-90.

Temple, R (1986), *The Genius of China: 3000 Years of Science Discovery and Invention*, Simon & Schuster Inc., New York.

Thompson, D (1999), *Design Analysis: Mathematical Modelling of Nonlinear Systems*, Cambridge University Press, New York.

Questions

1.1 What is mechanics modelling? Why is it so important to solve engineering problems? What is the difference between the mechanics model of an engineering problem and the engineering problem itself?

1.2 What are the major procedures to solve a stress and deformation problem?

1.3 What is the approach that we are going to take in this book? What are its features and merits?

1.4 Why should we understand the physical meaning of a solution obtained?

Problems

1.1 A government agency sponsored two projects in two different manufacturing companies for aeroplanes to study the machinability of a nickel-base alloy, Rene 41, and a cobalt-base alloy, HS25. Both studies are based on the tool life criterion of machinability against cutting speed. For the same tool material, one company found that the nickel alloy was 50% better, while the other reported that the cobalt alloy was 50% better. When the two reports are forwarded to you, the project supervisor in the governmental agency, what would you examine first to find out the causes of the discrepancy?

1.2 In the traditional application of cold-formed parts in automotive and aircraft manufacturing, a design is considered to be complete and acceptable when the most highly stressed regions are analysed to ensure that the structure using that material is safe. Nowadays, however, it is becoming increasingly more important that the design is on the basis of minimal weight and avoidance of strategic materials so that parts are manufactured from the most economical materials and have the most favourable ratio of strength to stress at all locations. State the stages in such an optimal design process in which deformation and stress analysis plays an important role.

1.3 In addition to designing a mechanical device to perform its function reliably, one should also design for ease of producibility and minimal cost or weight. The design must therefore be an iterative process encompassing some distinct phases, such as those specified in the flowchart **Fig. P1.1**. The engineer moves through the flowchart like a computer through its program, always evaluating whether the conditional clauses (functional requirements) are satisfied, and branching and repeating steps where appropriate. In the design phases listed, what are the stages that most need the knowledge of solid mechanics?

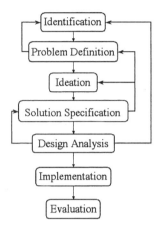

Figure P1.1 Engineering design as an iterative process.

1.4 Find two examples in your daily life that involve the application of solid mechanics.

Chapter 2

BASIC ASSUMPTIONS

To investigate an engineering problem efficiently and correctly, an engineer must know how to model it without losing its major characteristics. This chapter introduces the basic assumptions to apply to the establishment of the elementary theory of solid mechanics and discusses their rationale and limitations. It is important to understand that all the analyses and formulations in the theoretical framework to be studied in this text rely on the basic assumptions discussed here. In any practical application of the theory, therefore, one should always make sure that the conditions of these assumptions have been satisfied.

(a) ceramic components   (b) elements made by machining

(c) dislocations in alumina   (d) grains in alumina

With the naked eye, engineering materials look uniform, continuous and homogeneous, as shown in figures (a) and (b) above. However, they are not so under microscopes, as revealed by figures (c) and (d). In our deformation analysis of a component, should we consider the details of the material's microstructure.

2.1 SCALE OF ANALYSIS

The most important factors that govern the quantity of deformation and the process of motion of a component are the material properties of the component, its shape and the type and history of external loading and constraining. As we have discussed in Chapter 1, the central role of solid mechanics is to find the solution to the deformation and motion of a component by proper mechanics modelling that captures the characteristics of the problem with necessary accuracy but overcomes the mathematical and physical difficulties in obtaining the solution. To do so, we must have a good understanding of structures of materials, the behaviour of materials when subjected to external loading and constraining, and the rationale of mechanics modelling.

A wide variety of materials used in applications are called 'engineering materials' and can be broadly classified as metals and alloys, polymers, ceramics and glass, and composites. Differences among the classes of materials involve chemical bondings and micro-structures, leading to different mechanical behaviour and relative advantages and disadvantages among the classes. For example, the strong chemical bonding in ceramics and glass imparts mechanical strength and stiffness, and also temperature and corrosion resistance, but causes brittle behaviour. In contrast, many polymers are relatively weakly bonded between chain molecules, in which case the material has low strength and stiffness and is susceptible to creep deformation. How to quantitatively describe the behaviour of various materials and thus to make smart use of them in engineering practice is one of the major tasks of solid mechanics.

An important fact that must be borne in mind in the application of solid mechanics is the size scale of interest in engineering. There is a span of ten orders of magnitude in size from the scale of a metre down to the scale of an atom, which is around 10^{-10} m. This situation and various intermediate size scales of interest are indicated in **Fig. 2.1**. It is important to note that at any given size scale of analysis, a further understanding of the material behaviour can be sought by looking at what happens at a smaller scale. For instance, the behaviour of a machine, vehicle, or structure (Scale 10^0 m) is explained by the behaviour of its component parts, and the behaviour of the component parts can in turn be explained using smaller-scale test specimens of the materials (scales 10^{-1} m to 10^{-2} m). Similarly, the macroscopic behaviour of a material is explained by the behaviour of crystal grains, defects in crystal, polymer chains, and other microstructural features that exist in the size range of 10^{-3} m to 10^{-9} m. Thus knowledge of behaviour over the entire range of sizes from 10^0 m to 10^{-10} m contributes to understanding and predicting the performance of machines, vehicles, and structures. With the recent development of information technology, micro-machine technology and nanotechnology, engineering analysis has covered almost all the size scales shown in **Fig. 2.1**. For example, to design a micro-sliding system, one must understand the tribological behaviour of the encountering parts under a smaller scale, that is on the nanometre or atomic scale, see **Fig. 2.2**. However, our focus in this text will be on the scales 10^{-3} m to 10^0 m and the solutions we will achieve will be for design, manufacture and application of ordinary machines and structures.

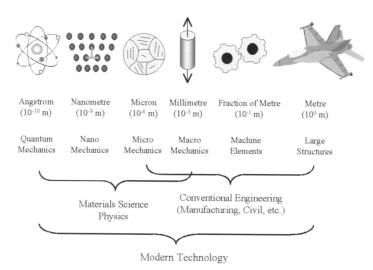

Figure 2.1 Size scales in mechanics analysis.

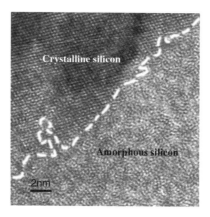

Figure 2.2 Atomic structural change in mono-crystalline silicon after a sliding of alumina particles on its surface. Each spot in the photo indicates a silicon atom.

2.2 BASIC ASSUMPTIONS IN ELEMENTARY SOLID MECHANICS

As mentioned, the behaviour of solid materials depends on their microstructures. However, to take into account all the details of their microstructures in a general macroscopic deformation analysis would be extremely difficult both physically and mathematically and thus unfeasible. It is therefore necessary to limit our attention to a realistic regime. To this end, we need to introduce the following assumptions for all the analyses and discussions in this introductory text.

2.2.1 Continuity

First, let us assume that all solids are continuous media, which means that there is not any vacancy in a solid material. This is certainly not true in practice. As we know, any material is composed of atoms. It is not continuous from one atom to another. In other words, there are gaps between any two atoms, as demonstrated by **Fig. 2.2**. Even at the scale of grains, materials are not continuous because there usually exist numerous pores between grains or micro-vessels, see for example **Fig. 2.3**.

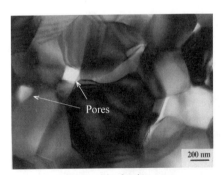
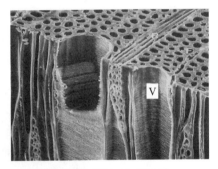

(a) Pores in alumina. (b) Vessels (V) in a hardwood.

Figure 2.3 Micro-discontinuity in engineering materials.

Fortunately, as illustrated in **Fig. 2.1**, the scales of engineering components and structures that we are going to investigate in this text, such as beams, plates, and machine elements, are much greater than the scales of the grains, gaps and pores in terms of the microstructure of materials. Furthermore, any macroscopically small engineering components contain huge numbers of grains, see *e.g.*, **Fig. 2.4**. From the point of view of statistics, the effect of the micro-discontinuity in a material on its macroscopic behaviour could be averaged when the solid object under consideration is much greater than the size of the discontinuity. Thus in our engineering analysis, the assumption of continuity is reasonable.

Because of this assumption, all physical quantities concerning the deformation of a solid, such as displacements, strains and stresses can be described by continuous functions of coordinates. Hence, we can conveniently use calculus of mathematics as a powerful tool in our stress and deformation analysis.

Figure 2.4 Microstructure of steel 4140 after quenching, ordinary martensite grains (Zhang and Zarudi, 2000).

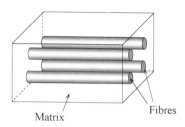

Figure 2.5 Long fibre-reinforced composites.

2.2.2 Homogeneity and Isotropy

As just mentioned, solid materials are microscopically neither homogeneous nor isotropic. For example, metals are composed of grains. Any individual grain is not isotropic because its atomic structure is strongly dependent on directions, or in other words, a grain itself is anisotropic (see the atomic structure of silicon shown in **Fig. 2.2** for instance). However, because the size of the grains is very small compared with the mechanical component subjected to deformation and because of the random distribution and orientation of the grains in the material (see **Fig. 2.4**), the anisotropy of individual grains would be averaged out macroscopically. Thus the assumption of homogeneity and isotropy can be applied to most engineering metals on the scale we are interested in. More clearly, a homogeneous and isotropic material means that the material constituents *at any point* and *in any direction* are the same. It then indicates that the properties of the material at any point and in any direction are the same. Hence, with the assumption of homogeneity and isotropy, our analysis will be much simplified and mathematical difficulties reduced.

However, there are a number of materials to which the above assumption cannot apply. A component made of a composite reinforced by unidirectional long fibres is a good example in this regard, as illustrated in **Fig. 2.5**. The fibres, *e.g.*, carbon fibres, have very different properties from the matrix material, *e.g.*, epoxy, and also impose extremely directional properties throughout the composite. Thus such a composite cannot be considered as either a homogeneous or an isotropic material as a whole. For these types of materials, a more complex theory that considers anisotropy and inhomogeneity must be used, which is beyond the scope of this text.

2.2.3 Small Deformation

We assume that the deformation of a solid is very small compared with its geometrical dimensions, such that we do not need to consider the geometrical changes of the solid during deformation. This is reasonable because in many cases we always try to avoid large deformation. For instance, the teeth of gears during power transmission must be as small as possible to avoid inefficient, instable and inaccurate transmission. Hence, we can discuss the deformation of the solid based on its initial dimensions, *i.e.*, its dimensions before deformation. This will simplify to a great extent our formulation in deformation analysis. As we will see later in the text, whenever the deformation is small, we are able to neglect all the small quantities beyond the second order and make the relations between displacements and strains linear so that the superposition principle applies in the regime of elastic deformation. This greatly facilitates our solutions.

2.2.4 Absence of Initial Stresses

This assumption indicates that the solid under consideration is free of initial stresses, that is, before the application of external forces, temperature changes, etc., the stresses at any point in the solid are zero. It means that the stresses obtained by deformation analysis are induced purely by the external forces applied, temperature changes that have occurred, etc. If a solid possesses initial stresses, then it must be particularly considered in the deformation analysis.

The above assumptions are the basic assumptions in the elementary theory of solid mechanics, in which 'continuity, homogeneity, isotropy and no initial stress' are physical but 'small deformation' is geometrical. In the engineering theory of mechanics of materials in junior courses, more assumptions have been applied in order to simplify the analytical procedures, but in turn they limit more the applicability of the engineering theory and the accuracy achievable by the theory.

As we will see in later chapters, the use of the basic assumptions not only overcomes the physical and mathematical difficulties in solving solid mechanics problems, but more importantly, produces reliable results in the scales of analysis that we are interested in.

References

DeGarmo, EP, Black, JT and Kohser, RA (1997), *Materials and Processes in Manufacturing*, Prentice-Hall, Inc., Englewood Cliffs, NJ.

Shackelford, JF (1996), *Introduction to Materials Science for Engineers*, 4th edition, Prentice-Hall, Inc., Englewood Cliffs, NJ.

Young, JF, Mindess, S, Gray, RJ and Bentur, A (1998), *The Science and Technology of Civil Engineering Materials*, Prentice-Hall, Inc., Englewood Cliffs, NJ.

Zarudi, I and Zhang, L (1999), Structural changes in mono-crystalline silicon subjected to indentation – experimental findings, *Tribology International*, **32**, 701-712.

Zarudi, I, Zhang, L and Mai, YW (1996), Subsurface damage in alumina induced by single-point scratching, *Journal of Materials Science*, **31**, 905-914.

Zhang, L and Zarudi, I (2000), Steel surface treatment by grinding, *Patent*, PQ7858.

Important Concepts

Continuity   Homogeneity
Isotropy   Small deformation
Initial stresses   Scale of analysis

Questions

2.1 What is the scope of solid mechanics?
2.2 What is the difference between engineering mechanics of materials and solid mechanics?
2.3 Why do we need to consider size scales in solving an engineering problem?
2.4 What are the advantages of using each of the four basic assumptions?

Problems

2.1 An annular plate made of mild steel with the dimensions shown in **Fig. P2.1** is certainly discontinuous from point A to B because it is a hollow component. Does this mean that it violates the assumption of continuity so that the theory of solid mechanics to be established based on the assumption will not be applicable to the deformation analysis of the plate?

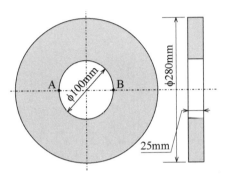

Figure P2.1 An annular plate made of mild steel.

2.2 Welding is a very common technique in engineering practice to join two parts permanently to build up a larger structure. However, when using an arc welding process, to assemble a vehicle frame as shown in **Fig. P2.2**, for example, heating and cooling during the welding will change the microstructure of the material in the vicinity of the weld. As a result, the material properties in the neighbourhood of the weld will be different from those in the base material. Thus in the local zone of the weld, the material is neither homogeneous nor isotropic. In the stress and deformation analysis of the vehicle frame, when can we still use solid mechanics theory based on the assumption of homogeneity and isotropy?

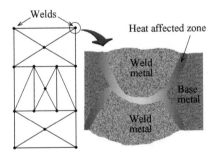

Figure P2.2 A vehicle frame made by arc welding. Left: the frame configuration; Right: the schematic of the usual microstructure of the material at a weld.

2.3 Portland cement concrete is a common material for load-bearing components of building construction such as columns, beams and slabs. It is basically a particulate composite material consisting of a single continuous phase binder, the portland cement, and a single discontinuous particulate phase, the aggregate. When we are interested in the bending deformation of a beam made of a portland cement concrete, can we approximately assume that it is an isotropic and homogeneous beam? The beam dimension is in the order of a metre and the average diameter of the aggregates is in the order of a centimetre.

Chapter 3

STRESS

To investigate the deformation of a solid, an important thing to understand is how to describe the stresses and strains at any point in any direction in the solid. This chapter first introduces the concept of stress at a point and its properties and then discusses the conditions that control the stress variation in the solid, *i.e.*, the equations of motion. The material in this chapter is of primary importance to the handling of the subject.

The above picture shows an attempt to study the mechanics of a cantilever beam loaded by a weight, which is the famous test carried out by Galileo, a pioneer in solid mechanics in the nineteenth century (Timoshenko, 1953).

3.1 FORCES

3.1.1 Type of Forces

When considering a mechanical element, a structure or a solid body in general, we can divide the forces in and on the body into internal and external ones. An *internal force* is that due to the material interaction inside the body through atomic attraction or repulsion. An *external force*, however, is that applied to the body due to the action of another body or field, such as the magnetic field. The understanding of these two types of forces is essential to stress analysis.

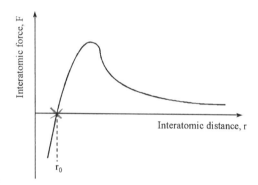

Figure 3.1 Variation of interatomic force F with the distance between two atoms, where $r = r_0$ is the equilibrium position at which the interatomic force is zero.

3.1.2 Internal Forces

Generally, atoms in a solid attract each other over long distances but repel each other at very short distances. In the absence of any external forces, atoms in the solid vibrate around their equilibrium positions where attractions and repulsions balance. **Fig. 3.1** is a schematic representation of the force, in a model with only two atoms, required to move one atom away from another from its equilibrium position. In reality, the situation is more complex because a solid contains a huge number of atoms that all interact with each other. Furthermore, the exact shape of the interaction curve, like that shown in **Fig. 3.1**, depends on the nature of the atomic bond of the solid (*e.g.*, ionic, covalent, or metallic). When a solid is in equilibrium, at any instant, the resultant force of all atomic interactions must be zero. Otherwise, the solid cannot be in equilibrium.

3.1.3 External Forces

When we take a careful look at the external forces acting on a mechanical component, we find that there are different types of such forces. For instance, the case of a pressure vessel under an internal pressure is that the inner surface of the vessel is subjected to a uniform pressure (**Fig. 3.2a**). The Eiffel Tower on a windy day will suffer from wind pressure (**Fig. 3.2b**). The side of a dam of a reservoir facing the water is subjected to water pressure (**Fig. 3.2c**). We often feel the weight of a steel component of a machine, such as a gear. It is due to gravity force, which, different from the water or wind pressure that is acting on the surface of a body, is acting throughout the body of the gear. Another example of such force is centrifugal force. When we drive at a relatively high speed to turn around a street corner, we will feel that a force is pulling us outwards. We can also imagine when a disk is rotating at a high speed, such as the grinding wheel shown in **Fig. 3.2d**, a centrifugal force is acting at any point throughout the wheel in the radial direction.

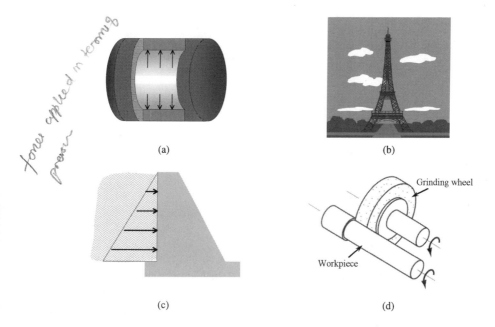

Figure 3.2 Various types of external forces applied to structures.

In many other cases, an external force on a surface of a mechanical element acts on a surface area that is very much smaller than the overall surface dimension of the element. The area effect in these cases is often negligible in analysing the overall deformation of the element. For convenience, we can consider such a small area as a mathematical point and call the force acting on it a *concentrated force*. The dimension of a concentrated force is N (Newton), or an equivalent unit. There are many practical examples. For instance, as shown in **Fig. 3.3**, when a man is standing on a very long flexible bridge and we are only interested in the overall deflection of the bridge, the force on the bridge surface due to the man's weight can be regarded as a concentrated force, since the contact area between the feet and the bridge surface is very much smaller than the upper surface area of the bridge. The above discussion indicates that whether an external load can be treated as a concentrated force or not depends totally on the size of the loading area relative to the characteristic dimension of the element to study and the accuracy to achieve by the analysis. This can be understood more deeply after the study of Saint-Venant's principle in Chapter 6.

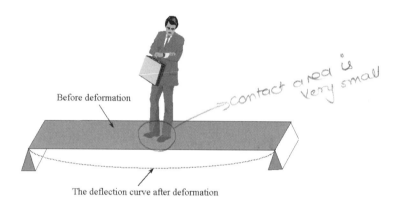

Figure 3.3 Bridge bending due to the man's weight.

The above examples show that we have at least three types of external forces on a solid, *i.e.*, *body forces*, such as gravitational force and magnetic force that act throughout the material points of the element, *surface forces*, such as water pressure that acts on any surface point of the element, and *concentrated forces* denoting an idealised case of a surface force on an extremely small area.

We can easily see that the magnitude and direction of surface or body forces at different points in a body or on its surface are different. For instance, the gravitational force is always towards the earth centre. The water pressure on the dam surface is always perpendicular to the dam surface but the magnitude increases with the depth from the water surface. To describe the direction and magnitude of such forces strictly and conveniently, let us introduce the concept of stress.

3.2 STRESS

3.2.1 Mathematical Definition

We have learnt in mechanics subjects, such as strength of materials, that a solid, such as a beam, subjected to some external forces will deform and accordingly generate stresses in the solid. We have also understood the analysis of deformation of a solid under a simple stress state, such as simple tension of a bar and pure bending of a beam. To characterise the deformation, we have learnt the concepts of stresses and strains. However, because of the importance of these concepts, we shall introduce here the strict definition of stress and discuss how to carry out stress analysis when a solid is under a complex stress state.

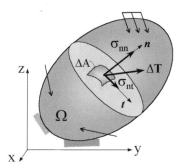

Figure 3.4 A solid under external loading.

Let us imagine a closed surface A within a solid Ω, as shown in **Fig. 3.4**. We would like to know the interaction between the material exterior to the surface A and that in the interior.

Consider a small surface element of area ΔA on our imagined surface A. Draw a unit vector n normal to ΔA, with its direction outwards from the interior of ΔA. Then, we distinguish the two sides of ΔA according to the direction of n. Consider the part of the material lying on the positive side of the normal, which exerts a force $\Delta \mathbf{T}$ on the inner part surrounded by the imagined surface ΔA. This force, $\Delta \mathbf{T}$, evidently, is a function of the area ΔA and its orientation.

Assume that as ΔA tends to zero, the ratio $\Delta \mathbf{T}/\Delta A$ tends to a definite limit and that the moment of forces acting on the surface ΔA about any point with the area vanishes in the limit:

$$\lim_{\Delta A \to 0} \frac{\Delta \mathbf{T}}{\Delta A} = \frac{d\mathbf{T}}{dA} = \overset{n}{\mathbf{T}}, \qquad (3\text{-}1)$$

where a superscript n is introduced to denote the direction of the normal $\boldsymbol{n}$ of the surface ΔA. The limiting vector $\overset{n}{\mathbf{T}}$, called the *stress vector*, or *traction*, represents the force per unit area acting on the imagined surface due to the interaction between the material exterior and that in the interior. Hence, if we follow the SI unit system, the dimension of $\overset{n}{\mathbf{T}}$ is N/m^2, or Pa (Pascal).

Stress vector $\overset{n}{\mathbf{T}}$ can be certainly resolved into two components, σ_{nn}, which is in the normal direction of ΔA, *i.e.*, along a unit vector $\boldsymbol{n}$, and σ_{nt}, which is in a tangential direction of ΔA, *i.e.*, along the unit vector $\boldsymbol{t}$ that is perpendicular to $\boldsymbol{n}$, as illustrated in **Fig. 3.4**. The first subscript of σ_{nn} or σ_{nt} indicates the external normal direction of the surface ΔA and the second shows the direction of the stress component. We usually call σ_{nn} *normal stress*, or *direct stress*, and σ_{nt} *shear stress*, or *shearing stress*. Thus mathematically, we have the relationship of

$$\overset{n}{\mathbf{T}} = \sigma_{nn}\boldsymbol{n} + \sigma_{nt}\boldsymbol{t}. \tag{3-2}$$

3.2.2 Physical Definition

The above mathematical definition of stress is based on the continuity assumption. Physically, it will become invalid if our analysis is on the nanometre scale because, as discussed in Chapter 2, a solid at such a small scale can no longer be treated as a continuum, *i.e.*, the continuity assumption cannot apply. In this circumstance, we have to define a stress vector on an element with a finite area that contains a sufficient number of atoms. In other words, we cannot take the limit of Eq. (3-1) and the stress is now an average one over a reasonably small area ΔA. Details of stress analysis on the nanometre and atomic scales can be found in the paper by Zhang and Tanaka (1999).

3.3 COMMONLY USED NOTATIONS FOR STRESSES

Obviously, the elementary notation described above is not sufficiently flexible and convenient for use in general, because the direction of surface ΔA can change and there are infinite tangential directions on the surface. Thus the normal stress, σ_{nn}, always varies with the direction change of $\boldsymbol{n}$, and the shear stress, σ_{nt}, can be in any tangential direction of the surface. On the other hand, a structural/mechanical element can be of very complex shape. It means that to conveniently analyse the stresses in a solid, or a structural/mechanical element, we must devise some more convenient scheme.

We have understood, based on the above definition, that the state of stress at any point will be determined if we specify, in direction and in magnitude, the stresses which act on all faces of an infinitesimal block of material situated at that point. Hence, we need to find a convenient notation scheme that distinguishes the direction of each face and the directions of the stresses acting on each face.

We can easily recall that we often solve engineering problems under a reference coordinate system, for instance, a Cartesian coordinate system, xyz as shown in **Fig. 3.4**. Clearly, it will be convenient to discuss stresses at a point of interest P on an infinitesimal plane through P with its external normal, n, in the direction of that of a reference coordinate, because in this case, we can resolve the stress vector on the plane, $\overset{n}{T}$, into the directions of the coordinate system. For example, as shown in **Fig. 3.5**, if we consider an infinitesimal plane through point P in the direction of coordinate z (n is coincident with z), then $\overset{n}{T}$, which can be denoted by $\overset{z}{T}$ in this special case, can be resolved into a normal stress component, σ_{zz}, and two shear stress components, σ_{zx} and σ_{zy}. As before, the first suffix of a stress component indicates the direction of the plane and the second denotes the direction of the stress component.

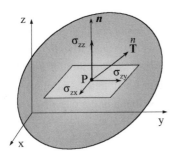

Figure 3.5 Resolving a stress vector.

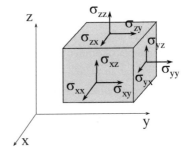

Figure 3.6 Resolving stress on planes in all the positive and negative coordinate directions.

Similarly, we can have other infinitesimal planes through point P in the directions of other coordinates, *i.e.*, in the directions of negative z, positive and negative x and positive and negative y. Thus in total, we have six such planes *through the same point* of interest in a solid. For the sake of convenient presentation, we use an infinitesimal cube formed by the six infinitesimal planes mentioned above, as illustrated in **Fig. 3.6**. On each such plane, we have one normal stress component and two shear stress components. If these stress components are arranged in the form of a square matrix, we obtain

	Direction of stress components		
	x	y	z
Plane normal to x	σ_{xx}	σ_{xy}	σ_{xz}
Plane normal to y	σ_{yx}	σ_{yy}	σ_{yz}
Plane normal to z	σ_{zx}	σ_{zy}	σ_{zz}

From the above, we know that the stress state at point P should be expressed by nine stress components. In engineering practice for stress analysis, we use several representations for the stress components. Referring to Cartesian coordinate system xyz, we shall write them as

$$\begin{pmatrix} \sigma_{xx} & \sigma_{xy} & \sigma_{xz} \\ \sigma_{yx} & \sigma_{yy} & \sigma_{yz} \\ \sigma_{zx} & \sigma_{zy} & \sigma_{zz} \end{pmatrix}, \qquad (3\text{-}3)$$

or, in a simplified notation, using σ for a direct stress and τ for a shear stress,

$$\begin{pmatrix} \sigma_{x} & \tau_{xy} & \tau_{xz} \\ \tau_{yx} & \sigma_{y} & \tau_{yz} \\ \tau_{zx} & \tau_{zy} & \sigma_{z} \end{pmatrix}. \qquad (3\text{-}4)$$

For theoretical development, it is sometimes convenient to refer to $x_1 x_2 x_3$ as a Cartesian system instead of xyz and then write a matrix representation of stresses as

$$\begin{pmatrix} \sigma_{11} & \sigma_{12} & \sigma_{13} \\ \sigma_{21} & \sigma_{22} & \sigma_{23} \\ \sigma_{31} & \sigma_{32} & \sigma_{33} \end{pmatrix}. \qquad (3\text{-}5)$$

In all cases the first suffix refers to the normal to the face or plane being considered, and the second suffix gives the direction of the resolved force component. In normal stresses, both suffixes are the same, *e.g.*, σ_{xx} and σ_{yy}. For convenience, it is also common in literature to use a bold letter σ to denote the above stress matrix or use σ_{ij}, where i = x, y, z and j = x, y, z, to represent stress components.

Sometimes, because of the special geometry of a solid structure, it is more convenient to use a polar coordinate system, (r, θ, z), as shown in **Fig. 3.7**. In this case, a convenient way of stress description should also be associated with the polar coordinate system. Thus the stress matrix becomes

$$\begin{pmatrix} \sigma_{rr} & \sigma_{r\theta} & \sigma_{rz} \\ \sigma_{\theta r} & \sigma_{\theta\theta} & \sigma_{\theta z} \\ \sigma_{zr} & \sigma_{z\theta} & \sigma_{zz} \end{pmatrix}. \tag{3-6}$$

It can be shown mathematically that stress is *a tensor of rank two* and therefore all the properties of a tensor, such as its convenient transformation with respect to the change of coordinate systems, apply to stress transformation. However, we prefer not to introduce tensor analysis into this elementary solid mechanics for engineers since the purpose of this book is to facilitate our understanding of mechanics of materials and emphasise the physical meaning of mechanics quantities. The reader who is interested in a more advanced mechanics formulation can always find many other relevant books, *e.g.*, those by Fung (1965) and Flugge (1972).

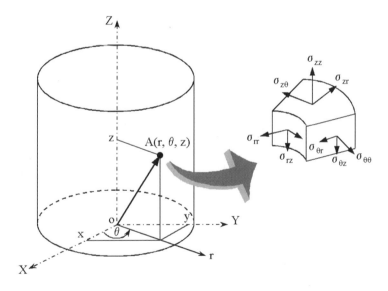

Figure 3.7 Stress description under a polar coordinate system. Left: a point, A, in a circular cylinder. Right: stress components at point A under the polar coordinate system.

3.4 SIGN OF STRESSES

As mentioned before, a stress has its direction. It is therefore necessary to define positive and negative directions of stresses for convenient applications. In the following, we will introduce the commonly accepted definition in engineering practice and literature.

3.4.1 The Positive Direction of Normal Stresses

Consider normal stress σ_{kk} on an infinitesimal plane ΔA_k whose external normal is k, where k can be x, y, or z. We define that σ_{kk} is positive if its direction is along the positive direction of k.

The above definition is clear but let us take σ_{xx} as a particular example to find out its positive direction with different orientations of surfaces. Following the general definition above, the normal stress σ_{xx} is positive if its direction is along the positive direction of the external normal of the surface, see **Fig. 3.8**. It is worthwhile emphasising that according to the above definition, the sign of a normal stress on an infinitesimal plane is independent of the directions of the reference coordinate system used.

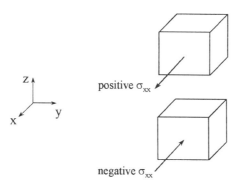

Figure 3.8 Sign of a normal stress.

In practice, we often hear the terms 'tensile' and 'compressive' stresses. As we can see according to the above sign definition, a tensile stress means a positive normal stress and a compressive stress is a negative normal stress. We will use the terms alternatively without any indication later in the text.

3.4.2 The Positive Direction of Shear Stresses

Let us define the positive direction of the shear stress σ_{kl} in the infinitesimal plane ΔA_k, where l is a tangential direction of ΔA_k. If the external normal, k, of ΔA_k has the same direction as a coordinate axis k (k = x, y, or z in a Cartesian coordinate system, but k = r, θ, or z in a polar coordinate system), then the positive σ_{kl} should have the same direction as coordinate axis l. Otherwise, the positive direction of σ_{kl} should be reversed.

Similarly, we take σ_{xy} as an example to demonstrate the definition of positive shear stresses. As shown in the upper half of **Fig. 3.9**, if the external normal of the surface has the same direction as coordinate axis x, the positive σ_{xy} must have the same direction as the coordinate direction y. However, in the case shown in the lower part of **Fig. 3.9**, the external normal of the surface is in the negative direction of x. Hence, the positive direction of σ_{xy} on the surface in this case should be in the negative direction of axis y. In all cases shown in **Figs. 3.6** and **3.7** we have drawn positive stresses.

The above sign definition of stress components is important to the modelling of a mechanics problem in describing its stress boundary conditions, is the key to a sound understanding of deformation and is also a common 'language' for technical communications in the engineering world.

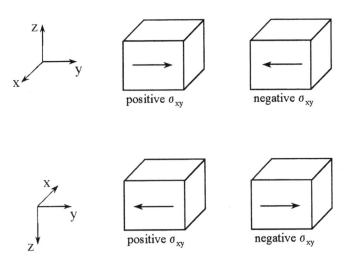

Figure 3.9 Sign of a shear stress.

Example 3.1 The stress states at two different points in a machine component were measured to be

$$A: \begin{pmatrix} 16 & 18 & 0 \\ 18 & 17 & -15 \\ 0 & -15 & 19 \end{pmatrix}, \quad B: \begin{pmatrix} -19 & 20 & 0 \\ 20 & -25 & 0 \\ 0 & 0 & 20 \end{pmatrix},$$

respectively (units of MN/m^2). Express the stress states graphically on infinitesimal elements.

Solution: The graphical presentations of these stress states are simple and are shown in **Fig. E3.1**, where the sign of a stress component is indicated by its arrow direction.

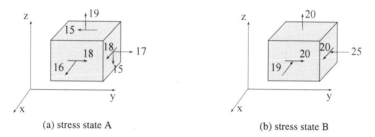

(a) stress state A (b) stress state B

Figure E3.1 The graphic presentation of stress states on infinitesimal elements (units of MN/m^2).

3.5 SYMMETRY OF STRESS MATRIX

From the above discussion we have understood that to describe the stress state at a point in a solid, we can use the stress matrix with nine stress components. A natural question is therefore: 'Are they all independent?' Or in other words, 'Can we use a smaller number of stress components to facilitate the description of a stress state?'

To answer the question, let us consider the solid Ω again that is in equilibrium under a set of external forces, see **Fig. 3.4**. As Ω is in equilibrium, *any element in it must also be in equilibrium*. Imagine that if we cut an infinitesimal element from Ω, with side lengths Δx, Δy and Δz, by applying a set of equivalent stresses on the element surfaces, as shown by the left part of **Fig. 3.10**, this element should be in equilibrium under the equivalent surface stresses. Let us check the equilibrium conditions of the element and see what conclusions can be drawn.

We check the equilibrium of the element in xy-plane first. The projection of all the stresses in xy-plane gives rise to the stress state in a plane element illustrated in the right half of **Fig. 3.10**. By taking moment about the centre O and assuming that an anticlockwise moment is positive, we obtain

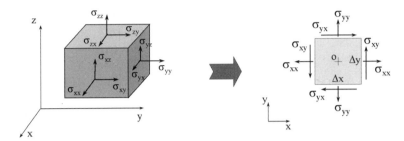

Figure 3.10 The equilibrium of an element with equivalent stresses.

$$\sigma_{xy}(\Delta y)(\Delta z)\left\{\frac{1}{2}(\Delta x)\right\} + \sigma_{xy}(\Delta y)(\Delta z)\left\{\frac{1}{2}(\Delta x)\right\}$$
$$-\sigma_{yx}(\Delta x)(\Delta z)\left\{\frac{1}{2}(\Delta y)\right\} - \sigma_{yx}(\Delta x)(\Delta z)\left\{\frac{1}{2}(\Delta y)\right\} \quad (3\text{-}7)$$
$$= (\text{moment of inertia})(\text{acceleration}).$$

This gives rise to

$$\sigma_{xy} - \sigma_{yx} = \left\{(\Delta x)^2 + (\Delta y)^2\right\} f(\rho, a), \quad (3\text{-}8)$$

where f is a finite function of the density of the solid material, ρ, and the angular acceleration, a, of the element. As we are discussing a stress state at a point, (Δx) and (Δy) must vanish. We therefore have

$$\lim_{\substack{\Delta x \to 0 \\ \Delta y \to 0}} (\sigma_{xy} - \sigma_{yx}) = 0 \quad (3\text{-}9)$$

which means that

$$\sigma_{xy} = \sigma_{yx}. \quad (3\text{-}10)$$

Similarly, by checking the equilibrium conditions in yz- and xz-planes, we can find that

$$\sigma_{xz} = \sigma_{zx}, \quad \sigma_{yz} = \sigma_{zy}. \quad (3\text{-}11)$$

Thus we have shown that the stress matrix is symmetrical, *i.e.*, $\sigma_{ij} = \sigma_{ji}$ (i, j = x, y, z). There are only six independent stress components that need to be used to describe the stress state at a point.

3.6 STRESS TRANSFORMATION

In the previous sections, we have shown how to describe the stress state at a point by resolving the stress vector $\overset{n}{T}$ onto infinitesimal planes perpendicular to the coordinate directions of a reference Cartesian system xyz. We can easily understand that the selection of the Cartesian system is artificial. We use the xyz system but some others may use x'y'x' system that is different from xyz, as illustrated in **Fig. 3.11**. We can immediately realise that the directions and magnitudes of the stresses at a point described under xyz are different from those at the same point described under x'y'z', although the stress state at the same point must be the same. This indicates that there must exist a transformation relationship between the stress descriptions for the same point under different coordinate systems.

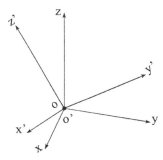

Figure 3.11 Two different Cartesian coordinate systems with relative rotations.

We often have cases in which components with special shapes work under special loading conditions. The gears in **Fig. 3.12** are such an example. All the external forces on the gear are acting within its plane. Thus to facilitate our understanding of stress transformation between two coordinate systems, without losing engineering importance, we can first consider simpler cases like the gear where all non-zero stresses are in a plane. To generalise the analysis, let us consider a thin plate of *arbitrary profile* whose thickness is much less than its other dimensions and that all external loads are on the side surface but act in the plane of the plate (xy-plane) only, as illustrated in **Fig. 3.13**.

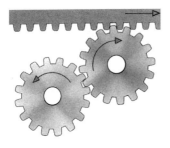
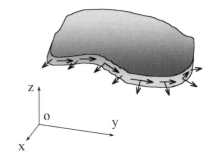

Figure 3.12 Gears subjected to an in-plane loading.

Figure 3.13 An arbitrary plate subjected to an in-plane loading.

3.6.1 Stress in Any Direction: Two-Dimensional Stress States

The thickness direction of the plate is in that of coordinate z so that its top and bottom surfaces are in parallel to the xy-coordinate plane. Because these surfaces are stress-free and the plate is very thin, we have

$$\sigma_{zz} = \sigma_{zx} = \sigma_{zy} = 0 \tag{3-12}$$

throughout the plate. The only non-vanishing stress components at a point in the plate are the two normal stresses, σ_{xx} and σ_{yy}, and a shear stress, σ_{xy}. Hence, if we consider an infinitesimal element at a point in the plate in the xy-plane that has a *unit thickness* in z-direction, the stresses on the element can be described by the two-dimensional state shown in **Fig. 3.14a**, or by the following simplified matrix

$$\begin{pmatrix} \sigma_{xx} & \sigma_{xy} \\ \sigma_{yx} & \sigma_{yy} \end{pmatrix}. \tag{3-13}$$

If we cut off part of the element with an angle θ and consider the stresses of the left triangle, we should apply equivalently a normal stress σ_{nn} and a tangential stress σ_{nt} on the cut surface to replace the action of the removed part of the element, where *n* is the unit external normal of the cut surface that has an included angle θ with the positive direction of x-axis, see **Fig. 3.14b**. Since θ can vary, if we are able to establish the relationship between the stresses on the inclined surface, *i.e.*, σ_{nn} and σ_{nt}, with those in coordinate directions, *i.e.*, σ_{xx}, σ_{yy}, and σ_{xy}, we obtain stresses in any directions.

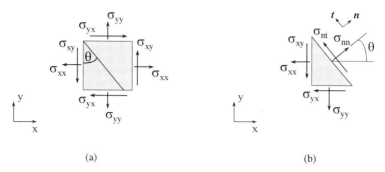

Figure 3.14 Stresses on a plane element.

As discussed before, since the whole plate is in equilibrium under the external loading, any element in the plate is also in equilibrium. Thus if we check the equilibrium of the triangle element of **Fig. 3.14b**, the resultant forces on the element in any direction must vanish. Hence the resultant forces in *n* and *t* directions being zero lead to

$$\sigma_{nn} = \sigma_{xx}\cos^2\theta + \sigma_{yy}\sin^2\theta + 2\sigma_{xy}\cos\theta\sin\theta \qquad (3\text{-}14a)$$

or

$$\sigma_{nn} = \frac{1}{2}(\sigma_{xx}+\sigma_{yy}) + \frac{1}{2}(\sigma_{xx}-\sigma_{yy})\cos 2\theta + \sigma_{xy}\sin 2\theta \qquad (3\text{-}14b)$$

and

$$\sigma_{nt} = (\sigma_{yy}-\sigma_{xx})\cos\theta\sin\theta + \sigma_{xy}(\cos^2\theta - \sin^2\theta) \qquad (3\text{-}15a)$$

or

$$\sigma_{nt} = \frac{1}{2}(\sigma_{yy}-\sigma_{xx})\sin 2\theta + \sigma_{xy}\cos 2\theta. \qquad (3\text{-}15b)$$

With the aid of the above equations, we can calculate the normal and shear stresses at a point in any direction if we know the stresses, σ_{xx}, σ_{yy}, and σ_{xy}, at the point. These relations can also be used to find the stresses when a coordinate system rotates.

3.6.2 Stresses with Coordinate Axis Rotation

Let us continue to consider the two-dimensional stress state discussed above. Now we would like to know how to calculate the stresses in a coordinate system ox'y' whose x'-axis has a relative angle θ with x-axis, as shown in **Fig. 3.15**.

Comparing **Fig. 3.15** with **Fig. 3.14b**, we immediately find that x' is coincide with n. Thus $\sigma_{x'x'}$ equals σ_{nn} of Eq. (3-14) and $\sigma_{y'y'}$ equals σ_{nn} when θ is replaced by 90°+θ. Since y'-axis is in the direction of t, $\sigma_{x'y'}$ equals σ_{nt} of Eq. (3-15). In mathematical forms, hence,

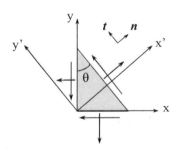

Figure 3.15 Coordinate rotation in the xy-plane.

$$\sigma_{x'x'} = \sigma_{xx}\cos^2\theta + \sigma_{yy}\sin^2\theta + 2\sigma_{xy}\cos\theta\sin\theta, \quad (3\text{-}16a)$$

$$\sigma_{y'y'} = \sigma_{xx}\sin^2\theta + \sigma_{yy}\cos^2\theta - 2\sigma_{xy}\cos\theta\sin\theta \quad (3\text{-}16b)$$

and

$$\sigma_{x'y'} = \frac{1}{2}(\sigma_{yy} - \sigma_{xx})\sin 2\theta + \sigma_{xy}\cos 2\theta. \quad (3\text{-}16c)$$

The above equations show us a very interesting phenomenon. When we calculate the summation of $\sigma_{x'x'}$ and $\sigma_{y'y'}$ using Eq. (3-16), we find that

$$\sigma_{x'x'} + \sigma_{y'y'} = \sigma_{xx} + \sigma_{yy}. \quad (3\text{-}17)$$

It means that the summation of the two normal stress components is independent of the rotation of the coordinate system. As we will find out later, this is the nature of stress and is also true in a general three-dimensional stress state where ($\sigma_{xx} + \sigma_{yy} + \sigma_{zz}$) keeps a constant when the coordinate system rotates. We will discuss it in detail later.

3.6.3 Principal Stresses and Maximum Shear Stress

In the design and assessment of many engineering structures, a very important step is to find out the maximum and minimum stresses and their directions under working conditions so that proper reinforcement of materials can be used in these directions to enhance the strength of the structure. The following are two examples in designing plastic and composite pressure vessels.

Design 1: A long, cylindrical pressure vessel with closed ends, subjected to an inner pressure p, is to be made by rolling a plastic strip into a helix and making a continuous fused joint, as illustrated in **Fig. 3.16a**. The thickness of the strip is t (t is much smaller than D, where D is the outer diameter of the vessel). It is desired that the fused joint will be subjected to a tensile stress only 80% of the maximum in the parent plastic strip. What angle α should be used in production?

Design 2: Lightweight pressure vessels, **Fig. 3.16b**, often use glass filaments for resisting tensile forces and use epoxy resin as a binder. What should be the angle of winding, α, of the filaments such that the tensile forces in the filaments are equal?

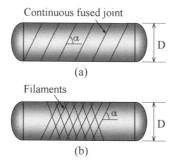

Figure 3.16 A design considerations of plastic and composite vessels.

Can we use the equations of stresses in any directions obtained, *i.e.*, Eqs. (3-14) to (3-15), to answer the questions above? To facilitate the solution, let us first introduce principal stresses and their directions.

Since the normal stress σ_{nn} is a function of θ, it must take its extrema when its first derivative with respect to θ is equal to zero, that is

$$\frac{d\sigma_{nn}}{d\theta} = 0. \tag{3-18}$$

The substitution of Eq. (3-14a) into the left-hand side of the above condition leads to

$$(\sigma_{yy} - \sigma_{xx})\sin 2\theta + 2\sigma_{xy}\cos 2\theta = 0, \tag{3-19}$$

which gives rise to

$$\tan 2\theta = \frac{2\sigma_{xy}}{\sigma_{xx} - \sigma_{yy}}. \tag{3-20}$$

The above process means that at the θ determined by Eq. (3-20), normal stress σ_{nn} reaches its maximum or minimum. We call the maximum and minimum σ_{nn} the *principal stresses*, their corresponding directions the *principal directions* and the planes that the principal stresses act the *principal planes*.

It is important to note that the left-hand side of Eq. (3-19) is just double that of the shear stress σ_{nt} given by Eq. (3-15b). This indicates that on the principal planes, shear stresses are all zero. In other words, σ_{nn} *reaches its extrema* (principal stresses) *on the planes with* $\sigma_{nt} = 0$. This can be regarded as *the physical definition of principal stresses*.

When substituting Eq. (3-20) into Eq. (3-14a), we obtain the principal stresses as

$$\left.\begin{matrix}\sigma_1\\\sigma_3\end{matrix}\right\} = \frac{1}{2}(\sigma_{xx} + \sigma_{yy}) \pm \sqrt{\left(\frac{\sigma_{xx} - \sigma_{yy}}{2}\right)^2 + \sigma_{xy}^2}, \tag{3-21}$$

where σ_1 is the maximum principal stress and σ_3 the minimum principal stress. In three-dimensional cases, as we will see in section **3.6.6**, there will be an intermediate principal stress, σ_2.

With a similar consideration, we can find the maximum and minimum shear stresses and their directions. However, we have known that on the principal planes shear stresses vanish. Thus for convenience, we can select a coordinate system oxy and let x-axis be in principal direction 1 and y-axis in principal direction 3, as shown in **Fig. 3.17**, such that shear stress σ_{xy} vanishes. Equation (3-15) thus becomes

$$\sigma_{nt} = -\frac{1}{2}(\sigma_1 - \sigma_3)\sin 2\theta \tag{3-22}$$

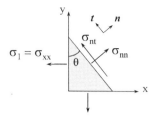

Figure 3.17 The stress state when coordinate axes x and y are in principal directions.

and the extremum condition $d\sigma_{nt}/d\theta = 0$ brings about $\cos 2\theta = 0$. Hence, the maximum and minimum shear stresses, τ_{max} and τ_{min}, are in the directions of 45° and 135° with respect to the principal direction 1 and can be obtained by using Eq. (3-22) with $\theta = 45°$ and 135°, *i.e.*,

$$\left.\begin{array}{c}\tau_{max}\\ \tau_{min}\end{array}\right\} = \pm\frac{1}{2}(\sigma_1 - \sigma_3). \qquad (3\text{-}23a,b)$$

Clearly, when the principal stresses and directions at a point are determined, the extrema of shear stresses and their directions are also determined.

Example 3.2 Now we can go back to answer the question in the design of our pressure vessels presented in **Fig. 3.16**. Let us take Design 1 as an example. Since $t \ll D$, we can approximately consider that stresses through the vessel thickness are uniform.[3.1] Because the vessel under the inner pressure is in equilibrium, any part in the vessel is also in equilibrium. Thus, as shown in **Fig. E3.2**, the equilibrium of the vessel parts in the x- and y-directions gives rise to the stresses at point A as

$$\sigma_{xx} = \frac{pR_i}{2t}, \quad \sigma_{yy} = \frac{pR_i}{t} \text{ and } \sigma_{xy} = 0,$$

where $R_i = (D - 2t)/2$ is the inner radius of the vessel. Because the shear stress σ_{xy} is zero, x and y directions are the principal directions so that $\sigma_1 = \sigma_{yy}$ and $\sigma_2 = \sigma_{xx}$. Now we need to find a direction, *n*, in which the normal stress σ_{nn} takes 80% of σ_{yy} that is the maximum tensile stress in the vessel according to the above equation. Using Eqs. (3-14b), we get

---

[3.1] If t is not much smaller than D, we have to use the elasticity theory to be established later in Chapter 6 to solve the problem. The approximation made here will become inappropriate.

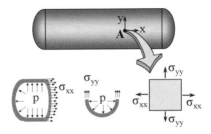

Figure E3.2 Stresses in a thin-walled pressure vessel.

$$\sigma_{nn} = 0.8\,\sigma_{yy} = \frac{1}{2}(\sigma_{xx} + \sigma_{yy}) + \frac{1}{2}(\sigma_{yy} - \sigma_{xx})\cos 2\theta$$

which gives rise to $\cos 2\theta = 0.2$ and $\theta = 39.23°$. Thus in the direction of $\alpha = 39.23°$, the tensile stress takes 80% of the maximum normal stress in the parent plastic strip.

Example 3.3 At a particular point in a structure, an external loading results in the measurement of a stress state described by

$$\begin{pmatrix} -10 & -10 & 0 \\ -10 & 20 & 0 \\ 0 & 0 & 0 \end{pmatrix} \text{MN/m}^2.$$

Find the principal stresses, maximum shear stress and their direction cosines with respect to x-axis at this point.

Solution: According to the given stress state, we have known that $\sigma_{xx} = -10$ MN/m^2, $\sigma_{yy} = 20$ MN/m^2 and $\sigma_{xy} = -10$ MN/m^2. Thus Eq. (3-21) gives rise to $\sigma_1 = 23.03$ MN/m^2 and $\sigma_3 = -13.03$ MN/m^2. To find the principal directions, let us use Eq. (3-20), which leads to $\tan 2\theta = 0.6667$ and thus $2\theta = 33.69°$ or $213.69°$. The Mohr's circle of the given stress state can be drawn easily as shown in **Fig.E3.3**, which shows that the direction of the first principal stress σ_1 is $\theta_1 = 106.85°$ and that of σ_3 is $\theta_3 = 16.85°$. (Mohr's circle studied in strength of materials or mechanics of materials is an important concept. To facilitate the review, it is briefly summarised in Appendix A of this book.) The corresponding direction cosines of the two principal stresses are therefore $\cos\theta_1 = -0.2899$ and $\cos\theta_3 = 0.9571$.

The maximum shear stress is determined by Eq. (3-23a), which brings about, for the present stress state, $\tau_{max} = 18.03$ MN/m^2. As we proved previously, the maximum stress is always in the direction of $45°$ with respect to σ_1, which is also shown by the above Mohr's circle, thus $\theta|\tau_{max} = 45° + \theta_1 = 151.85°$ and its direction cosine is $\cos(\theta|\tau_{max}) = -0.8817$.

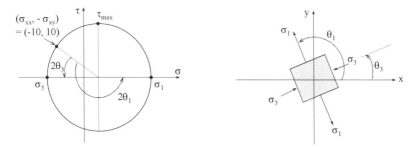

Figure E3.3 The Mohr's circle and the principal direction with respect to the x-axis.

3.6.4 Principal Stresses as Eigenvalues of the Stress Matrix

We have experienced the lengthy process of deriving the expressions of principal stresses in the above two-dimensional stress states. We can imagine that in a case with general three-dimensional stresses it will become more complicated. Thus if we can find a more straightforward way, the difficulties in understanding the properties of three-dimensional stress states may be overcome.

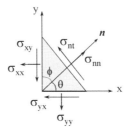

Figure 3.18 The equilibrium of an infinitesimal triangular element.

Let us consider again the triangular element in **Fig. 3.14b**. For convenience, as shown in **Fig. 3.18**, we use ϕ to denote the included angle between n and y-axis. The equilibrium of the element in x- and y-directions thus brings about

$$\begin{cases} \sigma_{xx} \cos\theta + \sigma_{xy} \cos\phi = \sigma_{nn} \cos\theta - \sigma_{nt} \cos\phi, \\ \sigma_{xy} \cos\theta + \sigma_{yy} \cos\phi = \sigma_{nn} \cos\phi + \sigma_{nt} \cos\theta. \end{cases}$$

Now, if n is a principal direction and σ_{nn} is a principal stress, then according to our previous discussion σ_{nt} vanishes. Denoting this principal stress by σ, we can rewrite the above two equations as

$$\begin{cases} \sigma_{xx}\cos\theta + \sigma_{xy}\cos\phi = \sigma\cos\theta, \\ \sigma_{xy}\cos\theta + \sigma_{yy}\cos\phi = \sigma\cos\phi. \end{cases}$$

If we further write them into a matrix form, we get

$$\begin{Bmatrix} \sigma_{xx} & \sigma_{xy} \\ \sigma_{xy} & \sigma_{yy} \end{Bmatrix} \begin{Bmatrix} l \\ m \end{Bmatrix} = \sigma \begin{Bmatrix} l \\ m \end{Bmatrix}, \tag{3-24a}$$

where l and m are the direction cosines of the principal direction, n, with respect to x- and y-axes, *i.e.*, $l = \cos\theta$ and $m = \cos\phi$. The above equation represents a standard eigenvalue problem,

$$\sigma X = \sigma X \tag{3-24b}$$

where σ is the stress matrix and X is the direction cosine vector, *i.e.*,

$$\sigma = \begin{Bmatrix} \sigma_{xx} & \sigma_{xy} \\ \sigma_{xy} & \sigma_{yy} \end{Bmatrix}, \quad X = \begin{Bmatrix} l \\ m \end{Bmatrix}.$$

If we rewrite Eq. (3-24) into

$$\begin{Bmatrix} \sigma_{xx} - \sigma & \sigma_{xy} \\ \sigma_{xy} & \sigma_{yy} - \sigma \end{Bmatrix} \begin{Bmatrix} l \\ m \end{Bmatrix} = \begin{Bmatrix} 0 \\ 0 \end{Bmatrix}, \tag{3-25}$$

we can easily see that it has a set of non-vanishing solutions *if and only if* the determinant of its coefficient matrix vanishes, that is

$$\begin{vmatrix} \sigma_{xx} - \sigma & \sigma_{xy} \\ \sigma_{xy} & \sigma_{yy} - \sigma \end{vmatrix} = 0, \tag{3-26}$$

which leads to

$$\sigma^2 - (\sigma_{xx} + \sigma_{yy})\sigma + (\sigma_{xx}\sigma_{yy} - \sigma_{xy}^2) = 0. \tag{3-27}$$

This quadratic equation has the solutions of

$$\left.\begin{matrix}\sigma_1\\\sigma_3\end{matrix}\right\} = \frac{1}{2}(\sigma_{xx}+\sigma_{yy}) \pm \sqrt{\left(\frac{\sigma_{xx}-\sigma_{yy}}{2}\right)^2 + \sigma_{xy}^2}, \quad (3\text{-}28)$$

which is exactly Eq. (3-21). Thus *mathematically principal stresses are the eigenvalues of the stress matrix while the corresponding eigenvector is the direction cosine vector*. The principal direction, say that of σ_1, can also be determined when σ in Eq. (3-25) is substituted by σ_1 from Eq. (3-28).

The above conclusion offers an alternative but convenient way to determine principal stresses and their direction cosines for three-dimensional stress states, because Eq. (3-24) represents a standard eigenvalue problem without any dimensional limitation. By keeping these in mind, we are now able to discuss stress transformation and principal stresses for general three-dimensional stress states.

3.6.5 Stress in Any Direction: Three-Dimensional Stress States

Similar to the two-dimensional analysis, to obtain stresses in any direction in a three-dimensional state, let us consider an infinitesimal tetrahedron formed by three surfaces parallel to the coordinate planes and an inclined surface abc normal to the unit vector n, see **Fig. 3.19**.

Let the area of the surface abc be ΔA, then the areas of the other three surfaces, *i.e.*, acp, bcp and abp, are respectively

$$\Delta A_x = \Delta A l, \quad \Delta A_y = \Delta A m, \quad \Delta A_z = \Delta A n, \quad (3\text{-}29)$$

as shown in **Fig. 3.20**, where $l = \cos(n, x)$, $m = \cos(n, y)$ and $n = \cos(n, z)$ are the direction cosines of the external normal, n, of surface abc. In the above expressions, (n, k) means the included angle between n and the coordinate axis k (k = x, y, z). If the stress vector on surface abc is $\overset{n}{T}$ and its components in x-, y- and z-directions are σ_{nx}, σ_{ny} and σ_{nz}, as illustrated in **Fig. 3.21**, the equilibrium of the element in the x-direction gives rise to the following equation:

$$(\Delta A)\sigma_{nx} = (\Delta A_x)\sigma_{xx} + (\Delta A_y)\sigma_{yx} + (\Delta A_z)\sigma_{zx}.$$

Using Eq. (3-29), the above equation can be simplified to

$$\sigma_{nx} = l\sigma_{xx} + m\sigma_{yx} + n\sigma_{zx}. \quad (3\text{-}30a)$$

Similarly, the equilibrium of the element in y and z-directions leads to

$$\sigma_{ny} = l\,\sigma_{xy} + m\,\sigma_{yy} + n\,\sigma_{yz} \qquad (3\text{-}30\text{b})$$

$$\sigma_{nz} = l\,\sigma_{xz} + m\,\sigma_{yz} + n\,\sigma_{zz} \qquad (3\text{-}30\text{c})$$

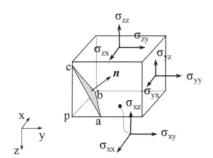

Figure 3.19 The tetrahedron element.

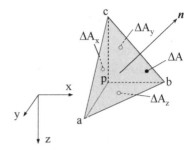

Figure 3.20 Surface areas of the tetrahedron element.

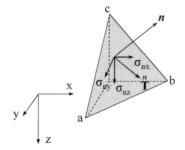

Figure 3.21 Stresses on surface abc of the tetrahedron element.

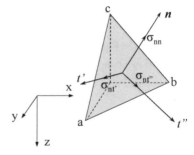

Figure 3.22 Stresses on plane abc in normal and tangential directions.

With the aid of Eq. (3-30), we can now find the stresses when a coordinate system rotates. Similar to the case in two-dimensional stress transformation, let the three axes of the rotated axes be in n, t' and t'', as shown in **Fig. 3.22**, where t' and t'' are two perpendicular tangential directions in plane abc with direction cosines being (l', m', n') and (l'', m'', n''), respectively.

Let us establish the relationship between the normal and tangential stresses on the new coordinate surfaces, *i.e.*, we try to find σ_{nn}, $\sigma_{nt'}$, $\sigma_{nt''}$, $\sigma_{t't'}$, $\sigma_{t't''}$ and $\sigma_{t''t''}$ in terms of the six stress components in the old coordinate directions, σ_{xx}, σ_{xy}, σ_{xz}, σ_{yy}, σ_{yz} and σ_{zz}.

The process is simple because we just need to resolve all the stress components on surface abc into *n*, *t'* and *t"*, respectively. For example, resolving stresses into the direction of *n* leads to

$$\sigma_{nn} = l\sigma_{nx} + m\sigma_{ny} + n\sigma_{nz}.$$

Using Eq. (3-30), the above relation can be rewritten as

$$\sigma_{nn} = l(l\sigma_{xx} + m\sigma_{yx} + n\sigma_{zx}) + m(l\sigma_{xy} + m\sigma_{yy} + n\sigma_{yz}) + \\ + n(l\sigma_{xz} + m\sigma_{yz} + n\sigma_{zz}),$$

or by expanding, it becomes

$$\sigma_{nn} = l^2 \sigma_{xx} + m^2 \sigma_{yy} + n^2 \sigma_{zz} + 2lm\sigma_{xy} + 2mn\sigma_{zy} + 2nl\sigma_{zx}. \quad (3\text{-}31)$$

In a similar way, we can obtain the other five stresses with respect to the new coordinate system, whose expressions are similar to Eq. (3-31). If we write all the six transformed expressions into a matrix form, they become

$$\begin{pmatrix} \sigma_{nn} & \sigma_{nt'} & \sigma_{nt"} \\ \sigma_{t'n} & \sigma_{t't'} & \sigma_{t't"} \\ \sigma_{t"n} & \sigma_{t"t'} & \sigma_{t"t"} \end{pmatrix} = \begin{pmatrix} l & m & n \\ l' & m' & n' \\ l" & m" & n" \end{pmatrix} \begin{pmatrix} \sigma_{xx} & \sigma_{xy} & \sigma_{xz} \\ \sigma_{yx} & \sigma_{yy} & \sigma_{yz} \\ \sigma_{zx} & \sigma_{zy} & \sigma_{zz} \end{pmatrix} \begin{pmatrix} l & m & n \\ l' & m' & n' \\ l" & m" & n" \end{pmatrix}^T, \quad (3\text{-}32)$$

where *l'*, *m'*, *n'* are the direction cosines of *t'* and *l"*, *m"*, *n"* are those of *t"*. In a more compact form, Eq. (3-32) can be written as

$$\sigma_{new} = R\,\sigma_{old}\,R^T, \quad (3\text{-}33)$$

where

$$\sigma_{new} = \begin{pmatrix} \sigma_{nn} & \sigma_{nt'} & \sigma_{nt"} \\ \sigma_{t'n} & \sigma_{t't'} & \sigma_{t't"} \\ \sigma_{t"n} & \sigma_{t"t'} & \sigma_{t"t"} \end{pmatrix}, \sigma_{old} = \begin{pmatrix} \sigma_{xx} & \sigma_{xy} & \sigma_{xz} \\ \sigma_{yx} & \sigma_{yy} & \sigma_{yz} \\ \sigma_{zx} & \sigma_{zy} & \sigma_{zz} \end{pmatrix} \quad (3\text{-}34)$$

and

$$\mathbf{R} = \begin{pmatrix} l & m & n \\ l' & m' & n' \\ l'' & m'' & n'' \end{pmatrix}, \tag{3-35}$$

where **R** is called the *transformation matrix*. In fact, if we consider stress as a tensor as mentioned early in this chapter, then the transformation equation (3-33) is straightforward mathematically. In the above, however, we achieved an understanding of the mechanics at the cost of a lengthy derivation.

3.6.6 Principal Stresses in Three Dimensions

We have realised the importance of principal stresses when discussing two-dimensional stress states. Since most engineering elements work under three-dimensional stress states, it is necessary to work out the way to find principal stresses at a point in a solid in a general three-dimensional stress field.

We have known that on a principal plane, shear stress vanishes and principal stresses at a point are the eigenvalues of the stress matrix. Thus in the following we can use a simple way to determine principal stresses and their directions.

The stress matrix in a general stress state is in the form of Eq. (3-3). Thus according to Eq. (3-24), the eigenvalue equation for our three-dimensional stress matrix is

$$\begin{pmatrix} \sigma_{xx} - \sigma & \sigma_{xy} & \sigma_{xz} \\ \sigma_{yx} & \sigma_{yy} - \sigma & \sigma_{yz} \\ \sigma_{zx} & \sigma_{zy} & \sigma_{zz} - \sigma \end{pmatrix} \begin{pmatrix} l \\ m \\ n \end{pmatrix} = \begin{pmatrix} 0 \\ 0 \\ 0 \end{pmatrix}. \tag{3-36}$$

To have a non-vanishing solution of the eigenvector $(l, m, n)^T$, the determinant of the coefficient matrix of Eq. (3-36) must be zero, *i.e.*,

$$\begin{vmatrix} \sigma_{xx} - \sigma & \sigma_{xy} & \sigma_{xz} \\ \sigma_{yx} & \sigma_{yy} - \sigma & \sigma_{yz} \\ \sigma_{zx} & \sigma_{zy} & \sigma_{zz} - \sigma \end{vmatrix} = 0,$$

which gives rise to

$$\sigma^3 - I_1^\sigma \sigma^2 + I_2^\sigma \sigma - I_3^\sigma = 0, \tag{3-37}$$

where the coefficients are

$$I_1^\sigma = \sigma_{xx} + \sigma_{yy} + \sigma_{zz},$$

$$I_2^\sigma = \begin{vmatrix} \sigma_{yy} & \sigma_{yz} \\ \sigma_{zy} & \sigma_{zz} \end{vmatrix} + \begin{vmatrix} \sigma_{xx} & \sigma_{xz} \\ \sigma_{zx} & \sigma_{zz} \end{vmatrix} + \begin{vmatrix} \sigma_{xx} & \sigma_{xy} \\ \sigma_{yx} & \sigma_{yy} \end{vmatrix}$$

$$= \sigma_{xx}\sigma_{yy} + \sigma_{yy}\sigma_{zz} + \sigma_{zz}\sigma_{xx} - (\sigma_{xy})^2 - (\sigma_{yz})^2 - (\sigma_{zx})^2, \quad (3\text{-}38\text{a, b, c})$$

$$I_3^\sigma = \begin{vmatrix} \sigma_{xx} & \sigma_{xy} & \sigma_{xz} \\ \sigma_{yx} & \sigma_{yy} & \sigma_{yz} \\ \sigma_{zx} & \sigma_{zy} & \sigma_{zz} \end{vmatrix}$$

$$= \sigma_{xx}\sigma_{yy}\sigma_{zz} + 2\sigma_{xy}\sigma_{yz}\sigma_{zx} - \sigma_{xy}(\sigma_{yz})^2 - \sigma_{yy}(\sigma_{zx})^2 - \sigma_{zz}(\sigma_{xy})^2.$$

Hence, the principal stresses are the three real roots of Eq. (3-37). Similar to the situation in two-dimensional cases of obtaining Eqs. (3-21) to (3-23), we use symbol σ_1 to stand for the maximum principal stress, σ_2 to denote the intermediate principal stress and σ_3 to indicate the minimum principal stress, *i.e.*, $\sigma_1 \geq \sigma_2 \geq \sigma_3$. We will follow this rule throughout the book without further indication.

When the principal stresses are obtained by solving Eq. (3-37) and when $\sigma_1 \neq \sigma_2 \neq \sigma_3$, the direction cosines l, m and n corresponding to a specific principal stress, say σ_i (i = 1, 2, 3), can be determined by any two of the equations in Eq. (3-36) by replacing σ with σ_i together with the geometrical relationship

$$(l_i)^2 + (m_i)^2 + (n_i)^2 = 1 \quad (i = 1, 2, 3). \tag{3-39}$$

Equation (3-39) is the relationship that must be followed by the direction cosines of a unit vector. For example, to determine the direction cosines l_1, m_1 and n_1 of the first principal stress σ_1, we can replace the σ by σ_1 in any two equations in Eq. (3-36), say the first two, together with Eq. (3-39) by taking i = 1 to obtain the solution. Hence, l_1, m_1 and n_1 are given by

$$(\sigma_{xx} - \sigma_1)l_1 + \sigma_{xy} m_1 + \sigma_{xz} n_1 = 0,$$
$$\sigma_{yx} l_1 + (\sigma_{yy} - \sigma_1)m_1 + \sigma_{yz} n_1 = 0, \tag{3-40}$$
$$(l_1)^2 + (m_1)^2 + (n_1)^2 = 1.$$

The unused third equation of Eq. (3-36) is often for checking the correctness of the solution obtained.

When two of the principal stresses are equal, *e.g.*, $\sigma_1 = \sigma_2 \neq \sigma_3$, any pair of orthogonal directions in the $\sigma_1\sigma_2$-plane, which is perpendicular to σ_3, can be the direction of σ_1 or σ_2. If all the three principal stresses are equal, *i.e.*, $\sigma_1 = \sigma_2 = \sigma_3$, any orthogonal directions are the principal directions. The proof of the above can be found in Fung (1965).

When a solid is subjected to a set of external loads, the principal stresses at any point in the solid are physically determined. Thus their magnitudes will not be affected by the coordinate system used for stress analysis. In other words, the magnitude of principal stresses, σ_1, σ_2 and σ_3, does not vary with the rotation of a reference frame. This means that the values of the coefficients in Eq. (3-37), *i.e.*, I_1^σ, I_2^σ and I_3^σ, must not vary with the rotation of the coordinate system. Hence, I_1^σ is called *the first stress invariant*, I_2^σ *the second stress invariant* and I_3^σ *the third stress invariant*. They are important quantities in solid mechanics as we will see later. In two-dimensional cases, the first stress invariant has been presented in Eq. (3-17). As we can see now, Eq. (3-17) is just a special case of I_1^σ when one of the normal stresses, σ_{zz}, vanishes. Based on the above physical consideration, we can easily understand that the three stress invariants are also determined by

$$\begin{aligned} I_1^\sigma &= \sigma_1 + \sigma_2 + \sigma_3, \\ I_2^\sigma &= \sigma_1\sigma_2 + \sigma_2\sigma_3 + \sigma_3\sigma_1, \\ I_3^\sigma &= \sigma_1\sigma_2\sigma_3 \end{aligned} \qquad (3\text{-}41a, b, c)$$

when the principal stresses are known.

Example 3.4 At a point in a structure subjected to three-dimensional loading, the stress state was measured to be

$$\begin{pmatrix} 50 & -20 & 0 \\ -20 & 80 & 60 \\ 0 & 60 & -70 \end{pmatrix} \text{MN/m}^2.$$

Find the principal stresses and their direction cosines.

Solution: Using Eq. (3-38), it is easy to obtain

$$I_1^\sigma = 60, \ I_2^\sigma = -9100, \ I_3^\sigma = -432000.$$

Thus Eq. (3-37) becomes

$$\sigma^3 - 60\sigma^2 - 9100\sigma + 43200 = 0.$$

The principal stresses are the three real roots of the above equation, which are

$$\sigma_1 = 107.3 \text{ MN/m}^2, \quad \sigma_2 = 44.1 \text{ MN/m}^2, \quad \sigma_3 = -91.4 \text{ MN/m}^2.$$

To obtain the direction cosine of a principal direction, say that of σ_1, substitute the value of σ_1 into Eq. (3-40). This gives rise to

$$l_1 = 0.314, \quad m_1 = -0.900, \quad n_1 = -0.303.$$

Similarly, when the values of σ_2 and σ_3 are used in Eq. (3-40) respectively, we get

$$l_2 = 0.948, \quad m_2 = 0.282, \quad n_2 = 0.146.$$
$$l_3 = -0.048, \quad m_3 = 0.337, \quad n_3 = -0.940.$$

3.6.7 Maximum Shear Stress in Three Dimensions

In a similar way to that for two-dimensional cases, we can find the maximum shear stress at a point under three-dimensional stresses. For convenience, let the coordinate directions of **Fig. 3.21** be the principal directions, i.e., $\sigma_1 = \sigma_{xx}$, $\sigma_2 = \sigma_{yy}$ and $\sigma_3 = \sigma_{zz}$. Thus Eq. (3-30) becomes

$$\sigma_{nx} = l\sigma_1, \quad \sigma_{ny} = m\sigma_2, \quad \sigma_{nz} = n\sigma_3.$$

Hence, the square of the magnitude of the total stress on the plane abc, $\overset{n}{T}$, is

$$\begin{aligned} T^2 &= (\sigma_{nx})^2 + (\sigma_{ny})^2 + (\sigma_{nz})^2 \\ &= l^2(\sigma_1)^2 + m^2(\sigma_2)^2 + n^2(\sigma_3)^2. \end{aligned} \quad (3\text{-}42)$$

On the other hand, according to Eq. (3-31), we know that the square of the normal stress on abc is

$$(\sigma_{nn})^2 = \left(l^2\sigma_1 + m^2\sigma_2 + n^2\sigma_3\right)^2.$$

Therefore, the square of the total shear stress on the plane is

$$(\sigma_{nt})^2 = T^2 - (\sigma_{nn})^2$$
$$= l^2(\sigma_1)^2 + m^2(\sigma_2)^2 + n^2(\sigma_3)^2 - (l^2\sigma_1 + m^2\sigma_2 + n^2\sigma_3)^2.$$

Thus conditions[3.2]

$$\frac{d\sigma_{nt}}{dl} = 0 \text{ and } \frac{d\sigma_{nt}}{dm} = 0 \qquad (3\text{-}43)$$

give rise to the extrema of shear stress as

$$\sigma_{nt}\big|_{extrema} = \begin{cases} \pm\frac{1}{2}(\sigma_1 - \sigma_2), \\ \pm\frac{1}{2}(\sigma_2 - \sigma_3), \\ \pm\frac{1}{2}(\sigma_1 - \sigma_3). \end{cases} \qquad (3\text{-}44a, b, c)$$

Their corresponding direction cosines are listed in **Table 3.1**. When compared with the cosines of principal directions listed in **Table 3.2**, we can see clearly that the direction of an extreme shear stress bisects the angle between two principal axes. Since $\sigma_1 \geq \sigma_2 \geq \sigma_3$, the maximum shear stress is

$$\sigma_{nt}\big|_{max} = \frac{1}{2}(\sigma_1 - \sigma_3). \qquad (3\text{-}45)$$

The direction of the maximum shear stress bisects the principal directions 1 and 3.

Table 3.1 The direction cosines of extreme shear stresses

l	0	$\pm 1/\sqrt{2}$	$\pm 1/\sqrt{2}$
m	$\pm 1/\sqrt{2}$	0	$\pm 1/\sqrt{2}$
n	$\pm 1/\sqrt{2}$	$\pm 1/\sqrt{2}$	0

[3.2] Note that only two of the direction cosines, l, m and n, are independent variables, because they are related by Eq. (3-39).

Table 3.2 The direction cosines of principal stresses

l	0	0	±1
m	0	±1	0
n	±1	0	0

Many structures and mechanical components, such as those made of low carbon steel, fail under shear stresses. The determination of maximum shear stress and its directions is of primary importance to design and strength analysis in engineering. We will discuss the failure theory and some practical examples in Chapter 9 of the book.

Example 3.5 A plate of 20 mm thickness carries an outward normal force of 1 MN, which is uniformly distributed, acting outwards on its edge faces as illustrated in **Fig. E3.4**. The lateral faces ABCD and A'B'C'D' are not subjected to forces. If AB = 1 m and BC = 2 m, determine the state of stress and the maximum shear stress, τ_{max}. Find the limits within which the forces in the y-direction could vary without affecting the magnitude of τ_{max}.

Solution: Areas of ABB'A' and CC'D'D are each 0.02 m² so that the normal stress in the x-direction in the plate is

$$\sigma_{xx} = 1 \text{ MN}/0.02 \text{ m}^2 = 50 \text{ MN/m}^2.$$

Similarly,

$$\sigma_{yy} = 25 \text{ MN/m}^2.$$

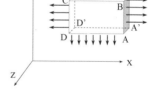

Figure E3.4

Since there are no lateral forces, $\sigma_{zz} = 0$ and thus the state of stress throughout the plate is

$$\begin{pmatrix} 50 & 0 & 0 \\ 0 & 25 & 0 \\ 0 & 0 & 0 \end{pmatrix}.$$

The principal stresses are $\sigma_1 = \sigma_{xx} = 50$ MN/m², $\sigma_2 = \sigma_{yy} = 25$ MN/m² and $\sigma_3 = \sigma_{zz} = 0$ because there is no shear stress on any of the plate surfaces. The maximum shear stress in the plate can be calculated by Eq. (3-45), *i.e.*,

$$\tau_{max} = (\sigma_1 - \sigma_3)/2 = (50 - 0)/2 = 25 \text{ MN/m}^2.$$

If τ_{max} is not to vary then σ_{yy} must remain the intermediate principal stress. Therefore the limits are $0 \le \sigma_{yy} \le 50$ MN/m². Hence, if the forces on surfaces CBC'B' and ADD'A' vary within the limits of zero and 2 MN, the maximum shear stress in the plate does not change.

3.7 EQUATIONS OF MOTION AND EQUILIBRIUM

We have understood many new concepts in the last few sections, such as stresses, their sign definitions, the variation rule with the rotation coordinate system and the physical meaning of principal stresses and their determination. However, hitherto we have only considered stress analysis *at a point* in a solid. We will naturally ask ourselves a question: 'How do stresses vary from one point to another in a solid?' As we know, for example, the stresses at different points in a solid, such as in the dam of a reservoir under water pressure (**Fig. 3.2c**), in a grinding wheel under centrifugal force (**Fig. 3.2d**) and in a bridge subjected to transverse bending (**Fig. 3.3**), will generally change in intensity. To answer the question, we must investigate the conditions that control the way in which stress components vary from one point to another.

The requirement that the laws of motion must be obeyed gives us the means for determining how the stresses vary from point to point. Let us consider first an infinitesimal element of material in a solid of dimensions Δx, Δy and Δz, having its surfaces parallel to the coordinate planes, as shown in **Fig. 3.23**, where $\Delta(\cdots)$ stands for an infinitesimal increment of quantity $(\cdots)$. On the coordinate planes, we have the following stresses:

(1) At $x = 0$, the stresses on plane ABCD are σ_{xx}, σ_{xy} and σ_{xz}.
(2) At $y = 0$, the stresses on plane AEHD are σ_{yy}, σ_{yx} and σ_{yz}.
(3) At $z = 0$, the stresses on plane CDHG are σ_{zz}, σ_{zx} and σ_{zy}.

On the other three planes, the stresses are different, *i.e.*,

(i) At $x = \Delta x$, the stresses on plane EFGH are $\sigma_{xx} + \Delta\sigma_{xx}$, $\sigma_{xy} + \Delta\sigma_{xy}$ and $\sigma_{xz} + \Delta\sigma_{xz}$.
(ii) At $y = \Delta y$, the stresses on plane CBFG are $\sigma_{yy} + \Delta\sigma_{yy}$, $\sigma_{yx} + \Delta\sigma_{yx}$ and $\sigma_{yz} + \Delta\sigma_{yz}$.
(iii) At $z = \Delta z$, the stresses on plane ABFE are $\sigma_{zz} + \Delta\sigma_{zz}$, $\sigma_{zx} + \Delta\sigma_{zx}$ and $\sigma_{zy} + \Delta\sigma_{zy}$.

In addition to the above stresses on the surfaces, the element may also be subjected to a body force. Assume that the body force per unit volume is $\rho\mathbf{f}$, where ρ is the material density at the element. The three components of the body force in coordinate directions are ρf_x, ρf_y and ρf_z.

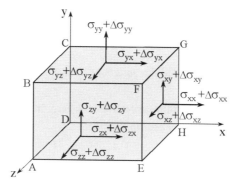

Figure 3.23 Variation of stresses moving from one point to another in a solid body.

Now consider the motion of the element in x-direction. Only the stresses acting in the x-direction can have an effect on the element motion in the direction. Thus according to the law of motion, in x-direction, we have

$$\{(\sigma_{xx}+\Delta\sigma_{xx})\Delta y\Delta z - \sigma_{xx}\Delta y\Delta z\} + \{(\sigma_{xz}+\Delta\sigma_{xz})\Delta x\Delta y - \sigma_{xz}\Delta x\Delta y\} + \\ + \{(\sigma_{xy}+\Delta\sigma_{xy})\Delta x\Delta z - \sigma_{xy}\Delta x\Delta z\} + \rho f_x \Delta x\Delta y\Delta z = a_x \rho \Delta x\Delta y\Delta z,$$

where a_x is the acceleration of the element in x-direction. In writing the above equation, we have used one of the basic assumptions that the material of the solid is homogeneous and thus the density of the material is a constant over the element. The above equation can be simplified to

$$\frac{\Delta\sigma_{xx}}{\Delta x} + \frac{\Delta\sigma_{xy}}{\Delta y} + \frac{\Delta\sigma_{xz}}{\Delta z} + \rho f_x = \rho a_x.$$

Since we have assumed that the material is continuous, proceeding to the limit as the infinitesimal quantities $\Delta(\cdots)$ tends to zero, we obtain

$$\frac{\partial\sigma_{xx}}{\partial x} + \frac{\partial\sigma_{xy}}{\partial y} + \frac{\partial\sigma_{xz}}{\partial z} + \rho f_x = \rho a_x. \tag{3-46a}$$

Similarly, in y- and z-directions, we have

$$\frac{\partial\sigma_{yx}}{\partial x} + \frac{\partial\sigma_{yy}}{\partial y} + \frac{\partial\sigma_{yz}}{\partial z} + \rho f_y = \rho a_y, \tag{3-46b}$$

$$\frac{\partial \sigma_{zx}}{\partial x} + \frac{\partial \sigma_{zy}}{\partial y} + \frac{\partial \sigma_{zz}}{\partial z} + \rho f_z = \rho a_z. \qquad (3\text{-}46c)$$

Equations (3-46a, b, c) are called the *equations of motion* that specify the conditions of stress variation in a solid.

If a solid is in static equilibrium, acceleration is zero. In this case, the equations of motion become the *equations of equilibrium*, i.e.,

$$\begin{cases} \dfrac{\partial \sigma_{xx}}{\partial x} + \dfrac{\partial \sigma_{xy}}{\partial y} + \dfrac{\partial \sigma_{xz}}{\partial z} + \rho f_x = 0, \\[4pt] \dfrac{\partial \sigma_{yx}}{\partial x} + \dfrac{\partial \sigma_{yy}}{\partial y} + \dfrac{\partial \sigma_{yz}}{\partial z} + \rho f_y = 0, \\[4pt] \dfrac{\partial \sigma_{zx}}{\partial x} + \dfrac{\partial \sigma_{zy}}{\partial y} + \dfrac{\partial \sigma_{zz}}{\partial z} + \rho f_z = 0. \end{cases} \qquad (3\text{-}47a, b, c)$$

Further, if body forces are negligible,[3.3] the equations of equilibrium reduce to

$$\begin{cases} \dfrac{\partial \sigma_{xx}}{\partial x} + \dfrac{\partial \sigma_{xy}}{\partial y} + \dfrac{\partial \sigma_{xz}}{\partial z} = 0, \\[4pt] \dfrac{\partial \sigma_{yx}}{\partial x} + \dfrac{\partial \sigma_{yy}}{\partial y} + \dfrac{\partial \sigma_{yz}}{\partial z} = 0, \\[4pt] \dfrac{\partial \sigma_{zx}}{\partial x} + \dfrac{\partial \sigma_{zy}}{\partial y} + \dfrac{\partial \sigma_{zz}}{\partial z} = 0. \end{cases} \qquad (3\text{-}48a, b, c)$$

In two-dimensional stress states, if all the non-vanishing stresses are in xy-plane as we discussed before, the equations of equilibrium are simplified to

$$\begin{cases} \dfrac{\partial \sigma_{xx}}{\partial x} + \dfrac{\partial \sigma_{xy}}{\partial y} = 0, \\[4pt] \dfrac{\partial \sigma_{yx}}{\partial x} + \dfrac{\partial \sigma_{yy}}{\partial y} = 0. \end{cases} \qquad (3\text{-}49a, b)$$

In some cases, such as the pressure vessel problem shown in **Fig. 3.2a**, the polar coordinate system, rθz (see **Fig. 3.7**), is more convenient for analysis because of the

---

[3.3] In many engineering components, the only body force is quite often that due to gravity. In many cases, its effect is negligible.

geometry of the special components. A direct coordinate transformation[3.4] from Eq. (3-46) gives rise to

$$\begin{cases} \dfrac{\partial \sigma_{rr}}{\partial r} + \dfrac{1}{r}\dfrac{\partial \sigma_{r\theta}}{\partial \theta} + \dfrac{\partial \sigma_{rz}}{\partial z} + \dfrac{\sigma_{rr} - \sigma_{\theta\theta}}{r} + \rho f_r = \rho\, a_r, \\[6pt] \dfrac{\partial \sigma_{r\theta}}{\partial r} + \dfrac{1}{r}\dfrac{\partial \sigma_{\theta\theta}}{\partial \theta} + \dfrac{\partial \sigma_{\theta z}}{\partial z} + \dfrac{2\sigma_{r\theta}}{r} + \rho f_\theta = \rho\, a_\theta, \\[6pt] \dfrac{\partial \sigma_{rz}}{\partial r} + \dfrac{1}{r}\dfrac{\partial \sigma_{\theta z}}{\partial \theta} + \dfrac{\partial \sigma_{zz}}{\partial z} + \dfrac{\sigma_{rz}}{r} + \rho f_z = \rho\, a_z. \end{cases} \quad (3\text{-}50\text{a, b, c})$$

These are the equations of motion in the polar coordinate system. Similarly, for cases with in-plane stress states, in rθ-plane say, the above equations become

$$\begin{cases} \dfrac{\partial \sigma_{rr}}{\partial r} + \dfrac{1}{r}\dfrac{\partial \sigma_{r\theta}}{\partial \theta} + \dfrac{\sigma_{rr} - \sigma_{\theta\theta}}{r} + \rho f_r = \rho\, a_r, \\[6pt] \dfrac{\partial \sigma_{r\theta}}{\partial r} + \dfrac{1}{r}\dfrac{\partial \sigma_{\theta\theta}}{\partial \theta} + \dfrac{2\sigma_{r\theta}}{r} + \rho f_\theta = \rho\, a_\theta. \end{cases} \quad (3\text{-}51\text{a, b})$$

Before finishing this chapter, we must mention the name of a French scientist, Augustin Cauchy (1789-1857), who made a great contribution to the development of the theory of elasticity, including the concepts of stress and strain,[3.5] stress transformation Eq. (3-30), the concepts of principal stresses, principal strains and their directions, the equations of equilibrium Eq. (3-47), and the formulation of the linear stress-strain relationship that is commonly called the generalised Hooke's law.[3.6]

Example 3.6 A structure subjected to a set of external loads is in static equilibrium. The engineer who is responsible for the safety of the structure would like to examine the stress distribution throughout the structure so that its performance can be monitored with confidence. He somehow developed the following stress field and wants to know if it is a possible stress distribution. The body forces can be ignored.

$$\sigma_{xx} = axy, \quad \sigma_{xy} = \frac{a}{2}(b^2 - y^2) + cz, \quad \sigma_{xz} = -cy,$$

$$\sigma_{yy} = \sigma_{zz} = \sigma_{yz} = 0,$$

[3.4] Details of the coordinate transformation can be found in any mathematics texts or in many books on elasticity, such as that by Barber (1992).
[3.5] We shall discuss the strain-displacement relationship in Chapter 4.
[3.6] The generalised Hooke's law will be discussed in Chapter 5.

where a, b and c are constants.

Solution: The stress distribution specifies the stress variation from one point to the other in the structure. If the given distribution is a possible stress field for the structure under static equilibrium, it must satisfy the equations of equilibrium, Eq. (3-48), because the body forces are negligible in the present case. The substitution of the above stress field into Eq. (3-48) shows that all three equations are satisfied. Hence, it is a possible stress distribution.

Example 3.7 A thin rectangular plate, as shown in **Fig. E3.5**, is in static equilibrium and body forces are negligible. The stress field in the plate has been found to be

$$\sigma_{xx} = qxy, \quad \sigma_{yy} = 0, \quad \sigma_{xy} = c\left(\frac{h^2}{4} - y^2\right),$$

$$\sigma_{zz} = \sigma_{zx} = \sigma_{zy} = 0,$$

where q is a known positive constant and c is an unknown constant. Determine c and illustrate graphically the stresses on the plate edges.

Solution: The variation of the stresses throughout the plate must satisfy the equations of equilibrium. However, the satisfaction of Eq. (3-48) requires that

$$qy - 2cy = 0.$$

Since y cannot be always zero, we must have c = q/2.

Now let us plot the stress distribution around the plate edges. On the edge of x = 0 and $-h/2 \le y \le h/2$, the only non-vanishing stress is

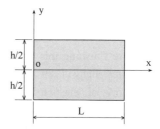

Figure E3.5

$$\sigma_{xy} = \frac{q}{2}\left(\frac{h^2}{4} - y^2\right),$$

which is always positive because the magnitude of y is always less or equal to h/2. On the edge of x = L and $-h/2 \le y \le h/2$, the non-zero stresses are

$$\sigma_{xx} = qLy, \quad \sigma_{xy} = \frac{q}{2}\left(\frac{h^2}{4} - y^2\right).$$

It is clear that on this edge, σ_{xy} is positive but the sign of σ_{xx} depends on that of y. Finally, on edges of y = -h/2 and h/2 with $0 \le x \le L$, both σ_{yy} and σ_{yx} are zero. Based on the above, the normal and shear stresses on the plate edges can be drawn individually as shown in **Fig. E3.6**.

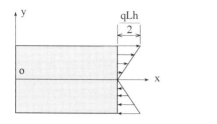
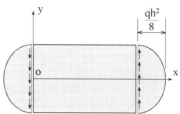

(a) distribution of normal stress σ_{xx}　　　(b) distribution of shear stress σ_{xy}

Figure E3.6 Stresses on the plate edges.

References

Barber, JR (1992), *Elasticity*, Kluwer Academic Publishers, Dordrecht.
Chou, PC (1967), *Elasticity: Tensor, Dyadic and Engineering Approaches*, Van Nostrand, Princeton, NJ.
Flugge, W (1972), *Tensor Analysis and Continuum Mechanics*, Springer-Verlag, Berlin.
Fung, YC (1965), *Foundations of Solid Mechanics*, Prentice-Hall, Inc., Englewood Cliffs, NJ.
Hibbeler, RC (1997), *Mechanics of Materials*, Prentice-Hall, Inc., Englewood Cliffs, NJ, pp.199.
Timoshenko, SP (1953), *History of Strength of Materials*, McGraw-Hill, New York.
Young, JF, Mindess, S, Gray, RJ and Bentur, A (1998), *The Science and Technology of Civil Engineering Materials*, Prentice Hall Inc, New Jersey.
Zhang, L and Tanaka, H (1999), On the mechanics and physics in the nano-indentation of silicon monocrystals, *JSME International Journal*, **A42** (4), 546-559.

Important Concepts

Stress vector	Stress matrix
Symmetry of stress	Sign of stresses
Principal stresses	Principal directions and planes
Stress in any direction	Maximum shear stress
Stress invariant	Stress transformation
Equations of motion	Equations of equilibrium

Questions

3.1 Why can we define a stress vector at a mathematical point in a solid?
3.2 Why do we need six independent stress components to describe the stress-state at a point?
3.3 How do we define the signs of normal and shear stresses?
3.4 What is the difference between the mathematical and physical definitions of stress? What is the dimension of stress?
3.5 What is the physical meaning of a principal stress and what is its meaning mathematically?
3.6 How do we determine principal stresses and their directions?
3.7 What are the relationships between extreme shear stresses and principal stresses? What are the relationships between their directions?
3.8 Why do we have stress invariants?
3.9 What are the conditions that control the variation of stresses from one point to another in a solid?
3.10 What is the difference between the stresses on the element in **Fig. 3.6** and those on the element in **Fig. 3.23**? Why are they different?
3.11 In stress analysis, what basic assumptions introduced in Chapter 2 have been used?

Problems

3.1 Illustrate graphically the following stress states (units MN/m^2):

(a) $\begin{pmatrix} 0 & 2 & 4 \\ 2 & 4 & 0 \\ 4 & 0 & 2 \end{pmatrix}$, (b) $\begin{pmatrix} 2 & 1 & 2 \\ 1 & -1 & 0 \\ 2 & 0 & -2 \end{pmatrix}$, (c) $\begin{pmatrix} 2 & -2 & 0 \\ -2 & 4 & 0 \\ 0 & 0 & 5 \end{pmatrix}$, (d) $\begin{pmatrix} 2 & -3 & 0 \\ -3 & 4 & 0 \\ 0 & 0 & 0 \end{pmatrix}$.

3.2 Write down the stress matrices for the states of stress illustrated in **Fig. P3.1**. Take the numerical magnitude of each non-zero stress component to be unity. (Note that only the stress components on the positive surfaces of the infinitesimal elements are illustrated in the figure.)

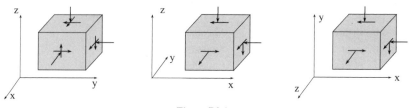

Figure P3.1

3.3 Transform the following stress matrices for changes of axis of (a) 60°, (b) 45° and (c) 20° (units MN/m²). Determine the orientation of the principal axes and deduce the principal stresses for the stress state described in (c).

(a)
$$\begin{pmatrix} 100 & 0 & 0 \\ 0 & -100 & 0 \\ 0 & 0 & 50 \end{pmatrix},$$

(b)
$$\begin{pmatrix} 100 & 0 & 0 \\ 0 & 0 & 0 \\ 0 & 0 & 50 \end{pmatrix},$$

(c)
$$\begin{pmatrix} -40 & -20 & 0 \\ -20 & 60 & 0 \\ 0 & 0 & -50 \end{pmatrix}.$$

3.4 Using an experimental measurement technique, an engineer obtained the stresses at points (a) and (b) of a loaded structure as follows:

(a)
$$\begin{pmatrix} 10 & 20 & 0 \\ 22 & -100 & 1.2 \\ 0 & 0.9 & 50 \end{pmatrix},$$

(b)
$$\begin{pmatrix} 1.9 & 3.1 & 0 \\ 3.2 & 0 & 10 \\ 0 & 8 & 15 \end{pmatrix},$$

Can you find any problem due to his measurement bearing in mind the property that a stress matrix must have?

3.5 A rectangular thin plate in the yz-plane is under a uniformly distributed stress around its edge, as shown in **Fig. P3.2**. The front and back surfaces of the plate are free from stress. Find the stress components on the plane A-A.

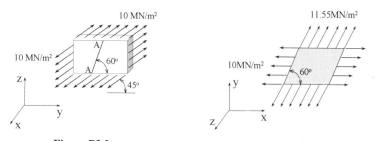

Figure P3.2   Figure P3.3

3.6 A thin plate in the xy-plane is under a uniform state of stress as shown in **Fig. P3.3**. (i) Among all planes that are normal to the xy-plane, find the one on which the maximum shear stress acts and the magnitude of this maximum shear stress. (ii) Find the greatest of all shear stresses on any plane (three-dimensional).

3.7 The thin plate shown in **Fig. P3.4** is under uniform shear stress and that in **Fig. P3.5** is subjected to uniform normal stress around edges. Find the principal stresses and their directions.

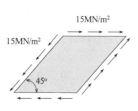

Figure P3.4

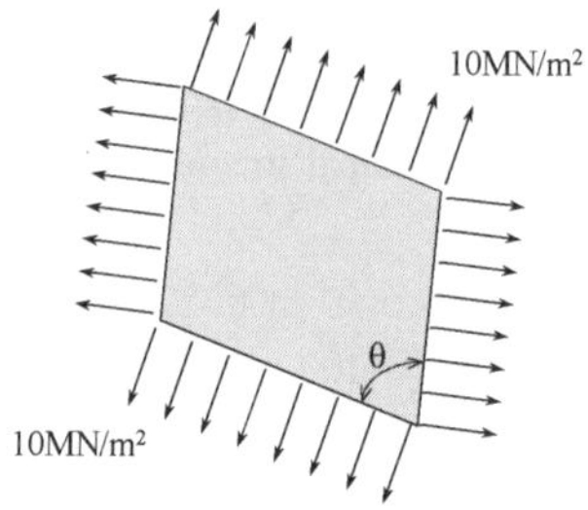

Figure P3.5

3.8 It is known that at a surface point of a solid circular cylinder subjected to tension, bending and torsion, the stresses are $\sigma_{rr} = \sigma_{rz} = \sigma_{r\theta} = \sigma_{\theta\theta} = 0$, $\sigma_{zz} = a$, $\sigma_{z\theta} = 2a$, where a is a known constant. Find the principal stresses at the point.

3.9 A punch tool is loaded by a hydraulic piston, as shown in **Fig. P3.6**, with a liquid pressure of 10 MN/m². If the tool diameter is 10 mm and that of the cylinder is 50 mm, what is the maximum shear stress in the tool?

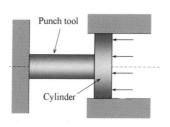

Figure P3.6

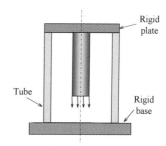

Figure P3.7

3.10 Fig. P3.7 shows a half-sectional elevation of a tube resting on a plane and closed by a rigid plate from which a rod is suspended along the tube axis. The lower end of the rod is subjected to a uniformly distributed normal force. Find the maximum permissible value of this load if the shear and compressive stresses in both the rod and tube are not to exceed 100 MN/m² and 80 MN/m² respectively. The tube and rod have cross-sectional areas of 2000 mm² and 1000 mm² respectively.

3.11 A pulley that is 0.6 m in diameter is mounted on a 50 mm circular shaft. Between two bearings that are 2.4 m apart, the pulley is 0.9 m from one of the bearings. Power is transmitted through the 3-foot portion of the shaft such that the taut side of the belt has a tension of 1.5 kN and the slack side 450 N. Calculate the maximum tensile, compressive and shear stresses in the shaft.

3.12 Show that the following quantities are invariants for a two-dimensional state of stress with $\sigma_{zz} = \sigma_{zx} = \sigma_{zy} = 0$:

(a) $\sigma_{x'x'} + \sigma_{y'y'}$ (b) $\sigma_{x'x'} \sigma_{y'y'} - (\sigma_{x'y'})^2$.

3.13 Show that stress components under the Cartesian and Polar coordinate systems have the following relationships:

$$\sigma_{xx} = \sigma_{rr}\cos^2\theta + \sigma_{\theta\theta}\sin^2\theta - 2\sigma_{r\theta}\sin\theta\cos\theta,$$

$$\sigma_{yy} = \sigma_{rr}\sin^2\theta + \sigma_{\theta\theta}\cos^2\theta + 2\sigma_{r\theta}\sin\theta\cos\theta,$$

$$\sigma_{zz} = \sigma_{zz},$$

$$\sigma_{xy} = \sigma_{rr}\sin\theta\cos\theta - \sigma_{\theta\theta}\sin\theta\cos\theta + \sigma_{r\theta}\left(\cos^2\theta - \sin^2\theta\right),$$

$$\sigma_{yz} = \sigma_{\theta z}\cos\theta + \sigma_{zr}\sin\theta,$$

$$\sigma_{yz} = -\sigma_{\theta z}\sin\theta + \sigma_{zr}\cos\theta.$$

3.14 The stresses in the pressure vessel illustrated in **Fig. 3.2a** can be obtained as

$$\sigma_{rr} = A + \frac{B}{r^2}, \quad \sigma_{\theta\theta} = A - \frac{B}{r^2}, \quad \sigma_{zz} = C, \quad \sigma_{r\theta} = \sigma_{\theta z} = \sigma_{zr} = 0,$$

where A, B and C are constants. Find the stresses in the pressure vessel under the Cartesian coordinate system xyz.

3.15 Is the following stress distribution possible for a body in equilibrium?

$$\sigma_{xx} = 3x^2 + 3y^2 - z, \quad \sigma_{xy} = z - 6xy - 3/4,$$

$$\sigma_{yy} = 3y^2, \quad \sigma_{xz} = x + y - 3/2,$$

$$\sigma_{zz} = 3x + y - z + 5/4, \quad \sigma_{yz} = 0.$$

For the state of stress at the specific point $(x, y, z) = (0.5, 1, 0.75)$, determine the principal stresses and their direction cosines.

3.16 Given the following stress distribution,

$$\sigma_{xx} = 3x^2 + 4xy - 8y^2, \quad \sigma_{yy} = 2x^2 + xy + 3y^2,$$

$$\sigma_{xy} = -\frac{1}{2}x^2 - 6xy - 2y^2, \quad \sigma_{zz} = \sigma_{xz} = \sigma_{yz} = 0,$$

determine, in the absence of body forces, whether equilibrium exists.

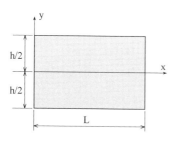

Figure P3.8

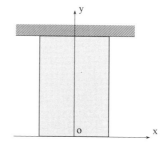

Figure P3.9

3.17 Is the following stress distribution, where q is a known constant, a possible stress field in the thin plate illustrated in **Fig. P3.8**? Determine constants c_1 and c_2 when body forces are negligible. Illustrate graphically the stresses on the plate edges.

$$\sigma_{xx} = qx^2y - \frac{2}{3}qy^3, \quad \sigma_{yy} = \frac{1}{3}qy^3 - c_1y + c_2,$$

$$\sigma_{xy} = -qxy^2 + c_1x, \quad \sigma_{zz} = \sigma_{xz} = \sigma_{yz} = 0.$$

3.18 A thin rectangular plate is suspended as illustrated in **Fig. P3.9**. The density of the plate material is ρ and the only external load applied on the plate is its own weight. Determine constants c_1 and c_2 in the following stress field that represents the stress distribution in the plate.

$$\sigma_{xx} = 0, \quad \sigma_{yy} = c_1y + c_2,$$

$$\sigma_{xy} = \sigma_{zz} = \sigma_{xz} = \sigma_{yz} = 0.$$

Chapter 4

DISPLACEMENT AND STRAIN

Like stress, displacement and strain are also fundamental concepts in solid mechanics. This chapter discusses the method used to describe these important mechanics quantities, investigates their relationships and examines the compatibility of deformation. We shall see that strain transformation follows exactly the same rule as that of stress.

In some applications we need to make use of deformation, but in some other cases we must avoid deformation-caused problems. The car body shown above was manufactured by introducing large plastic deformation using the technique of sheet metal forming. Car body panels after forming must be able to keep their shapes unchanged to guarantee a successful assembly.

In the last chapter, we discussed the stresses at a point in a solid subjected to external forces and obtained the equations of motion that specify the conditions of stress variation in the solid. In the meantime, however, we can easily see that the solid must also deform under the external forces. It is therefore necessary to know how to quantitatively describe deformation. This is the task of this chapter.

4.1 DISPLACEMENT

When a solid is subjected to external loading, the positions of material points in the solid will change, *i.e.*, displacement occurs. There are two types of displacements. The first corresponds to a rigid body rotation, or translation, or both. In this case, although all the points in the solid move, their relative positions do not vary and thus the shape of the solid is unchanged. We call this type of displacement *rigid body displacement*. On the other hand, if the relative positions of material points in the solid change during a motion, the shape of the solid must have changed. Thus the displacements are the result of *deformation* in the solid. In many cases, rigid body motion and deformation take place at the same time.

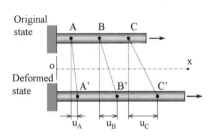
Figure 4.1 Displacement in a bar.

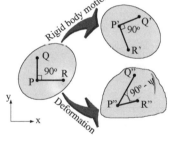
Figure 4.2 Type of displacements in a solid.

The example in **Fig. 4.1** shows both the original and deformed states of a uniform bar subjected to a longitudinal tension when the left end is fully clamped so that rigid body displacement does not appear. The points A, B and C on the bar in its undeformed state move to A', B' and C', respectively after deformation. Correspondingly, their displacements are u_A, u_B and u_C and as we can see easily they are unequal. *It is generally true that displacements in a solid subjected to deformation are non-uniform and are functions of coordinates*. In the uni-axial tension of the bar shown in **Fig. 4.1**,

displacement at any point in x-direction, u, is a function of coordinate x, *i.e.*, u = u(x). In a solid under two-dimensional deformation in the xy-plane, the displacement in the x-direction, u, and that in the y-direction, v, are the functions of x and y, *i.e.*, u = u(x, y) and v = v(x, y). Similarly, in the case with three-dimensional deformation under the coordinate system xyz, the displacements in x-, y- and z-directions, u, v and w, are generally the functions of x, y and z, and can be written as u = u(x, y, z), v = v (x, y, z) and w = w(x, y, z), respectively.

The in-plane deformation illustrated in **Fig. 4.2** can help us to understand more clearly the above two types of displacements in a solid, rigid body displacement and *deformation induced displacement*. Under a rigid body motion, points P, Q and R move to P', Q' and R'. Although displacements happen, no deformation occurs because the segment lengths of PQ and PR do not change, *i.e.*, PQ = P'Q' and PR = P'R', and the included angle ∠QPR keeps constant, *i.e.*, ∠QPR = ∠Q'P'R'. However, when deformation happens, both the segment lengths and the included angle vary. In the case shown in the lower part of **Fig. 4.2**, P"Q" becomes longer than PQ and P"R" becomes shorter than PR. The included angle of the segments experiences a reduction of ψ.

Since rigid body displacements do not introduce deformation in a body, they do not relate to stresses. For the sake of convenience, in all the cases to be discussed in this book we assume that displacements u, v and w are caused by deformation only. In other words, we have assumed that there are always enough constraints to prevent a body from moving as a rigid body.

4.2 STRAIN

As pointed out above, deformation in a solid will introduce shape change or distortion. More precisely, deformation changes the relative distance of two end points of a segment and the included angle of two segments in the solid. It is therefore necessary to introduce physical quantities, called *strains*, to provide a measure of the relative deformation of neighbouring points during a distortion process.

Figure 4.2 shows that to describe a strain state we must introduce two kinds of strains, a *direct strain* (or a *normal strain*), ε, to measure the relative change of distance between two neighbouring points and a *shear strain*, γ, to measure the relative change of the angle between two directions. When deformation is small, as we have specified in Chapter 2, the direct and shear strains can be defined as

$$\varepsilon = \frac{l - l_0}{l_0},$$

$$\gamma = \frac{1}{2}\tan\psi \cong \frac{1}{2}\psi,$$

(4-1a, b)

where l_0 is the length of a segment before deformation, *e.g.*, the length of PQ in **Fig. 4.2** and l is the length of the segment after deformation, *e.g.*, that of P"Q" in **Fig. 4.2**. Clearly when l is larger than l_0, *i.e.*, when the segment is subjected to elongation, the direct strain ε is positive. Otherwise, when the segment is under compression, l becomes shorter than l_0 and ε is negative. The sign of the shear strain, γ, is not implied by the definition itself. We thus define that when a shear deformation causes an angle reduction, *i.e.*, when $\angle Q"P"R"$ becomes smaller than $\angle QPR$, the corresponding shear strain is positive. Otherwise if $\angle Q"P"R"$ becomes larger, γ is negative.

The above concepts of direct and shear strains are straightforward.[4.1] However, we still have the following question to answer: How can we describe the strains at any point in any direction throughout a solid since the measurement of the relative distance change of any two points in a body is difficult? Practically, it is easier to measure the displacements at a point. When the displacements are all known, based on the definitions, strains can be calculated under a reference coordinate frame. To answer the above question, therefore, we must find the relationships between strains and displacements. Again, we will limit our attention to small deformation cases, *i.e.*, the displacements in a solid caused by deformation are much smaller than the minimum dimension of the solid.

4.3 STRAIN-DISPLACEMENT RELATIONS UNDER SMALL DEFORMATION

As mentioned above, when a body undergoes in-plane deformation in the xy-plane, the displacements at any point P in the body are $u = u(x, y)$ in the x-direction and $v = v(x, y)$ in the y-direction. As illustrated in **Fig. 4.3**, the displacement in the x-direction of an adjacent point R should be

$$u + \Delta u = u + \frac{\Delta u}{\Delta x}(\Delta x),$$

[4.1] The strain gauge method for strain measurement is directly based on the definition of ε. Details about the strain gauge technique can be found in relevant textbooks, such as those by Murray (1992) and Window and Holister (1982).

where the increment of displacement, $\Delta u = \dfrac{\Delta u}{\Delta x}(\Delta x)$, is due to the coordinate change from point P, which is at x, to point R, which is at x+Δx. When deformation is small, the effect of the angle between P'R' and PR on the length change of P'R' is negligible. Hence, according to Eq. (4-1a), the direct strain in segment PR due to deformation in the x-direction is

$$\varepsilon = \lim_{\Delta x \to 0}\left(\dfrac{P'R' - PR}{PR}\right) = \lim_{\Delta x \to 0}\left\{\dfrac{\left(\Delta x + u + \dfrac{\Delta u}{\Delta x}(\Delta x) - u\right) - \Delta x}{\Delta x}\right\} = \dfrac{\partial u}{\partial x}.$$

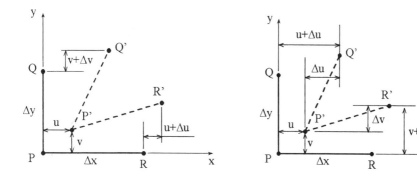

Figure 4.3 Displacements in two-dimensional deformation.

Figure 4.4 Angle change between two segments in the xy-plane.

Since this strain is in x-direction, we denote it as ε_{xx}, *i.e.*,

$$\varepsilon_{xx} = \dfrac{\partial u}{\partial x}. \qquad (4\text{-}2)$$

Similarly, if we consider the length change of segment PQ that is in the y-direction and note that the displacement of point Q in the y-direction is

$$v + \Delta v = v + \dfrac{\Delta v}{\Delta y}(\Delta y),$$

which is due to coordinate change from point P at y to point Q at y+Δy, we can easily find that the direct strain in the y-direction is

$$\varepsilon_{yy} = \frac{\partial v}{\partial y}. \tag{4-3}$$

Equations (4-2) and (4-3) show the relationship between direct strains in the xy-plane with the displacement components.

Now let us consider the distortion of the angle between segments PR and PQ during deformation. As shown in **Fig. 4.4**, the total angle change from ∠QPR to ∠Q'P'R' is

$$\frac{\Delta u}{\Delta y} + \frac{\Delta v}{\Delta x}$$

when deformation is small. (The angle change is also small.) According to the definition of shear strain given by Eq. (4-1b), we have

$$\gamma = \frac{1}{2} \lim_{\substack{\Delta x \to 0 \\ \Delta y \to 0}} \left(\frac{\Delta u}{\Delta y} + \frac{\Delta v}{\Delta x} \right) = \frac{1}{2} \left(\frac{\partial u}{\partial y} + \frac{\partial v}{\partial x} \right).$$

Since the shear strain is in the xy-plane, we denote it as ε_{xy}, *i.e.*,

$$\varepsilon_{xy} = \frac{1}{2} \left(\frac{\partial u}{\partial y} + \frac{\partial v}{\partial x} \right). \tag{4-4}$$

Thus to describe a plane deformation in the xy-plane, we need three independent strain components, ε_{xx}, ε_{yy} and ε_{xy}.

Under three-dimensional deformation, displacements at a point are generally functions of x, y and z, as discussed in section **4.1**. We can follow the same procedure above to analyse the length and angle changes of segments in the xz- and yz-planes individually. As a result, we obtain the following strain components:

$$\varepsilon_{zz} = \frac{\partial w}{\partial z}, \ \varepsilon_{yx} = \frac{1}{2}\left(\frac{\partial u}{\partial y} + \frac{\partial v}{\partial x}\right) = \varepsilon_{xy}, \ \varepsilon_{yz} = \frac{1}{2}\left(\frac{\partial w}{\partial y} + \frac{\partial v}{\partial z}\right) = \varepsilon_{zy}, \ \varepsilon_{zx} = \frac{1}{2}\left(\frac{\partial u}{\partial z} + \frac{\partial w}{\partial x}\right) = \varepsilon_{zx}.$$
(4-5)

In a matrix form, the strain state in three-dimension can be expressed as

$$\varepsilon = \begin{pmatrix} \varepsilon_{xx} & \varepsilon_{xy} & \varepsilon_{xz} \\ \varepsilon_{yx} & \varepsilon_{yy} & \varepsilon_{yz} \\ \varepsilon_{zx} & \varepsilon_{zy} & \varepsilon_{zz} \end{pmatrix}.$$
(4-6)

Equation (4-5) indicates that the strain matrix is symmetrical. Similar to the stresses discussed in Chapter 3, it can be shown mathematically that the strain of Eq. (4-6) is also a tensor of rank two (Fung, 1965). *Hence all the transformation rules obtained in Chapter 3 for stresses also apply to the transformation of strains provided that σ in those formulae is replaced by ε correspondingly.*

In summary, to describe the general strain state at a point under small deformation, we need six independent strain components, *i.e.*, three direct strains and three shear strains, that have the following relationships with the three displacements in coordinate directions:

$$\begin{cases} \varepsilon_{xx} = \dfrac{\partial u}{\partial x}, \\ \varepsilon_{yy} = \dfrac{\partial v}{\partial y}, \\ \varepsilon_{zz} = \dfrac{\partial w}{\partial z}, \\ \varepsilon_{xy} = \dfrac{1}{2}\left(\dfrac{\partial u}{\partial y} + \dfrac{\partial v}{\partial x}\right), \\ \varepsilon_{yz} = \dfrac{1}{2}\left(\dfrac{\partial w}{\partial y} + \dfrac{\partial v}{\partial z}\right), \\ \varepsilon_{zx} = \dfrac{1}{2}\left(\dfrac{\partial u}{\partial z} + \dfrac{\partial w}{\partial x}\right). \end{cases}$$
(4-7)

In the polar coordinate system rθz, the above relationships become

$$\begin{cases}
\varepsilon_{rr} = \dfrac{\partial u}{\partial r}, \\
\varepsilon_{\theta\theta} = \dfrac{1}{r}\dfrac{\partial v}{\partial \theta} + \dfrac{u}{r}, \\
\varepsilon_{zz} = \dfrac{\partial w}{\partial z}, \\
\varepsilon_{r\theta} = \dfrac{1}{2}\left(\dfrac{\partial v}{\partial r} + \dfrac{1}{r}\dfrac{\partial u}{\partial \theta} - \dfrac{v}{r}\right), \\
\varepsilon_{\theta z} = \dfrac{1}{2}\left(\dfrac{1}{r}\dfrac{\partial w}{\partial \theta} + \dfrac{\partial v}{\partial z}\right), \\
\varepsilon_{rz} = \dfrac{1}{2}\left(\dfrac{\partial u}{\partial z} + \dfrac{\partial w}{\partial r}\right).
\end{cases} \qquad (4\text{-}8)$$

In Eq. (4-8), we have assumed that u is the displacement in the r-direction, v is that in the θ-direction and w is that in the z-direction. Since the strain-displacement relations, Eqs. (4-7) and (4-8), are obtained based on the geometric change of an element in a solid subjected to deformation, they are often called *geometric equations* in the literature.

Example 4.1 A rectangular parallelepiped of infinitesimal dimensions, l_x, l_y and l_z is subjected to a state of small strain expressed by the strain matrix of Eq. (4-6). Find the relative volume change of the element during deformation.

Solution: The volume of the element before deformation is $V_0 = l_x l_y l_z$. After deformation, the side lengths in the x-, y- and z-directions will become $l_x(1 + \varepsilon_{xx})$, $l_y(1 + \varepsilon_{yy})$ and $l_z(1 + \varepsilon_{zz})$, respectively. Because deformation is small, the relative angle changes of the element surfaces can be ignored in calculating the volume. Thus the volume of the element after deformation becomes $V = V_0(1 + \varepsilon_{xx})(1 + \varepsilon_{yy})(1 + \varepsilon_{zz})$. If we ignore the products of strains when expanding the expression of V, which is reasonable because we are dealing with small deformation problems, we get $V = V_0 + V_0(\varepsilon_{xx} + \varepsilon_{yy} + \varepsilon_{zz})$. Hence, the relative volume change of the element due to deformation is

$$\Delta = \dfrac{V - V_0}{V_0} = \varepsilon_{xx} + \varepsilon_{yy} + \varepsilon_{zz}.$$

With the above formula, the relative volume change of an infinitesimal element at a point in a solid subjected to deformation can be easily obtained. For example, if the state of strain at a point in a structure is measured to be

$$\begin{pmatrix} 10 & 2 & 4 \\ 2 & -6 & 8 \\ 4 & 8 & 15 \end{pmatrix} \times 10^{-5},$$

then the relative volume change at the point, $\Delta = (10 - 6 + 15) \times 10^{-5} = 1.9 \times 10^{-4}$, is positive. This indicates that the volume of the element is increasing under the specified straining. However, if the state of strain is measured to be

$$\begin{pmatrix} -10 & 2 & 4 \\ 2 & -6 & 8 \\ 4 & 8 & 15 \end{pmatrix} \times 10^{-5},$$

the relative volume change at the point, $\Delta = (-10 - 6 + 15) \times 10^{-5} = -1 \times 10^{-5}$, becomes negative. In this case, the volume of the element is decreasing. Δ is often called the *dilatation* or *volume strain* of a material.

4.4 PRINCIPAL STRAINS AND THEIR DIRECTIONS

Since the transformation rule is the same for both strain and stress, it is natural that the principal strains, ε_1, ε_2 and ε_3, and their direction cosines, l, m and n, can be obtained by using the identical equations for principal stresses and their directions, Eq. (3-37) to Eq. (3-40), provided that σ in those equations are all replaced by ε. For example, principal strains are the real roots of equation

$$\varepsilon^3 - I_1^\varepsilon \varepsilon^2 + I_2^\varepsilon \varepsilon - I_3^\varepsilon = 0, \tag{4-9}$$

where I_1^ε, I_2^ε and I_3^ε are the first, second and third *strain invariants*. According to Eqs. (3-38) and (3-41), they can be expressed as

$$\begin{cases} I_1^\varepsilon = \varepsilon_{xx} + \varepsilon_{yy} + \varepsilon_{zz} \\ \quad = \varepsilon_1 + \varepsilon_2 + \varepsilon_3, \\ I_2^\varepsilon = \varepsilon_{xx}\varepsilon_{yy} + \varepsilon_{yy}\varepsilon_{zz} + \varepsilon_{zz}\varepsilon_{xx} - (\varepsilon_{xy})^2 - (\varepsilon_{yz})^2 - (\varepsilon_{zx})^2 \\ \quad = \varepsilon_1\varepsilon_2 + \varepsilon_2\varepsilon_3 + \varepsilon_3\varepsilon_1, \\ I_3^\varepsilon = \varepsilon_{xx}\varepsilon_{yy}\varepsilon_{zz} + 2\varepsilon_{xy}\varepsilon_{yz}\varepsilon_{zx} - \varepsilon_{xx}(\varepsilon_{yz})^2 - \varepsilon_{yy}(\varepsilon_{zx})^2 - \varepsilon_{zz}(\varepsilon_{xy})^2 \\ \quad = \varepsilon_1\varepsilon_2\varepsilon_3. \end{cases} \quad (4\text{-}10)$$

By comparing the expression of I_1^ε in Eq. (4-10) with that of Δ in Example 4.1, we find that the physical meaning of the first strain invariant, I_1^ε, is the relative volume change of an infinitesimal element at a point due to straining.

Mathematically, principal strains are the eigenvalues of the strain matrix, Eq. (4-6), and the corresponding principal direction cosines form the eigenvector. Physically, on the principal strain planes, shear strains vanish. Under a two-dimensional deformation in xy-plane, the strain matrix of Eq. (4-6) reduces to

$$\varepsilon = \begin{pmatrix} \varepsilon_{xx} & \varepsilon_{xy} \\ \varepsilon_{yx} & \varepsilon_{yy} \end{pmatrix} \quad (4\text{-}11)$$

Thus in this case, Eqs. (3-14) to (3-21) in stress analysis also apply to strain analysis when σ is replaced by ε. Similarly, the extrema of shear strains also have the same relationships with principal strains as those for stresses in Eqs. (3-44) and (3-45). Correspondingly, we also define $\varepsilon_1 \geq \varepsilon_2 \geq \varepsilon_3$.

Example 4.2 A thin plate is subjected to a set of external loads within the xy-plane as illustrated in **Fig. E4.1**. Using the strain gauge rosette shown in the figure, the direct strains at a point are measured to be $\varepsilon_A = -1 \times 10^{-4}$, $\varepsilon_B = 1 \times 10^{-4}$ and $\varepsilon_C = 1.8 \times 10^{-4}$. Find ε_{xx}, ε_{yy} and ε_{xy} at this point of the plate.

Solution: Since the rule of strain transformation is the same as that for stresses, using Eq. (3-14a) the direct strain in any direction *n* with an angle θ to the x-axis can be obtained as

$$\varepsilon_{nn} = \varepsilon_{xx}\cos^2\theta + \varepsilon_{yy}\sin^2\theta + 2\varepsilon_{xy}\cos\theta\sin\theta.$$

Based on the coordinate system in **Fig. E4.1**, it is clear that for strain gauges A, B and C, we have $\theta_A = 0°$,

$\theta_B = 120°$ and $\theta_C = 240°$, respectively. Thus the above transformation formula gives rise to

$$\varepsilon_A = \varepsilon_{xx}, \quad \varepsilon_B = \frac{1}{4}\varepsilon_{xx} + \frac{3}{4}\varepsilon_{yy} - \frac{\sqrt{3}}{2}\varepsilon_{xy}, \quad \varepsilon_C = \frac{1}{4}\varepsilon_{xx} + \frac{3}{4}\varepsilon_{yy} + \frac{\sqrt{3}}{2}\varepsilon_{xy}.$$

Hence,

$$\varepsilon_{xx} = \varepsilon_A = -1 \times 10^{-4},$$

$$\varepsilon_{yy} = 2(\varepsilon_B + \varepsilon_C) - \varepsilon_A = 2.6 \times 10^{-4},$$

$$\varepsilon_{xy} = \frac{\sqrt{3}}{3}(\varepsilon_C - \varepsilon_B) = 0.46 \times 10^{-4}.$$

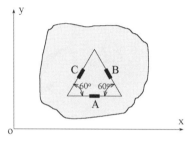

Figure E4.1

4.5 COMPATIBILITY OF STRAINS

The relationships among displacements and strains, Eq. (4-7) or (4-8), show that if we know displacements, we can calculate strains easily because it is only a process of differentiation. However, if we know the six strains how can we uniquely get the *three displacements* by integrating the *six partial differential equations*?

Mathematically, since there are six equations for three unknown functions, u, v and w, the system of Eqs. (4-7) will not generally have a single-valued solution, if the strains are arbitrarily assigned. We therefore must expect that a solution may exist only if the strains satisfy certain conditions.

Physically, as we understood in defining a strain, since strain components only determine the relative positions of points in a solid, and since any rigid-body motion corresponds to zero strain, we can see that the displacement solution of Eq. (4-7) can be determined only up to an arbitrary rigid body motion. Hence, if the strains vary without following a rule, we may end with a problem. For instance, because we are considering a continuous body that must remain continuous during the process of deformation, a continuous triangle (portion of material in the body) in the unstrained state must still be a closed and continuous shape during deformation so that displacements at any point are continuous and single-valued, as shown in **Fig. 4.5**. But if strains were specified arbitrarily, we may expect that by following the strain field starting from point *a*, we might end at point *c* and *d* either with overlapping of material or a gap. This cannot be avoided unless the specified strains along the edges of the triangle obey certain conditions. We call these conditions the *compatibility conditions of deformation*.

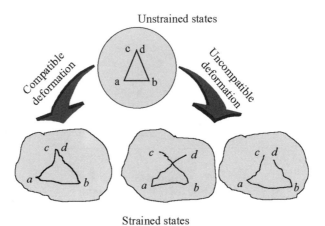

Figure 4.5 Deformation compatibility.

To facilitate our understanding, let us first discuss the compatibility conditions under two-dimensional deformation in the xy-plane. In this case, we have only three independent strain components, ε_{xx}, ε_{yy} and ε_{xy}, and two displacement components, u and v, which are all functions of coordinates x and y. We have three equations, *i.e.*,

$$\begin{cases} \varepsilon_{xx} = \dfrac{\partial u}{\partial x}, \\ \varepsilon_{yy} = \dfrac{\partial v}{\partial y}, \\ \varepsilon_{xy} = \dfrac{1}{2}\left(\dfrac{\partial u}{\partial y} + \dfrac{\partial v}{\partial x}\right), \end{cases} \quad (4\text{-}12a, b, c)$$

but two unknowns, u and v. Thus to obtain a single-valued solution of u and v, an additional relation must exist among strain components. If we differentiate the two sides of Eq. (4-12a) twice with respect to y, those of Eq. (4-12b) twice with respect to x, but those of Eq. (4-12c) once about x and once about y, we obtain

$$\begin{cases} \dfrac{\partial^2 \varepsilon_{xx}}{\partial y^2} = \dfrac{\partial^3 u}{\partial x \partial y^2}, \\ \dfrac{\partial^2 \varepsilon_{yy}}{\partial x^2} = \dfrac{\partial^3 v}{\partial x^2 \partial y}, \\ \dfrac{\partial^2 \varepsilon_{xy}}{\partial x \partial y} = \dfrac{1}{2}\left(\dfrac{\partial^3 v}{\partial x^2 \partial y} + \dfrac{\partial^3 u}{\partial x \partial y^2}\right). \end{cases}$$

It is then clear that

$$\frac{\partial^2 \varepsilon_{xx}}{\partial y^2} + \frac{\partial^2 \varepsilon_{yy}}{\partial x^2} = 2\frac{\partial^2 \varepsilon_{xy}}{\partial x \partial y}. \tag{4-13}$$

This is the compatibility condition that strain components under two-dimensional deformation in xy-plane must follow, which guarantees a continuous deformation in the xy-plane and provides a single-valued solution of u and v. We call Eq. (4-13) the *compatibility equation* of deformation in the xy-plane. In a polar coordinate system, it becomes

$$r\frac{\partial^2}{\partial r^2}(r\varepsilon_{\theta\theta}) - r\frac{\partial \varepsilon_{rr}}{\partial r} + \frac{\partial^2 \varepsilon_{rr}}{\partial \theta^2} = 2\frac{\partial^2 (r\varepsilon_{r\theta})}{\partial r \partial \theta}. \tag{4-14}$$

For a general three-dimensional deformation problem specified by Eq. (4-7), we can similarly obtain

$$\begin{cases} \dfrac{\partial^2 \varepsilon_{xx}}{\partial y\, \partial z} = \dfrac{\partial}{\partial x}\left(-\dfrac{\partial \varepsilon_{yz}}{\partial x} + \dfrac{\partial \varepsilon_{zx}}{\partial y} + \dfrac{\partial \varepsilon_{xy}}{\partial z} \right), \\[6pt] \dfrac{\partial^2 \varepsilon_{yy}}{\partial z\, \partial x} = \dfrac{\partial}{\partial y}\left(-\dfrac{\partial \varepsilon_{zx}}{\partial y} + \dfrac{\partial \varepsilon_{xy}}{\partial z} + \dfrac{\partial \varepsilon_{yz}}{\partial x} \right), \\[6pt] \dfrac{\partial^2 \varepsilon_{zz}}{\partial x\, \partial y} = \dfrac{\partial}{\partial z}\left(-\dfrac{\partial \varepsilon_{xy}}{\partial z} + \dfrac{\partial \varepsilon_{yz}}{\partial x} + \dfrac{\partial \varepsilon_{zx}}{\partial y} \right), \\[6pt] \dfrac{\partial^2 \varepsilon_{xx}}{\partial y^2} + \dfrac{\partial^2 \varepsilon_{yy}}{\partial x^2} = 2 \dfrac{\partial^2 \varepsilon_{xy}}{\partial x\, \partial y}, \\[6pt] \dfrac{\partial^2 \varepsilon_{yy}}{\partial z^2} + \dfrac{\partial^2 \varepsilon_{zz}}{\partial y^2} = 2 \dfrac{\partial^2 \varepsilon_{yz}}{\partial y\, \partial z}, \\[6pt] \dfrac{\partial^2 \varepsilon_{zz}}{\partial x^2} + \dfrac{\partial^2 \varepsilon_{xx}}{\partial z^2} = 2 \dfrac{\partial^2 \varepsilon_{zx}}{\partial z\, \partial x}. \end{cases} \quad (4\text{-}15)$$

These are the six compatibility equations that the six strain components must follow.

Normally the importance of the equations of motion obtained in Chapter 2 can be understood easily because they represent the relationships among stresses in a solid under dynamic equilibrium. However, if the importance of the above compatibility equation is ignored, it may often cause difficulties in solving problems in solid mechanics as we will see in later chapters. Mathematically, we can prove that the compatibility equations are the *necessary and sufficient conditions* for obtaining single-valued displacements when a body is simply connected.[4.2]

The concept of deformation compatibility has not only theoretical value but also direct practical applications. This can be partly understood through the following example.

[4.2] Practically, a simply connected body means that there is no hole and material discontinuity inside. Mathematically, it means that any simple closed contour drawn in the body can shrink continuously to a point without leaving the body. Otherwise, the body is called multiply connected. A multiply connected body will make our investigation more complicated. Further discussion in this regard can be found in more advanced textbooks.

Example 4.3 The details of strain distribution over a thin steel panel in a ship, which is subjected to an in-plane deformation, are important to the design of the inner structure of the ship. In their scaled-model testing, a design team used a huge number of strain gauges to measure the strain components on the panel surface, as illustrated in **Fig. E4.2**. With the aid of the multi-variable regression method, they obtained the following expressions for the strain distribution over the panel. However, when the team examined the applicability of the formulae, they found that the prediction accuracy varied from 5% to 30% over the limited points checked. Although the accuracy was still acceptable to the design the team was not sure if the formulae were so rational in terms of mechanics that their prediction on other unexamined points would be reasonable.

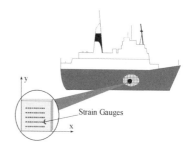

Figure E4.2 Strain field in a ship panel subjected to in-plane deformation.

$$\begin{cases} \varepsilon_{xx} = A_0 + A_1(x^2 + y^2) + x^4 + y^4, \\ \varepsilon_{yy} = B_0 + B_1(x^2 + y^2) + x^4 + y^4, \\ \varepsilon_{xy} = C_0 + C_1 xy(x^2 + y^2 + C_2). \end{cases} \quad (a)$$

Solution: The first question we need the answer to is the compatibility of the strain field. If the strains are incompatible, there must be something wrong in the process of measurement and regression. Since the deformation was only in the xy-plane, we can simply check whether Eq. (a) satisfies Eq. (4-13) or not. By substituting Eq. (a) into Eq. (4-13), we find that if the strains in (a) are compatible, the coefficients must obey the following conditions, *i.e.*,

$$\begin{cases} C_1 = 2, \\ A_1 + B_1 - 2C_2 = 0. \end{cases}$$

References

Brown, J (1973), *Introductory Solid Mechanics*, John Wiley & Sons, Dorking.
Chou, PC (1967), *Elasticity: Tensor, Dyadic and Engineering Approaches*, Van Nostrand, Princeton, NJ.
Fung, YC (1965), *Foundations of Solid Mechanics*, Prentice-Hall, Inc., Englewood Cliffs, NJ.
Koistinen, DP and Wang, N-M (1978), *Mechanics of Sheet Metal Forming*, Plenum Press, New York.

Murray, WM (1992), *The Bounded Electrical Resistance Strain Gauge: An Introduction*, Oxford University Press, Oxford.

Timoshenko, SP and Goodier, JN (1970), *Theory of Elasticity*, McGraw-Hill Book Company, Singapore.

Window, AL and Holister, GS (eds) (1982), *Strain Gauge Technology*, Applied Science, London.

Important Concepts

Displacement
Deformation
Shear strain
Strains in any direction
Principal strains
Strain invariants
Symmetry of strain matrix
Compatibility of deformation

Rigid-body motion
Direct strain
Sign of strains
Volume strain
Principal strain directions
Strain-displacement relations
Extreme shear strains
Similarity to stress analysis

Questions

4.1 How many types of displacements exist in the motion and deformation of a solid?

4.2 What is a rigid body motion and what is deformation?

4.3 Why is a rigid body displacement not related to stress?

4.4 What are the two basic strains that must be used to measure the distortion of a solid?

4.5 What is a positive direct strain and what is a positive shear strain?

4.6 What are the physical and mathematical meanings of principal strains?

4.7 Why do we have strain invariants?

4.8 What is the relationship between the transformation rules of stresses and those of strains? Why do they have such a relationship?

4.9 Why must strain components be compatible? How many compatibility equations are there for a general deformable body and how many for a plane deformation problem?

4.10 Why are strain-displacement relations also called geometric equations?

Problems

4.1 Give two examples in engineering practice for each of the following cases:
   (a) A component has a rigid body displacement only.
   (b) A component has deformation-induced displacement only.
   (c) A component has displacements caused by both deformation and rigid body motion.

4.2 By following a similar method to that used in section **4.3**, find the strain-displacement relationships given in Eq. (4-5).

4.3 It was found that the displacements in a structure follow the rule of $u = k(2x + y^2)$, $v = k(x^2 - 3y^2)$ and $w = 0$, where $k = 10^{-4}$.
   (a) Show graphically the distorted configuration of a two-dimensional element with sides dx and dy.
   (b) Find the principal strains and their direction cosines at point $(x, y, z) = (1, 1, 1)$ in the structure.

4.4 It is found that the displacements in a machine component are described by

$$u = \frac{z^2 + v(x^2 - y^2)}{2a}, \quad v = \frac{vxy}{a}, \quad w = -\frac{xz}{a},$$

where a is a constant. Find the strain components, principal strains and their direction cosines at point $(x, y, z) = (0.5, 1, 0.1)$.

4.5 A thin plate in the xy-plane is subjected to a set of edge forces within the plane, as illustrated in **Fig. P4.1**. Using the strain-gauge technique, one has measured that the direct strains at point A in directions 0°, 45° and 90° with respect to the positive x-direction are ε_0, ε_{45} and ε_{90}, respectively.
   (a) If the out-of-plane strains at the point are $\varepsilon_{yz} = \varepsilon_{xz} = 0$ and $\varepsilon_{zz} = -v(\varepsilon_{xx} + \varepsilon_{yy})/(1 - v)$ where v is a known constant, find strains ε_{xx}, ε_{yy} and ε_{xy} in terms of ε_0, ε_{45} and ε_{90}.
   (b) If the measured strains are $\varepsilon_0 = 0.0001$, $\varepsilon_{45} = 0.00005$ and $\varepsilon_{90} = 0.0001$, determine the principal strains and their directions at point A.

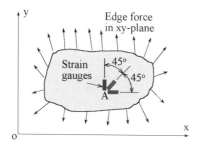
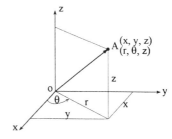

Figure P4.1   **Figure P4.2**

4.6 The relationship between a polar coordinate system and a Cartesian coordinate system is illustrated in **Fig. P4.2**. Show that the displacements and strains follow the following transformation rules:

$$u_r = u\cos\theta + v\sin\theta,$$
$$u_\theta = v\cos\theta - u\sin\theta,$$

and

$$\varepsilon_{rr} = \varepsilon_{xx}\cos^2\theta + \varepsilon_{yy}\sin^2\theta + 2\varepsilon_{xy}\cos\theta\sin\theta,$$
$$\varepsilon_{\theta\theta} = \varepsilon_{xx}\sin^2\theta + \varepsilon_{yy}\cos^2\theta - 2\varepsilon_{xx}\cos\theta\sin\theta,$$
$$\varepsilon_{r\theta} = -\varepsilon_{xx}\cos\theta\sin\theta + \varepsilon_{yy}\cos\theta\sin\theta + 2\varepsilon_{xx}\left(\cos^2\theta - \sin^2\theta\right),$$

where u_r and u_θ are displacements in r and θ directions, respectively.

4.7 Determine ε_n, ε_t and ε_{nt} if $\varepsilon_{xy} = 0.0016$ and $\varepsilon_{xx} = \varepsilon_{yy} = 0$ for the element in **Fig. P4.3**.

4.8 Find the change in length of diagonal $\overline{oA}$ of the plate in **Fig. P4.4** when it is subjected to the following uniform strain field: $\varepsilon_{xx} = 0.005$, $\varepsilon_{yy} = 0.0025$, $\varepsilon_{xy} = 0.002$, $\varepsilon_{zz} = \varepsilon_{zx} = \varepsilon_{zy} = 0$.

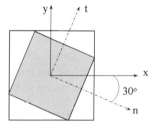

Figure P4.3   **Figure P4.4**

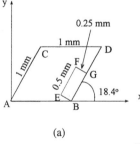   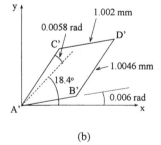

(a)   (b)

Figure P4.5

4.9 The parallelogram ABCD and rectangle BEFG are scribed on the surface of the plate shown in **Fig. P4.5a**. When loads are applied to the plate, the parallelogram deforms as illustrated in **Fig. P4.5b**. If the strain condition throughout the plate is uniform, *i.e.*, the strains are constants, calculate ε_{xy} and determine the deformed configuration of rectangle BEFG, evaluating the length and angular position of each line.

4.10 Explain the use of Mohr's circle of strain in determining the strain components for any direction in a two-dimensional element.

4.11 Use coordinate transformation between the polar and Cartesian coordinate systems to derive Eq. (4-8) from Eq. (4-7).

4.12 Are the following strain states possible ones in a continuous body subjected to deformation?

(a) $\varepsilon_{xx} = k(x^2 + y^2)z$, $\varepsilon_{yy} = ky^2z$, $\varepsilon_{xy} = kxyz$, $\varepsilon_{zz} = \varepsilon_{zx} = \varepsilon_{zy} = 0$,

(b) $\varepsilon_{xx} = k(x^2 + y^2)$, $\varepsilon_{yy} = ky^2$, $\varepsilon_{xy} = kxy$, $\varepsilon_{zz} = \varepsilon_{zx} = \varepsilon_{zy} = 0$,

(c) $\varepsilon_{xx} = axy^2$, $\varepsilon_{yy} = ax^2y$, $\varepsilon_{zz} = axy$

$\varepsilon_{xy} = 0$, $\varepsilon_{zx} = \frac{1}{2}(ax^2 + by^2)$, $\varepsilon_{zy} = \frac{1}{2}(az^2 + by)$,

where a, b and k are constants.

4.13 A prismatic bar with an elliptical cross-section $\left(\frac{x^2}{a^2} + \frac{y^2}{b^2} = 1\right)$ is subjected to torsion. It is found that the strains in the bar vary in the following way

$$\varepsilon_{zx} = -\frac{M_t y}{\pi ab^3 G}, \quad \varepsilon_{zy} = \frac{M_t x}{\pi a^3 b G},$$

$$\varepsilon_{xx} = \varepsilon_{yy} = \varepsilon_{zz} = \varepsilon_{xy} = 0,$$

where M_t is the torque, G is a material constant and a and b are half-lengths of the major and minor axes of the ellipse. Show that the strains are compatible.

4.14 A body is heated by a temperature increment of T(x, y, z). Show that when the strain components are

$$\varepsilon_{xx} = \varepsilon_{yy} = \varepsilon_{zz} = \alpha T,$$

$$\varepsilon_{xy} = \varepsilon_{yz} = \varepsilon_{zx} = 0,$$

where α is the thermal expansion coefficient of the material, T must be a linear function of coordinates x, y and z if the strains are compatible.

4.15 Show that under small deformation, if a material is incompressible, then the first strain invariant must vanish, *i.e.*, $I_1^\varepsilon = \varepsilon_{xx} + \varepsilon_{yy} + \varepsilon_{zz} = 0$.

4.16 A circular shaft is subjected to pure torsion, as illustrated in **Fig. P4.6**. The displacement in the shaft can be described by

$$u = -\theta yz + ay + bz + c,$$
$$v = \theta xz + ez - ax + f,$$
$$w = -bx - ey + k,$$

where a, b, c, e, f and k are undetermined constants but θ is a known constant. Use the following conditions, respectively, to find the undetermined constants.

(a) Point O is completely fixed.

(b) An infinitesimal line element dz on z-axis at point O is not allowed to rotate in the xOz and yOz planes.

(c) An infinitesimal line element dx on x-axis at point O is not allowed to rotate in the xOy plane.

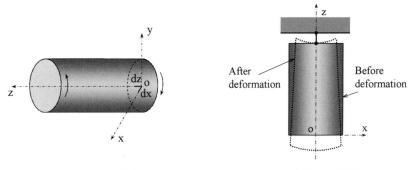

Figure P4.6   Figure P4.7

4.17 A prismatic bar suspended from the ceiling, as shown in **Fig. P4.7**, is subjected to its own weight. The suspension point is at the centre of gravity of the bar cross-section. The strain components can be obtained as

$$\varepsilon_{xx} = \varepsilon_{yy} = -\frac{\nu\rho g}{E} z, \quad \varepsilon_{zz} = \frac{\rho g}{E} z, \quad \varepsilon_{xy} = \varepsilon_{yz} = \varepsilon_{zx} = 0,$$

where ρ is the density of the bar material, g is the gravitational acceleration and ν and E are known material constants. Use the strain-displacement relations, Eq. (4-7), to find displacement u, v and w.

(*Hint*: At the suspension point (x, y, z) = (0, 0, *l*), where *l* is the bar length, w = 0. In addition, because of the symmetry of deformation, u = v = 0 and ∂u/∂y = 0 along the z-axis.)

(*Answer*: $u = -\frac{\nu\rho g}{E} xz$, $v = -\frac{\nu\rho g}{E} yz$ and $w = \frac{\rho g}{2E}\left\{z^2 + \nu\left(x^2 + y^2\right) - l^2\right\}$.)

4.18 Find displacement components corresponding to rigid body motion.

(*Hint*: For any rigid body motion, strains are zero throughout the body. Thus the displacements can be obtained by integrating Eq. (4-7) with vanishing strains.)

(*Answer*: $u = \omega_y z - \omega_z y + u_0$, $v = \omega_z x - \omega_x z + v_0$, $w = \omega_x y - \omega_y x + w_0$, where ω_x, ω_y and ω_z are constants and represent rigid body rotations around the x-, y- and z-axes, while u_0, v_0 and w_0 are constants but represent rigid body translations in the x-, y- and z-directions.)

Chapter 5

STRESS–STRAIN RELATIONSHIP

In the last two chapters, we studied stress and strain completely independently. In reality, however, they are related to each other. Our daily experience also shows that components made of different materials behave differently even when their dimensional structures and working conditions are exactly the same. (For example steel and plastic gears have obviously different strength and stiffness.) This indicates that stress and strain in a solid also depend on the mechanical properties of its material. All these are important to the design and material selection for machines and structures and will be discussed in this chapter.

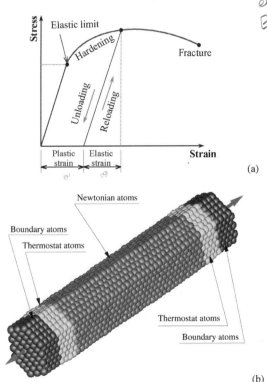

Uniaxial tension is a fundamental test for evaluating the mechanical behaviour of a material. (a) a typical stress-strain curve for a ductile material under tension. (b) Tension simulation of a nanowhisker of monocrystalline copper (Zhang et al., 1997).

5.1 MECHANISM OF STRESS-STRAIN BEHAVIOUR

Before going to find the relationships between stresses and strains, let us ask ourselves a question: 'Why does a material deform under external stresses?' Although we focus only on the deformation of engineering materials on and beyond the scale of microns in this text, it is beneficial to look at the atomic-scale mechanisms.

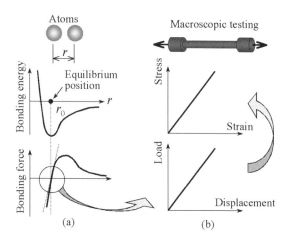

Figure 5.1 Mechanism of elastic deformation: from atomic scale to macroscopic scale.

As we understood in the elementary mechanics of materials, *elastic deformation* in a material under external load is temporary deformation that will fully recover when the load is removed. The elastic region in a stress-strain curve, as shown in figure (a) of the opening page of this chapter, is the initial linear portion below the elastic limit. Physically, the mechanism of elastic deformation is the stretching of atomic bonds. The fractional deformation of the material in its elastic region is so small that on the atomic scale the distance of neighbouring atoms only varies in the immediate vicinity of their equilibrium separation distance.[5.1] As can be seen in **Fig. 5.1a**, the force required to move one atom away from another is very nearly proportional to the atomic distance r. Thus macroscopically the relationship between stress and strain in a solid is also linear when the solid undergoes elastic deformation, as illustrated in **Fig. 5.1b**. In reality, a material contains a huge number of atoms that form anisotropic structures. The situation of course becomes much more complex because of the effect of neighbouring atoms and the three-

---

[5.1] The exact shape of the force-distance curve depends on the nature of the atomic bond of a specific material, such as ionic, covalent and metallic bonds. However, all bonds possess the general characteristic of the force-distance relationship shown in **Fig. 5.1**. Further details can be found in any textbook on materials science.

dimensional character of the material. However, the overall elastic property, such as Young's modulus, of a material is not very different from that on the atomic scale (Zhang *et al.*, 1997).

The fundamental mechanism of *plastic deformation* is the distortion and reformation of atomic bonds when the stress applied on the solid is large enough to create irreversible deformation. In this case, some atoms in the material have moved to new equilibrium positions, in which they have formed new bonds, with no tendency to return to their original positions. When this happens, we say that *yielding* of the material occurs. In many crystalline materials, like metals, the end of the elastic range and beginning of plastic behaviour occurs at the point when the stress-strain curve deviates from linearity, such as that shown in figure (a) on this chapter's opening page. The most common mechanism of plastic deformation for crystalline materials is slip, which, in metals, is associated with movement of dislocations (Callister, 1994). However, many other mechanisms exist and sometimes plastic deformation may be caused by a combination of several mechanisms, such as phase transformation and dislocations in monocrystalline silicon, as shown in **Fig. 5.2**.

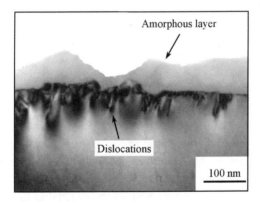

Figure 5.2 Plastic deformation in monocrystalline silicon with both dislocations and amorphous phase transformation. This is a cross-sectional view of a silicon after surface grinding with a diamond wheel.

In the analysis in this book, we will consider plastic deformation generally as a kind of macroscopic-irreversible deformation by ignoring its details on the atomic scale. In solid mechanics, it is called the *phenomenological method of study*.

Engineering components made of different materials perform differently even though their dimensional structures are exactly the same and they are working under the same environmental conditions. This is because different materials have different atomic bonds or micro-structures. Thus to make an optimal design under a given specification, we must know the relationship between stress and strain and its dependence upon the properties of a material.

5.2 GENERALISED HOOKE'S LAW

We have understood that physically stress and strain are related and in the elastic regime of deformation, stress and strain has a linear relationship. Macroscopically, such behaviour of solid materials was first discovered by a British scientist, Robert Hooke, in the seventeenth century. For the special case shown in **Fig. 5.3**, Hooke's finding can be briefly stated as 'The force, F, applied at the end along the rod axis is proportional to the axial elongation of the rod, δ.' That is

$$F = k\delta,$$

where k is a constant dependent on the properties of the rod material. It is called *Hooke's law*. If we divide both sides of the above equation by the area of the rod and then use $\varepsilon = \delta/l$ to replace δ, where l is the initial length of the rod and ε is its longitudinal strain according to the definition of direct strain given in Eq. (4-1a), Hooke's law can be rewritten as

$$\sigma = E\varepsilon, \tag{5-1}$$

where E is still a constant of course but is dependent on the properties of the rod material and is commonly known as *Young's modulus, elastic modulus*, or *modulus of elasticity*.

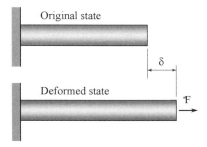

Figure 5.3 Relationship between the force and elongation.

In fact Hooke's law is a phenomenological expression of the atomic mechanism of elastic deformation in a solid. Since the relationship between the force and deformation is linear in this regime, the principle of superposition applies if the solid is subjected to more complex loading and deformation. Thus under general three-dimensional loading conditions, each of the six stress components must be linearly related to all the six strain components *via* superposition. Mathematically, by incorporating the contributions from all the strain components, we can extend the linear relation of Eq. (5-1) to

$$\begin{cases} \sigma_{xx} = C_{11}\varepsilon_{xx} + C_{12}\varepsilon_{yy} + C_{13}\varepsilon_{zz} + C_{14}\varepsilon_{xy} + C_{15}\varepsilon_{yz} + C_{16}\varepsilon_{zx}, \\ \sigma_{yy} = C_{21}\varepsilon_{xx} + C_{22}\varepsilon_{yy} + C_{23}\varepsilon_{zz} + C_{24}\varepsilon_{xy} + C_{25}\varepsilon_{yz} + C_{26}\varepsilon_{zx}, \\ \sigma_{zz} = C_{31}\varepsilon_{xx} + C_{32}\varepsilon_{yy} + C_{33}\varepsilon_{zz} + C_{34}\varepsilon_{xy} + C_{35}\varepsilon_{yz} + C_{36}\varepsilon_{zx}, \\ \sigma_{xy} = C_{41}\varepsilon_{xx} + C_{42}\varepsilon_{yy} + C_{43}\varepsilon_{zz} + C_{44}\varepsilon_{xy} + C_{45}\varepsilon_{yz} + C_{46}\varepsilon_{zx}, \\ \sigma_{yz} = C_{51}\varepsilon_{xx} + C_{52}\varepsilon_{yy} + C_{53}\varepsilon_{zz} + C_{54}\varepsilon_{xy} + C_{55}\varepsilon_{yz} + C_{56}\varepsilon_{zx}, \\ \sigma_{zx} = C_{61}\varepsilon_{xx} + C_{62}\varepsilon_{yy} + C_{63}\varepsilon_{zz} + C_{64}\varepsilon_{xy} + C_{65}\varepsilon_{yz} + C_{66}\varepsilon_{zx}, \end{cases} \quad (5\text{-}2\text{a})$$

or in matrix form,

$$\begin{Bmatrix} \sigma_{xx} \\ \sigma_{yy} \\ \sigma_{zz} \\ \sigma_{xy} \\ \sigma_{yz} \\ \sigma_{zx} \end{Bmatrix} = \begin{pmatrix} C_{11} & C_{12} & C_{13} & C_{14} & C_{15} & C_{16} \\ C_{21} & C_{22} & C_{23} & C_{24} & C_{25} & C_{26} \\ C_{31} & C_{32} & C_{33} & C_{34} & C_{35} & C_{36} \\ C_{41} & C_{42} & C_{43} & C_{44} & C_{45} & C_{46} \\ C_{51} & C_{52} & C_{53} & C_{54} & C_{55} & C_{56} \\ C_{61} & C_{62} & C_{63} & C_{64} & C_{65} & C_{66} \end{pmatrix} \begin{Bmatrix} \varepsilon_{xx} \\ \varepsilon_{yy} \\ \varepsilon_{zz} \\ \varepsilon_{xy} \\ \varepsilon_{yz} \\ \varepsilon_{zx} \end{Bmatrix}, \quad (5\text{-}2\text{b})$$

where C_{ij} (i = 1, ..., 6; j = 1, ..., 6) are constants of the material. It can be shown theoretically (Fung, 1965) that for any anisotropic material, the constant matrix **C** is symmetrical, *i.e.*, $C_{ij} = C_{ji}$, thus only 21 components of **C** are independent. In other words, even for the most anisotropic material, there are only 21 material constants that correlate strain with stress components. These material constants are different from material to material and must be determined by experiment. Equation (5-2) is a generalised form of Hooke's law, Eq. (5-1), and therefore is often called the *Generalised Hooke's Law*.

If a material is isotropic, then the stress-strain relations must not change by any transformation of coordinate axes. This leads to the conclusion that for an isotropic material, there only exist two material constants that are independent.[5.2] Since a great number of engineering materials can be considered to be isotropic, we shall focus on

---

[5.2] In the early stage of development of solid mechanics, many scientists thought that there existed only one independent material constant to define an isotropic body. It was George Green, a British scientist (1783-1841), who pointed out that there must be two independent constants (Timoshenko, 1983).

the stress-strain relationships of this type of materials[5.3] and try to find what the two material constants are and what they mean physically.

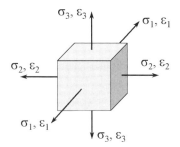

Figure 5.4 An element with surfaces in principal directions.

Consider an infinitesimal element of an isotropic material whose surfaces are all in principal stress and strain directions, as shown in **Fig. 5.4**. Thus on the element surfaces there are no shear stresses and shear strains. The generalised Hooke's law then becomes

$$\begin{cases} \sigma_1 = C_{11}\varepsilon_1 + C_{12}\varepsilon_2 + C_{13}\varepsilon_3, \\ \sigma_2 = C_{12}\varepsilon_1 + C_{22}\varepsilon_2 + C_{23}\varepsilon_3, \\ \sigma_3 = C_{13}\varepsilon_1 + C_{23}\varepsilon_2 + C_{33}\varepsilon_3, \end{cases} \qquad (5\text{-}3\text{a, b, c})$$

because $C_{ij} = C_{ji}$. Furthermore, since the material is isotropic, the magnitude of σ_1 should not change if we interchange ε_2 and ε_3. Hence $C_{12} = C_{13}$ in Eq. (5-3a). Based on the same argument, we must have $C_{12} = C_{23}$, $C_{13} = C_{23}$ and $C_{11} = C_{22} = C_{33}$, *i.e.*, we have only two independent constants. For convenience, let

$$\lambda = C_{12}, \quad \mu = \frac{1}{2}(C_{11} - C_{12}), \qquad (5\text{-}4)$$

then Eq. (5-3) becomes

$$\begin{cases} \sigma_1 = 2\mu\varepsilon_1 + \lambda I_1^\varepsilon, \\ \sigma_2 = 2\mu\varepsilon_2 + \lambda I_1^\varepsilon, \\ \sigma_3 = 2\mu\varepsilon_3 + \lambda I_1^\varepsilon, \end{cases} \qquad (5\text{-}5)$$

[5.3] Some other anisotropic properties, such as orthotropic and transversely isotropic properties, can be found in many other textbooks, *e.g.*, those by Love (1944) and Fung (1965).

where $I_1^\varepsilon = \varepsilon_1 + \varepsilon_2 + \varepsilon_3$ is the first strain invariant defined by Eq. (4-10) in Chapter 4. λ and μ are called *Lamé's constants* to recognise the discovery of these constants by a French engineer, Gabriel Lamé (1795-1890).

Equation (5-5) is the generalised Hooke's law for isotropic materials relating principal stresses and principal strains. Using the stress and strain transformation formulae obtained in Chapters 3 and 4, we can get the general relations between stress and strain components in any Cartesian coordinate system xyz. For example, the transformation of stresses and strains, e.g., Eq. (3-32), shows that the normal stress and direct strain in direction $\boldsymbol{n} = (l, m, n)$ are related to the stress components in a Cartesian coordinate system x'y'z' by

$$\sigma_{nn} = l^2 \sigma_{x'x'} + m^2 \sigma_{y'y'} + n^2 \sigma_{z'z'} + 2lm\,\sigma_{x'y'} + 2mn\,\sigma_{z'y'} + 2nl\,\sigma_{z'x'}$$

and

$$\varepsilon_{nn} = l^2 \varepsilon_{x'x'} + m^2 \varepsilon_{y'y'} + n^2 \varepsilon_{z'z'} + 2lm\,\varepsilon_{x'y'} + 2mn\,\varepsilon_{z'y'} + 2nl\,\varepsilon_{z'x'},$$

respectively. If x'-, y'- and z'-directions are the corresponding first, second and third principal directions, the shear stress and shear strain components vanish on these coordinate planes. Hence, if x-axis is in $\boldsymbol{n}$-direction, we have

$$\sigma_{nn} = \sigma_{xx},\ \sigma_{x'x'} = \sigma_1,\ \sigma_{y'y'} = \sigma_2,\ \sigma_{z'z'} = \sigma_3,\ \sigma_{x'y'} = \sigma_{x'z'} = \sigma_{y'z'} = 0,$$
$$\varepsilon_{nn} = \varepsilon_{xx},\ \varepsilon_{x'x'} = \varepsilon_1,\ \varepsilon_{y'y'} = \varepsilon_2,\ \varepsilon_{z'z'} = \varepsilon_3,\ \varepsilon_{x'y'} = \varepsilon_{x'z'} = \varepsilon_{y'z'} = 0.$$

Therefore,

$$\sigma_{xx} = l^2 \sigma_1 + m^2 \sigma_2 + n^2 \sigma_3$$

and

$$\varepsilon_{xx} = l^2 \varepsilon_1 + m^2 \varepsilon_2 + n^2 \varepsilon_3.$$

Then by using Eq. (5-5), we obtain

$$\begin{aligned}\sigma_{xx} &= l^2\left(2\mu\varepsilon_1 + \lambda I_1^\varepsilon\right) + m^2\left(2\mu\varepsilon_2 + \lambda I_1^\varepsilon\right) + n^2\left(2\mu\varepsilon_3 + \lambda I_1^\varepsilon\right) \\ &= 2\mu\left(l^2\varepsilon_1 + m^2\varepsilon_2 + n^2\varepsilon_3\right) + \lambda I_1^\varepsilon\left(l^2 + m^2 + n^2\right) \\ &= 2\mu\varepsilon_{xx} + \lambda I_1^\varepsilon,\end{aligned}$$

where we have applied the geometrical relationship $l^2 + m^2 + n^2 = 1$ in the derivation. Similarly, the transformation rule, *e.g.*, Eq. (3-32), also shows that the shear stress, $\sigma_{nt'}$, and shear strain, $\varepsilon_{nt'}$, on the plane with external normal $\boldsymbol{n} = (l, m, n)$ in direction $\boldsymbol{t'} = (l', m', n')$ are related to the stress and strain components in x'-, y'- and z'- coordinate planes by

$$\sigma_{nt'} = ll'\sigma_{x'x'} + mm'\sigma_{y'y'} + nn'\sigma_{z'z'} +$$
$$(ml' + lm')\sigma_{x'y'} + (mn' + nm')\sigma_{z'y'} + (nl' + ln')\sigma_{z'x'},$$
$$\varepsilon_{nt'} = ll'\varepsilon_{x'x'} + mm'\varepsilon_{y'y'} + nn'\varepsilon_{z'z'} +$$
$$(ml' + lm')\varepsilon_{x'y'} + (mn' + nm')\varepsilon_{z'y'} + (nl' + ln')\varepsilon_{z'x'}.$$

If $\boldsymbol{t'}$ is in the y-direction and the x'-, y'- and z'-directions are the first, second and third principal directions, respectively, the above relations reduce to

$$\sigma_{xy} = ll'\sigma_1 + mm'\sigma_2 + nn'\sigma_3,$$
$$\varepsilon_{xy} = ll'\varepsilon_1 + mm'\varepsilon_2 + nn'\varepsilon_3,$$

Following the same procedure as before, we have

$$\sigma_{xy} = ll'(2\mu\varepsilon_1 + \lambda I_1^\varepsilon) + mm'(2\mu\varepsilon_2 + \lambda I_1^\varepsilon) + nn'(2\mu\varepsilon_3 + \lambda I_1^\varepsilon)$$
$$= 2\mu(ll'\varepsilon_1 + mm'\varepsilon_2 + nn'\varepsilon_3) + \lambda I_1^\varepsilon(ll' + mm' + nn')$$
$$= 2\mu\varepsilon_{xy}.$$

In the above, we have applied the geometrical relation $ll' + mm' + nn' = 0$. As a result, the general relationships between stress and strain components can be written as

$$\begin{cases} \sigma_{xx} = 2\mu\varepsilon_{xx} + \lambda I_1^\varepsilon, \\ \sigma_{yy} = 2\mu\varepsilon_{yy} + \lambda I_1^\varepsilon, \\ \sigma_{zz} = 2\mu\varepsilon_{zz} + \lambda I_1^\varepsilon, \\ \sigma_{xy} = 2\mu\varepsilon_{xy}, \\ \sigma_{yz} = 2\mu\varepsilon_{yz}, \\ \sigma_{zx} = 2\mu\varepsilon_{zx}, \end{cases} \quad (5\text{-}6)$$

where I_1^ε can be expressed as $I_1^\varepsilon = \varepsilon_{xx} + \varepsilon_{yy} + \varepsilon_{zz}$ according to Eq. (4-10). The generalised Hooke's law in the form of Eq. (5-6) is similar to what Cauchy developed as mentioned in Chapter 2.

Under uniaxial tension in x-direction, we have

$$\sigma_{xx} \neq 0, \; \sigma_{yy} = \sigma_{zz} = \sigma_{xy} = \sigma_{yz} = \sigma_{zx} = 0.$$

Equation (5-6) thus leads to

$$\begin{cases} 2\mu\varepsilon_{xx} + \lambda I_1^\varepsilon = \sigma_{xx}, \\ 2\mu\varepsilon_{yy} + \lambda I_1^\varepsilon = 0, \\ 2\mu\varepsilon_{zz} + \lambda I_1^\varepsilon = 0. \end{cases} \quad \text{(a, b, c)}$$

The summation of the above three equations leads to

$$I_1^\varepsilon = \frac{1}{3\lambda + 2\mu}\sigma_{xx}. \quad \text{(d)}$$

The substitution of (d) into (a) then gives rise to

$$\sigma_{xx} = \frac{\mu(3\lambda + 2\mu)}{\lambda + \mu}\varepsilon_{xx}. \quad \text{(e)}$$

Comparing Eq. (e) with Eq. (5-1), we get

$$E = \frac{\mu(3\lambda + 2\mu)}{\lambda + \mu}. \quad (5\text{-}7)$$

Similarly, the substitution of Eq. (d) into Eqs. (b) and (c) brings about

$$\varepsilon_{yy} = \varepsilon_{zz} = -\frac{\lambda}{2\mu(3\lambda + 2\mu)}\sigma_{xx}. \quad \text{(f)}$$

Using Eq. (e), the above equation becomes

$$\varepsilon_{yy} = \varepsilon_{zz} = -\frac{\lambda}{2(\lambda + \mu)}\varepsilon_{xx}. \quad \text{(g)}$$

If we denote

$$\nu = \frac{\lambda}{2(\lambda+\mu)}, \tag{5-8}$$

then Eq. (g) becomes

$$\frac{\varepsilon_{yy}}{\varepsilon_{xx}} = \frac{\varepsilon_{zz}}{\varepsilon_{xx}} = -\nu. \tag{5-9}$$

The above equation states that the material constant ν is the ratio of transverse deformation to the axial deformation when the material is subjected to a uniaxial stress state. ν is called the *Poisson's ratio* of the material to acknowledge this important finding in material deformation by a French scientist, S D Poisson (1781-1840).

Lamé's constants, λ and μ, can also be expressed by Young's modulus, E, and Poisson's ratio, ν. Using Eqs. (5-7) and (5-8), we can easily find

$$\lambda = \frac{E\nu}{(1+\nu)(1-2\nu)}, \quad \mu = \frac{E}{2(1+\nu)}. \tag{5-10}$$

Thus the generalised Hooke's law, Eq. (5-6), can also be written in terms of Young's modulus and Poisson's ratio as

$$\begin{cases} \sigma_{xx} = \dfrac{E}{(1+\nu)(1-2\nu)}\left[(1-\nu)\varepsilon_{xx} + \nu\left(\varepsilon_{yy}+\varepsilon_{zz}\right)\right], \\ \sigma_{yy} = \dfrac{E}{(1+\nu)(1-2\nu)}\left[(1-\nu)\varepsilon_{yy} + \nu\left(\varepsilon_{zz}+\varepsilon_{xx}\right)\right], \\ \sigma_{zz} = \dfrac{E}{(1+\nu)(1-2\nu)}\left[(1-\nu)\varepsilon_{zz} + \nu\left(\varepsilon_{xx}+\varepsilon_{yy}\right)\right], \\ \sigma_{xy} = \dfrac{E}{1+\nu}\varepsilon_{xy}, \\ \sigma_{yz} = \dfrac{E}{1+\nu}\varepsilon_{yz}, \\ \sigma_{zx} = \dfrac{E}{1+\nu}\varepsilon_{zx}. \end{cases} \tag{5-11}$$

In many cases, we need to calculate strains when stresses are known. Since the above equations are all linear, it is easy to obtain

$$\begin{cases} \varepsilon_{xx} = \frac{1}{E}\left[\sigma_{xx} - \nu\left(\sigma_{yy} + \sigma_{zz}\right)\right], \\ \varepsilon_{yy} = \frac{1}{E}\left[\sigma_{yy} - \nu\left(\sigma_{zz} + \sigma_{xx}\right)\right], \\ \varepsilon_{zz} = \frac{1}{E}\left[\sigma_{zz} - \nu\left(\sigma_{xx} + \sigma_{yy}\right)\right], \\ \varepsilon_{xy} = \frac{(1+\nu)}{E}\sigma_{xy}, \\ \varepsilon_{yz} = \frac{(1+\nu)}{E}\sigma_{yz}, \\ \varepsilon_{zx} = \frac{(1+\nu)}{E}\sigma_{zx}. \end{cases} \quad (5\text{-}12)$$

In principal directions, shear stresses and strains vanish, the above equations reduce to

$$\begin{cases} \varepsilon_1 = \frac{1}{E}\left[\sigma_1 - \nu\left(\sigma_2 + \sigma_3\right)\right], \\ \varepsilon_2 = \frac{1}{E}\left[\sigma_2 - \nu\left(\sigma_1 + \sigma_3\right)\right], \\ \varepsilon_3 = \frac{1}{E}\left[\sigma_3 - \nu\left(\sigma_1 + \sigma_2\right)\right]. \end{cases} \quad (5\text{-}13)$$

Example 5.1 An alloy steel bar of 10 mm × 10 mm square cross-section and 400 mm length is subjected to an axial tensile force of 20 kN. Find the state of strain, the volume strain and the volume change of the bar. The material constants of the bar are E = 200 GPa and ν = 0.25.

Solution: If we take the longitudinal axis of the bar as the x-axis of a Cartesian coordinate system, the other two coordinate directions will be transverse, as shown in **Fig. E5.1**. Clearly, since the bar is under a uniform tension, the only non-zero stress component in this case is

$\sigma_{xx} = 20000/(0.01 \times 0.01) = 200$ MPa.

According to the generalised Hooke's law, Eq. (5-12), therefore, the strain components are

$\varepsilon_{xx} = \dfrac{\sigma_{xx}}{E} = \dfrac{200\,\text{MPa}}{200\,\text{GPa}} = 0.001$,

$\varepsilon_{yy} = \varepsilon_{yy} = -\nu\varepsilon_{xx} = -0.25 \times 0.001 = -0.00025$,

$\varepsilon_{xy} = \varepsilon_{yz} = \varepsilon_{zx} = 0$.

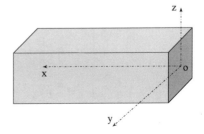

Figure E5.1

The volume strain Δ (see Example 4.1), *i.e.*, the first strain invariant I_1^ε, is

$$\Delta = \varepsilon_{xx} + \varepsilon_{yy} + \varepsilon_{zz} = 0.001 + (-0.00025) + (-0.00025) = 0.0005.$$

The total volume change of the bar is therefore

$$\delta V = V_0 \Delta = (0.01 \times 0.01) \times 0.4 \times 0.0005 = 2 \times 10^{-8} \text{ m}^3,$$

where V_0 is the initial volume of the bar.

Example 5.2 A thin cantilever beam is subjected to external stresses and a concentrated shear force in xy-plane, as shown in **Fig. E5.2**. Using the strain gauge technique, the strains at point A on the beam were measured to be $\varepsilon_{xx} = 0.1 \times 10^{-3}$, $\varepsilon_{xy} = 0.05 \times 10^{-3}$ and $\varepsilon_{yy} = 0.1 \times 10^{-3}$. The beam is made of mild steel, with Young's modulus 210 GPa and Poisson's ratio 0.3. Find the stress components, σ_{xx}, σ_{yy} and σ_{xy}.

Solution: The front and back surfaces of the beam are free from external stresses and loading is within the xy-plane. Thus on these surfaces, we have $\sigma_{zz} = \sigma_{zx} = \sigma_{zy} = 0$. Because the beam is very thin, it is reasonable to assume that the above three stress components vanish throughout the beam. Hence, the third equation of Eq. (5-11) leads to

$$\varepsilon_{zz} = \frac{-\nu}{1-\nu}(\varepsilon_{xx} + \varepsilon_{yy}).$$

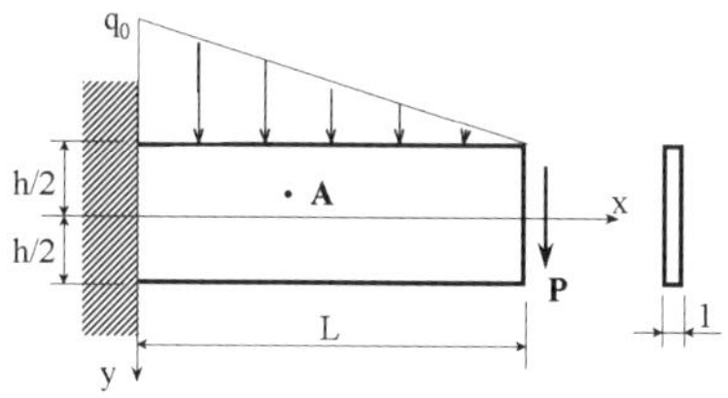

Figure E5.2

Using the above relationship and considering the first, second and fourth equations of Eq. (5-11), we have

$$\sigma_{xx} = \frac{E}{1-\nu^2}(\varepsilon_{xx} + \nu\varepsilon_{yy}),\ \sigma_{yy} = \frac{E}{1-\nu^2}(\varepsilon_{yy} + \nu\varepsilon_{xx}),\ \sigma_{xy} = \frac{E}{1+\nu}\varepsilon_{xy}.$$

Using the strains measured and the material constants known, we can easily obtain

$$\sigma_{xx} = \frac{210}{1-0.3^2}(0.1 + 0.3 \times 0.1) \times 10^{-3} = 30 \text{ MPa},$$

$$\sigma_{yy} = \frac{210}{1-0.3^2}(0.1 + 0.3 \times 0.1) \times 10^{-3} = 30 \text{ MPa},$$

$$\sigma_{xy} = \frac{210}{1+0.3} \times 0.05 \times 10^{-3} = 8.08 \text{ MPa}.$$

Example 5.3 The transverse deformation of a uniform solid cylinder is completely restricted by a rigid body, as illustrated in **Fig. E5.3**. The axial direction of the cylinder is loaded by a uniformly distributed normal stress q. If the cylinder material is isotropic, find the ratio of axial stress to axial strain. When the cylinder is made of a steel with Young's modulus equal to 210 GPa and Poisson's ratio 0.3, calculate the value of the ratio and compare it with that of Young's modulus.

Solution: Take the x-axis along the horizontal axis of the cylinder then the y- and z-axes are in the transverse direction. Because the transverse deformation of the cylinder is fully restricted by the rigid body, it is clear that all the transverse direct strains are zero. Under the Cartesian coordinate system used, the above conclusion means that $\varepsilon_{yy} = \varepsilon_{zz} = 0$.

On the other hand, we can see that on the two end surfaces of the cylinder, we have $\sigma_{xx} = -q$. The second and third equations of the generalised Hooke's law, Eq. (5-12), then bring about

$$\sigma_{yy} = \nu(\sigma_{zz} - q), \quad \sigma_{zz} = \nu(\sigma_{yy} - q),$$

which give rise to

$$\sigma_{yy} = \sigma_{zz} = \frac{-\nu}{1-\nu} q.$$

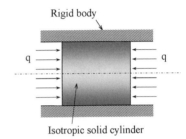

Isotropic solid cylinder

Figure E5.3

Thus the first equation of Eq. (5-12) results in the ratio of axial stress to axial strain, Ξ, *i.e.*,

$$\Xi = \frac{\sigma_{xx}}{\varepsilon_{xx}} = \frac{E(1-\nu)}{1-\nu-2\nu^2}.$$

This ratio is sometimes called *nominal elastic modulus* of a material. With the steel cylinder, we have

$$\Xi = \frac{210 \times (1-0.3)}{1-0.3-2\times 0.3^2} \approx 283 \text{ GPa}.$$

Compared with the Young's modulus of the material, E = 210 GPa, we find that Ξ is greater. In fact the analytical equation derived above shows that for a material with $\nu = 0$, $\Xi = E$; for that with $\nu = 0.5$, Ξ becomes infinite.

5.3   PHYSICAL INDICATIONS OF ELASTIC CONSTANTS

We have concluded that for an isotropic material, there only exist two independent elastic constants, Young's modulus, E, and Poisson's ratio, ν, or Lamé's constants, λ and μ, that define the stress-strain relationships of the material.[5.4] Let us now try to

---

[5.4] In many textbooks, particularly in technical papers, stress-strain relationships of a material are often called the *constitutive equations* of the material.

understand their physical indications so that we can make wise use of materials in the design and manufacture of machines and structures.[5.5]

5.3.1 Young's Modulus E

When a material is subjected to a uniaxial uniform tension or compression, for instance in a uniaxial tensile test, the stress state at a point in the material is

$$\sigma = \begin{pmatrix} \sigma_1 & 0 & 0 \\ 0 & 0 & 0 \\ 0 & 0 & 0 \end{pmatrix}, \qquad (h)$$

where the tensile direction is the first principal direction. Thus according to Eq. (5-13), or directly from Eq. (5-1), we get

$$E = \frac{\sigma_1}{\varepsilon_1}. \qquad (i)$$

Figure 5.5 A uniaxial tensile test.

This equation shows that Young's modulus E indicates the ability of the material against axial deformation.

Example 5.4 To design a stand structure for a heavy machine, as illustrated in **Fig. E5.4**, it is required that the four columns that support the platform of the stand deform as little as possible when the machine is operating. There are three materials available for making the columns, aluminium alloy 2014, low-carbon steel and high-carbon steel. Which material is the best in this application?

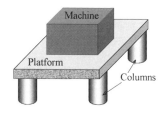

Figure E5.4

Solution: Each column of the stand is basically subjected to an axial compression when the platform is sufficiently stiff. To have less axial deformation, a better column material must have a larger elastic modulus E. Among the three materials available, the aluminium alloy possesses a much smaller E, 69GPa, than that of the carbon steels, 207GPa. Hence, it is not a good choice even without taking the cost into account. Low-carbon steel and high-carbon

---

[5.5] Applications of selection of materials in design can be found in Ashby (1992).

steel have the same value of elastic modulus but the former is cheaper. Thus in this application low-carbon steel is the most appropriate material available for making the columns.

5.3.2 Poisson's Ratio ν

Under a uniaxial tension again, Eq. (5-9) has stated that Poisson's ratio of a material is the ratio of transverse deformation to the axial deformation. If ν > 0, Eq. (5-9) means that an axial tension will cause a transverse shrinkage and *vice versa*. If ν = 0, then the material will not shrink or expand transversely under an axial tension or compression. However, if ν < 0, an axial tension will be accompanied by a transverse expansion or an axial compression will lead to a transverse shrinkage, which seems not possible to happen in practice. Theoretically, it has been proved that for any isotropic material,

$$-1 < \nu < \frac{1}{2}. \tag{5-14}$$

Although no isotropic materials with negative Poisson's ratios have been discovered so far, it is possible in engineering to make a material with ν < 0 when its micro-structure is carefully designed (*e.g.*, Almgren, 1985; Lakes, 1993). If fact, materials with negative Poisson's ratios are important to engineering applications. For example, if a fastener is made of such a material, there is no need to worry about losing its function when it is subjected to a tension perpendicular to its fastening direction.

Example 5.5 Wine bottles are made of glass that is a brittle material and can crack easily subjected to tensile stress. Is rubber a good material for wine bottle plugs?

Solution: For a cylindrical plug made of a material with a positive Poisson's ratio ν, an axial compression, which must occur in order to plug in, will cause a transverse expansion and thus increase its diameter. The greater ν the larger transverse expansion during axial compression and thus the larger transverse force to the bottleneck that may break it.

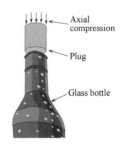

Figure E5.5

The Poisson's ratio of rubber can be as high as 0.499 that is almost the highest possible ν of all engineering materials. Hence, rubber is not a suitable material for plugs of wine bottles. The most commonly used material is corkwood that has ν ≈ 0.

5.3.3 Shear Modulus G

If a material is subjected to a pure shear in xy-plane, then

$$\sigma = \begin{pmatrix} 0 & \sigma_{xy} & 0 \\ \sigma_{xy} & 0 & 0 \\ 0 & 0 & 0 \end{pmatrix}.$$

The generalised Hooke's law, Eq. (5-12), gives rise to

$$\frac{\sigma_{xy}}{2\varepsilon_{xy}} = G, \tag{j}$$

where G is defined as

$$G = \frac{E}{2(1+\nu)} \tag{5-15}$$

and is called the *shear modulus* of the material, because Eq. (j) clearly indicates that G reflects the ability of the material to resist shear deformation. If ν approaches to -1, G approaches to infinite according to Eq. (5-15). This means that a material with $\nu = -1$ is impossible to shear, or in other words, the material is rigid in the sense of shear deformation. We must emphasise that the above conclusion is for isotropic materials and does not necessarily apply to those with anisotropic properties.

5.3.4 Bulk Modulus K

If a material is subjected to a hydrostatic stress state, then

$$\sigma = \begin{pmatrix} \sigma & 0 & 0 \\ 0 & \sigma & 0 \\ 0 & 0 & \sigma \end{pmatrix},$$

i.e., $\sigma_1 = \sigma_2 = \sigma_3 = \sigma$. The generalised Hooke's law, Eq. (5-13), brings about

$$\varepsilon_1 = \varepsilon_2 = \varepsilon_3 = \frac{1-2\nu}{E}\sigma.$$

Hence,

$$\begin{aligned}\sigma &= \frac{1}{3}(\sigma_1 + \sigma_2 + \sigma_3) \\ &= \frac{E}{3(1-2\nu)}(\varepsilon_1 + \varepsilon_2 + \varepsilon_3) \\ &= \frac{E}{3(1-2\nu)}I_1^\varepsilon \\ &= KI_1^\varepsilon,\end{aligned} \quad\quad (k)$$

where

$$K = \frac{E}{3(1-2\nu)}, \quad\quad (5\text{-}16)$$

and I_1^ε is the first strain invariant as we have been familiar with. As discussed in Example 4.1, physically, I_1^ε represents the relative volume change of an infinitesimal material element under small deformation. Hence, Eq. (k) means that constant K is actually a measure of volume change of a material when it is undergoing a hydrostatic stress. K defined by Eq. (5-16) is thus called the *bulk modulus* of the material. It is interesting to note that if $\nu = 0.5$, K becomes infinite, which indicates that if the Poisson's ratio of a material is 0.5 the material is volumetrically incompressible. In engineering practice, rubber is nearly incompressible since its Poisson's ratio is very close to 0.5.

Table 5.1 lists the general relationships among independent elastic constants that are alternatively used by various textbooks and technical papers. **Table 5.2** gives the properties of some commonly used engineering materials. For later use, the table also includes density, yield stress and coefficient of thermal expansion.

Physically, we know that to deform a body, the work done by external loads must be positive. This requires that Young's modulus E, shear modulus G $(= \mu)$ and bulk modulus K must all be positive.

The above material constants can be determined by experiment. For example, E and ν can be determined by a simple tensile test, G can be obtained by a thin tube torsion and K can be measured by uniform compression. Of course, when E and ν are known, the others can be calculated directly based on their relationships listed in **Table 5.1**.

It must be noted, however, that an isotropic material is an idealised model. Real materials are basically anisotropic. **Table 5.3** lists the stiffness coefficients of some materials with cubic atomic structure, where C_{ij} are the coefficients in Eq. (5-2) and

$(C_{11}-C_{12})/2C_{44}$ is called the *anisotropy ratio*. We can see that tungsten has a unit anisotropy ratio and thus is elastically isotropic.

Table 5.1 Relationships among elastic constants of isotropic materials

	E	ν	λ	G (= μ)	K
E, ν	E	ν	$\dfrac{E\nu}{(1+\nu)(1-2\nu)}$	$\dfrac{E}{2(1+\nu)}$	$\dfrac{E}{3(1-2\nu)}$
E, G	E	$\dfrac{E}{2G}-1$	$\dfrac{G(E-2G)}{3G-E}$	G	$\dfrac{GE}{3(3G-E)}$
E, K	E	$\dfrac{1}{2}-\dfrac{E}{6K}$	$\dfrac{3K(3K-E)}{9K-E}$	$\dfrac{3KE}{9K-E}$	K
E, λ	E	$\dfrac{2\lambda}{E+\lambda+R}$	λ	$\dfrac{E-3\lambda+R}{4}$	$\dfrac{E+3\lambda+R}{6}$
ν, λ	$\dfrac{\lambda(1+\nu)(1-2\nu)}{\nu}$	ν	λ	$\dfrac{\lambda(1-2\nu)}{2\nu}$	$\dfrac{\lambda(1+\nu)}{3\nu}$
ν, G	$2G(1+\nu)$	ν	$\dfrac{2G\nu}{1-2\nu}$	G	$\dfrac{2G(1+\nu)}{3(1-2\nu)}$
ν, K	$3K(1-2\nu)$	ν	$\dfrac{3K\nu}{1+\nu}$	$\dfrac{3K(1-2\nu)}{2(1+\nu)}$	K
λ, G	$\dfrac{G(3\lambda+2G)}{\lambda+G}$	$\dfrac{\lambda}{2(\lambda+G)}$	λ	G	$\dfrac{3\lambda+2G}{3}$
λ, K	$\dfrac{9K(K-\lambda)}{3K-\lambda}$	$\dfrac{\lambda}{3K-\lambda}$	λ	$\dfrac{3(K-\lambda)}{2}$	K
G, K	$\dfrac{9KG}{3K+G}$	$\dfrac{3K-2G}{2(3K+G)}$	$\dfrac{3K-2G}{3}$	G	K

In the above expressions, $R = \sqrt{E^2 + 9\lambda^2 + 2E}$

Table 5.2 Properties at room-temperature of some commonly used materials[5.6]

Material	Density, ρ (g/cm^3)	Elastic modulus, E (GPa)	Poisson's ratio, ν	Yield stress, Y (MPa)	Coefficient of thermal expansion, α ($^{\circ}$C^{-1}×10^{-6})
Ferrous Alloys					
Iron	7.87	207	0.29	130	11.8
Low-carbon steel	7.86	207	0.30	295	11.7
Medium-carbon steel	7.85	207	0.30	350	11.3
High-carbon steel	7.84	207	0.30	380	11.0
Stainless steels	7.77*	198*	0.30	276*	36.3*
Nonferrous Alloys					
Aluminium(>99.5)	2.71	69	0.33	17	23.6
Aluminium alloy 2014	2.80	72	0.33	97	22.5
Copper (99.95)	8.94	110	0.35	69	16.5
Magnesium (>99)	1.74	45	0.29	41	27.0
Nickel (>99)	8.9	207	0.31	138	13.3
Silver (>99)	10.49	76	0.37	55	19.0
Titanium (>99)	4.51	107	0.34	240	9.0
Ceramics					
Alumina (Al$_2$O$_3$)	3.97	393	0.27	-	8.8
Magnesia (MgO)	3.58	207	0.36	-	13.5
Zirconia (ZrO$_2$)	5.56	152	0.32	-	10.0
Fused silica (SiO$_2$)	2.2	75	0.16	-	0.5
Soda-lime glass	2.5	69	0.23	-	9.0
Silicon carbide (SiC)	3.22	414	0.19	-	4.7
Silicon nitride (Si$_3$N$_4$)	3.44	304	0.24	-	3.6
Titanium carbide (TiC)	4.92	462	-	-	7.4
Thermoplastic Polymers		Tensile modulus			
Polyethylene [-CH$_2$-CH$_2$-]$_n$ High density, 70-80% crystalline	0.952-0.965	1.07-1.09	-	-	60-110
Nylon 6,6, 30-40% crystalline	1.13-1.15	1.58-3.97	-	-	80
Thermoset Polymers					
Epoxy, complex network, amorphous	1.11-1.40	2.41	-	-	45-65
Polyester, complex network, amorphous	1.04-1.46	2.07-4.41	-	-	55-100
Elastomer		Modulus at 100% elongation			
Natural rubber	-	3.3-5.9×10^{-3}	0.499	-	-

[5.6] The values in the table are selected from the table given by Callister (1994). Those with asterisks have been averaged out.

Table 5.3 Stiffness coefficients for selected cubic materials.

Material	C_{11} (GPa)	C_{12} (GPa)	C_{44} (GPa)	Anisotropy ratio $(C_{11}-C_{12})/2C_{44}$
Metals				
Ag	124	93	46	0.34
Al	108	61	29	0.81
Au	186	157	42	0.35
Cu	168	121	75	0.31
α-Fe	233	124	117	0.47
Ni	247	147	125	0.40
W	501	198	151	1.00
Covalent solids				
Si	166	64	80	0.64
Diamond	1076	125	576	0.83
TiC	512	110	177	1.14
Ionic solids				
LiF	112	46	63	0.52
MgO	291	90	155	0.65
NaCl	49	13	13	1.38

Example 5.6 Two silicon carbide plates are joined together and loaded in the way illustrated in **Fig. E5.6**. In one case, the joining material is aluminium and in the other it is copper. In which case is the relative horizontal displacement of the plates larger? If such a joined structure is placed into an environment with high hydrostatic pressure, as illustrated in **Fig. E5.7**, which joining material will have larger volume shrinkage?

Solution: In the loading case in **Fig. E5.6**, the joining material is under a shear deformation. Thus the relative horizontal displacement of the two silicon carbide plates will be greater when the shear strain in the joining material is larger. According to the generalised Hooke's law, we can easily understand that a material with a smaller shear modulus, G, will have a larger shear strain when the shear stress applied is a constant. Using the material properties given in **Table 5.2**, we have

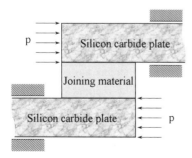

Figure E5.6 The joining under a shear stress.

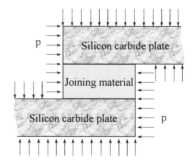

Figure E5.7 The joining under a pressure.

$$G_{Al} = \frac{E_{Al}}{2(1+v_{Al})} = \frac{69}{2(1+0.33)} \approx 25.9 \, \text{GPa}, \quad G_{Cu} = \frac{E_{Cu}}{2(1+v_{Cu})} = \frac{110}{2(1+0.35)} \approx 40.7 \, \text{GPa}.$$

Hence, the horizontal displacement with aluminium joining will be larger because $G_{Cu} > G_{Al}$.

When the joining is subjected to a hydrostatic pressure, as in the case of **Fig. E5.7**, more volume shrinkage will occur if the volume strain, I_1^ε, is larger. Equation (k) then indicates that when the hydrostatic stress is a constant, a material with a smaller bulk modulus, K, will have a larger volume strain. Using the material properties in **Table 5.2** again, we have

$$K_{Al} = \frac{E_{Al}}{3(1-2v_{Al})} = \frac{69}{3(1-2\times 0.33)} \approx 67.7 \, \text{GPa},$$

$$K_{Cu} = \frac{E_{Cu}}{3(1-2v_{Cu})} = \frac{110}{3(1-2\times 0.35)} \approx 122 \, \text{GPa}.$$

Clearly, the joining with aluminium will have greater volume shrinkage.

5.4   EFFECT OF TEMPERATURE CHANGE

As we understand in physics, temperature affects the vibration of atoms in a solid. An increase of temperature causes the atoms to vibrate more severely and makes the average atomic spacing greater. The accumulation of such spacing increments over a macroscopic distance in a solid produces a dimensional increase of the solid. Similarly, a decrease of temperature causes the spacing to decrease and induces a dimensional reduction (shrinkage) macroscopically. Such deformation is called *thermal deformation* and the corresponding strains are called *thermal strains*.

In an ideal isotropic material, the effect is the same in all directions. Thus a temperature change only introduces the variation of direct strains. In other words, thermal deformation does not produce shear strains in an isotropic material. It has been found that over a limited range of temperature, T, the thermal strains in a solid are approximately proportional to the temperature change, $\Delta T = T - T_0$, *i.e.*,

$$\varepsilon = \alpha (\Delta T), \tag{5-17}$$

where T_0 is a reference temperature at which thermal strains are considered to be zero and α is a material parameter, called *coefficient of thermal expansion*, with dimension $(^\circ C)^{-1}$. Thus when considering thermal strains, the generalised Hooke's law, Eq. (5-12), becomes

$$\begin{cases} \varepsilon_{xx} = \frac{1}{E}\left[\sigma_{xx} - \nu\left(\sigma_{yy} + \sigma_{zz}\right)\right] + \alpha(\Delta T), \\ \varepsilon_{yy} = \frac{1}{E}\left[\sigma_{yy} - \nu\left(\sigma_{xx} + \sigma_{zz}\right)\right] + \alpha(\Delta T), \\ \varepsilon_{zz} = \frac{1}{E}\left[\sigma_{zz} - \nu\left(\sigma_{xx} + \sigma_{yy}\right)\right] + \alpha(\Delta T), \\ \varepsilon_{xy} = \frac{1}{2G}\sigma_{xy}, \\ \varepsilon_{yz} = \frac{1}{2G}\sigma_{yz}, \\ \varepsilon_{zx} = \frac{1}{2G}\sigma_{zx}. \end{cases}$$ (5-18)

Thermal effects are generally greater at higher temperatures, *i.e.*, the coefficient of thermal expansion α is an increasing function of temperature. Young's modulus and Poisson's ratio of a material may also vary when temperature changes. All these need a more comprehensive analysis and are ignored in this book.[5.7] Generally speaking, a material with stronger atomic bonding has a smaller α. Accordingly, a material with a higher melting temperature has a smaller α. **Figure 5.6** shows the variation trend of coefficients of thermal expansion of some materials with respect to their melting temperatures. In an engineering design, if thermal expansion or shrinkage must be small, the material to be used must have a smaller coefficient of thermal expansion. **Table 5.2** has listed the coefficients of thermal expansion of some commonly used engineering materials.

Engineering components and structures often experience significant temperature change. Engine parts, turbine blades, chemical vessels, aircraft, castings, gears, bearings, rolls of rolling mills, reservoir dams, buildings, bridges, railways and road surfaces are all examples whose working environment experiences considerable temperature change. If the thermal deformation throughout such a solid is free, it will not result in any stress in the solid. Unfortunately, components and structures often have complex external constraints such that their thermal deformation is restrained. In addition, temperature change over a component or structure is not uniform in many cases, which in turn causes non-uniform deformation in the body. When this happens, the thermal deformation in part of the material that is subjected to a larger temperature change and thus strained more, is constrained by that part of material surrounding it that is subjected to a smaller temperature change and strained less. In all these cases, thermal deformation may create considerable stresses, called *thermal stresses*, that may become harmful to the components or structures. In

---

[5.7] More advanced discussions can be found in relevant monographs and technical papers on thermal stress analysis, *e.g.*, Kurpisz and Nowak (1995) and Hetnarski (1996).

Chapter 7, we shall discuss the general solution of thermal deformation induced stress problems.

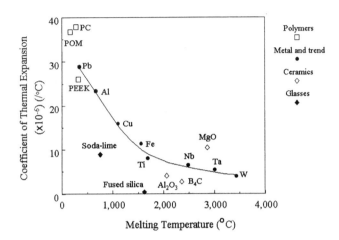

Figure 5.6 Thermal expansion coefficients of some materials at room temperature (Dowling, 1993).

Example 5.7 A prismatic bar restrained in the x-direction only, and free to expand in the y- and z-directions, as shown in **Fig. E5.8**, is subjected to a uniform temperature change ΔT. Show that the only non-vanishing stress and strain components due to the temperature change are

$$\sigma_{xx} = -E\alpha(\Delta T), \quad \varepsilon_{yy} = \varepsilon_{zz} = \alpha(\Delta T)(1+\nu).$$

Solution: Since the bar is restrained in the x-direction, $\varepsilon_{xx} = 0$. On the other hand, the surfaces of the bar normal to the y- and z-directions are stress free. This indicates that $\sigma_{yy} = \sigma_{zz} = 0$. Then the first equation of Eq. (5-18) gives rise to

$$\sigma_{xx} = -E\alpha(\Delta T).$$

The substitution of the above into the second and third equations of Eq. (5-18) leads to

$$\varepsilon_{yy} = \varepsilon_{zz} = \alpha(\Delta T)(1+\nu).$$

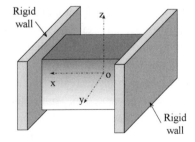

Figure E5.8

5.5 DEFORMATION BEYOND ELASTICITY

When the stress applied to a solid is sufficiently large, distortion and reformation of atomic bonds take place and as a result irreversible deformation appears. In this circumstance, the generalised Hooke's law is no longer valid and a new phenomenological rule, the theory of plasticity, must be established to relate strains with stresses. It is worthwhile to point out that stress and strain analyses carried out in Chapters 3 and 4 are independent of material properties. They are still valid even beyond the elastic limit of a material provided that the material is continuous, without initial stress and is under small deformation.

We shall discuss the mechanics of plasticity briefly in Chapter 9. Before that let us focus on the solution of modelling and elastic deformation problems.

References

Almgren, RF (1985), Isotropic three-dimensional structure with Poisson's ratio equals minus 1, *Journal of Elasticity* **15**, 427-430.

Ashby, MF (1992), *Materials Selection in Mechanical Design*, Pergamon Press, Oxford.

Callister, Jr, W (1994), *Materials Science and Engineering – An Introduction*, 3rd Edition, John Wiley & Sons, Inc., New York.

Chou, PC (1967), *Elasticity: Tensor, Dyadic and Engineering Approaches*, Van Nostrand, Princeton, NJ.

Dowling, NE (1993), *Mechanical Behaviour of Materials*, Prentice-Hall, Inc., Englewood Cliffs, NJ, pp.109.

Fung, YC (1965), *Foundations of Solid Mechanics*, Prentice-Hall, Inc., Englewood Cliffs, NJ.

Hetnarski, RB (ed.) (1996), *Thermal Stresses IV*, Elsevier, Amsterdam.

Kurpisz, K and Nowak, AJ (1995), *Inverse Thermal Problems*, Computational Mechanics Publications, Southampton.

Lakes, RS (1993), Design considerations for materials with negative Poisson's ratios, *Transactions of the ASME Journal of Mechanical Design*, **115**, 696-700.

Love, AEH (1944), *A Treatise on the Mathematical Theory of Elasticity*, Dover, New York.

Shackelford, JF (1996), *Introduction to Materials Science for Engineers*, 4th Edition, Prentice-Hall, Inc., Englewood Cliffs, NJ.

Timoshenko, SP (1983), *History of Strength of Materials*, Dover, New York.

Zhang, L and Tanaka, H (1998), Atomic scale deformation in silicon monocrystals induced by two-body and three-body contact sliding, *Tribology International*, **31**, 425-433.

Zhang, L, Tanaka, H and Gupta, P (1998), Some theoretical aspects in deforming nano-whiskers of copper monocrystals, in: *Proceedings of the International Symposium on Designing, Processing and Properties of Advanced Engineering Materials*, edited by T Kobayashi, M Umemoto and M Morinaga, Japan Society for the Promotion of Science, Toyohashi, Japan, pp.727-732.

Important Concepts

Elastic deformation
Young's modulus
Poisson's ratio
Independent elastic constants
Principle of superposition
Coefficient of thermal expansion
Thermal stress

Plastic deformation
Shear modulus
Bulk modulus
Hooke's law
Generalised Hooke's law
Thermal deformation
Thermal strain

Questions

5.1 Why do we need to establish the stress-strain relationships? Why do we have linear stress-strain relationships in the elastic deformation regime?

5.2 What are the two independent elastic constants of an isotropic material?

5.3 What are the physical meanings of the following material constants:
(a) Young's modulus, (b) Poisson's ratio, (c) shear modulus, (d) bulk modulus?
Can you give an example of material selection in design or manufacture of a machine or a structure that makes use of the physical inherence of each of the above material constants?

5.4 What is the principle of superposition? Why can we apply the principle to obtain the generalised Hooke's law according to Hooke's law?

5.5 Why can a temperature change cause thermal deformation?

5.6 Can we always use a constant coefficient of thermal expansion α? Why do some materials have higher α and some have lower α?

5.7 For an isotropic material, why does a temperature change only produce direct strains?

5.8 Must a thermal deformation create thermal stresses in a component or structure? Why does non-uniform heating in a solid induce thermal stresses?

Problems

5.1 A thin rectangular plate is subjected to a set of in-plane forces on its edges. The top and bottom surfaces are stress-free. With the aid of the strain gauge technique, as illustrated in **Fig. P5.1**, the direct strains at surface point A in three directions were measured to be $\varepsilon_{AB} = 0.0001$, $\varepsilon_{AC} = 0.0004$ and $\varepsilon_{AD} = 0.0006$. The properties of the plate material are E = 210 GPa and $\nu = 0.3$. Determine the principal stresses and maximum shear stress at point A. Find the directions of these stresses.

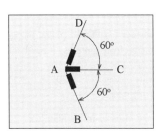
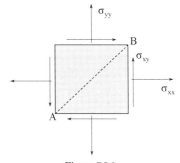

Figure P5.1   Figure P5.2

5.2 A square alumina plate is loaded as shown in **Fig. P5.2**, where $\sigma_{xx} = \sigma_{yy} = \sigma_{xy} = 10\text{MPa}$ and all the other stress components are zero. The material properties of alumina are given in **Table 5.2**. Find the change in the length of the diagonal AB.

5.3 A rubber cube is placed inside a steel box with exactly the same dimension, as illustrated in **Fig. P5.3**. A steel plate is then applied on the top surface of the rubber cube with a uniform pressure p. Find the interaction stresses between the cube and the box surfaces. Determine the volume strain of the cube and its total volume change. The friction between the rubber and steel surfaces is negligible in the present case. (*Hint*: Compared with rubber, the steel plate and box can be considered as rigid bodies in the present case.)

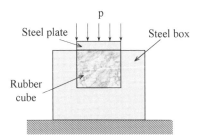
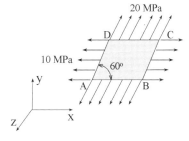

Figure P5.3   Figure P5.4

5.4 A thin plate of high-carbon steel is under stresses as shown in **Fig.P5.4**. The plate is in the xy-plane and the out-of-plane stresses are zero. The initial side length of the plate is 3 mm. The material properties are listed in **Table 5.2**. Determine (1) the change in length of edge DC due to deformation, (2) the maximum shear strain in the plane of the plate, and (3) the principal strains and their directions.

5.5 The thin plate shown in **Fig. P3.4** is made of aluminium alloy 2014. If the initial length of its horizontal sides is 5 mm, find (1) their length after deformation and (2) the principal strains and their directions.

5.6 The bar of Example 5.1 is subjected to an axial compressive force of 12 kN and a lateral fluid pressure p. (1) Find the change in length of the bar if p is made great enough to prevent any lateral strain. Determine the value of p. (2) Find the change in length when p = 0.

5.7 An arbitrarily shaped thin plate is loaded on its edges within its plane, the xy-plane. The plate surfaces parallel to the xy-plane are free from stresses. It has been found that the in-plane displacements are

$$u = -p\frac{1-v}{E}x, \quad v = -p\frac{1-v}{E}y,$$

where p is a known constant. Find the strains and stresses in the plate.

5.8 Determine the slope of the σ_{xx}~ε_{xx} curve of a material under the following stress state:

$$\sigma_{xx} = 2\sigma_{yy} = 3\sigma_{zz}.$$

5.9 Prove some of the relationships between elastic constants listed in **Table 5.1**, *e.g.*,

$$\lambda = \frac{3K(3K-E)}{9K-E}, \quad K = \frac{GE}{3(3G-E)}.$$

5.10 Determine the stress and strain components if the bar in Example 5.4 is restrained in the x- and y-directions but is free to expand in the z-direction.

5.11 Repeat the preceding problem if the bar is restrained in all directions.

5.12 Show that if thermal strains are considered, stresses can be determined by

$$\begin{cases} \sigma_{xx} = \frac{E}{(1+v)(1-2v)}\left[(1-v)\varepsilon_{xx} + v\left(\varepsilon_{yy} + \varepsilon_{zz}\right)\right] - \frac{E}{1-2v}\alpha T, \\ \sigma_{yy} = \frac{E}{(1+v)(1-2v)}\left[(1-v)\varepsilon_{yy} + v\left(\varepsilon_{zz} + \varepsilon_{xx}\right)\right] - \frac{E}{1-2v}\alpha T, \\ \sigma_{zz} = \frac{E}{(1+v)(1-2v)}\left[(1-v)\varepsilon_{zz} + v\left(\varepsilon_{xx} + \varepsilon_{yy}\right)\right] - \frac{E}{1-2v}\alpha T, \\ \sigma_{xy} = \frac{E}{1+v}\varepsilon_{xy}, \\ \sigma_{yz} = \frac{E}{1+v}\varepsilon_{yz}, \\ \sigma_{zx} = \frac{E}{1+v}\varepsilon_{zx}. \end{cases}$$

5.13 Show that the equilibrium equations can be written in terms of displacement components as

$$\begin{cases} (\lambda+\mu)\frac{\partial I_1^\varepsilon}{\partial x} + \mu\nabla^2 u + \rho f_x = 0, \\ (\lambda+\mu)\frac{\partial I_1^\varepsilon}{\partial y} + \mu\nabla^2 v + \rho f_y = 0, \\ (\lambda+\mu)\frac{\partial I_1^\varepsilon}{\partial z} + \mu\nabla^2 w + \rho f_z = 0. \end{cases}$$

(*Hint*: Make use of the generalised Hooke's law and the strain-displacement relations.)

5.14 Show that the compatibility equations can be written in terms of stress components as

$$\begin{cases} (1+\nu)\nabla^2\sigma_{xx} + \dfrac{\partial^2 I_1^\sigma}{\partial x^2} = -\dfrac{1+\nu}{1-\nu}\left\{(2-\nu)\dfrac{\partial(\rho f_x)}{\partial x} + \nu\dfrac{\partial(\rho f_y)}{\partial y} + \nu\dfrac{\partial(\rho f_z)}{\partial z}\right\}, \\[6pt]
(1+\nu)\nabla^2\sigma_{yy} + \dfrac{\partial^2 I_1^\sigma}{\partial y^2} = -\dfrac{1+\nu}{1-\nu}\left\{\nu\dfrac{\partial(\rho f_x)}{\partial x} + (2-\nu)\dfrac{\partial(\rho f_y)}{\partial y} + \nu\dfrac{\partial(\rho f_z)}{\partial z}\right\}, \\[6pt]
(1+\nu)\nabla^2\sigma_{zz} + \dfrac{\partial^2 I_1^\sigma}{\partial z^2} = -\dfrac{1+\nu}{1-\nu}\left\{\nu\dfrac{\partial(\rho f_x)}{\partial x} + \nu\dfrac{\partial(\rho f_y)}{\partial y} + (2-\nu)\dfrac{\partial(\rho f_z)}{\partial z}\right\}, \\[6pt]
(1+\nu)\nabla^2\sigma_{xy} + \dfrac{\partial^2 I_1^\sigma}{\partial x \partial y} = -(1+\nu)\left\{\dfrac{\partial(\rho f_y)}{\partial x} + \dfrac{\partial(\rho f_x)}{\partial y}\right\}, \\[6pt]
(1+\nu)\nabla^2\sigma_{yz} + \dfrac{\partial^2 I_1^\sigma}{\partial y \partial z} = -(1+\nu)\left\{\dfrac{\partial(\rho f_z)}{\partial y} + \dfrac{\partial(\rho f_y)}{\partial z}\right\}, \\[6pt]
(1+\nu)\nabla^2\sigma_{zx} + \dfrac{\partial^2 I_1^\sigma}{\partial z \partial x} = -(1+\nu)\left\{\dfrac{\partial(\rho f_x)}{\partial z} + \dfrac{\partial(\rho f_z)}{\partial x}\right\}. \end{cases}$$

Chapter 6

MODELLING AND SOLUTION

In any mechanics analysis, it is always the first key step to convert an engineering problem to a mechanics model that not only captures the major characteristics of the original but also makes the analysis feasible. With an incorrect model, we can never obtain a correct solution to the problem. The second necessary step is then to solve the mechanics model efficiently and accurately. This chapter discusses some fundamental issues in mechanics modelling and introduces some basic skills of solution.

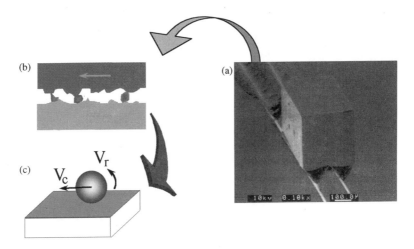

Damage in materials of two counter parts of a sliding mechanism, such as the silicon micro-slider shown in figure (a) (Courtesy of Prof. Nobuyuki Moronuki, Tokyo Metropolitan University), may easily take place especially when the sliding involves some particles between the two surfaces, as shown in figure (b) above. Understanding the deformation caused by the particles is therefore of primary importance to the design of sliding mechanisms. Figure (c) is a simple mechanics model for generating fundamental understanding, which considers the interaction of a single particle with silicon and simplifies the particle into a sphere with independent translation velocity V_c and rotational velocity V_r.

6.1 DIFFICULTIES

It seems that we have got all the necessary knowledge of stress and strain analyses and have understood the elastic stress-strain relationships. Are we now able to solve solid mechanics problems?

Recalling the mechanics quantities introduced in the previous chapters, we are to obtain the solution of fifteen unknown variables, that is, six stress components (σ_{xx}, σ_{yy}, σ_{zz}, σ_{xy}, σ_{yz}, σ_{zx}), six strain components (ε_{xx}, ε_{yy}, ε_{zz}, ε_{xy}, ε_{yz}, ε_{zx}) and three displacement components (u, v, w). In the meantime, we have also got fifteen equations, that is, three equations of motion, Eq. (3-46), six geometric equations, Eq. (4-7), and six stress-strain relations, Eq. (5-11). (These equations together with the compatibility equation, Eq. (4-15), are often called *basic equations* in elasticity.) Thus in principle we can solve the fifteen linear equations mathematically for the fifteen unknowns when the external loading and constraining conditions of a mechanics problem are properly specified.

Unfortunately, it is not actually so straightforward. First, we need to know *how to correctly translate a physical engineering problem to its mechanics model* that not only captures the major characteristics of the original engineering problem but also makes the solution mathematically feasible. This process is very important and is called *mechanics modelling*, because with an incorrect model we can never obtain correct results even if the model is solved accurately. Secondly, it is easy to imagine the mathematical difficulties to encounter in solving fifteen equations simultaneously. When a correct mechanics model is available, therefore, it is necessary to develop some methods that can facilitate the solution of the model.

In the following, we shall demonstrate the modelling and solution skills through a number of practical examples.

6.2 MECHANICS MODELLING

6.2.1 Displacement and Stress Boundary Conditions

To guarantee the dimensional accuracy and surface quality of turned workpieces, the cutting tool of a lathe, as shown in **Fig. 6.1**, must be sufficiently stiff to avoid vibration and tool-tip deflection. When we want to know whether the tool is good enough for a turning operation, we need to analyse its possible deformation when subjected to the turning forces.

To study the deformation of the cutting tool, we first need to understand the loading and supporting conditions of the cutting tool. The cutting force inserted by the workpiece and the holding of the tool shank by the tool post or the square turret of the lathe are all *known conditions* that are applied externally on the boundary surfaces of the cutting tool. These are commonly called *boundary conditions*. Since the cutting force, F, is acting through the interaction zone between the tool and workpiece[6.1] and the tool shank is held tightly in a turning operation, the problem can be translated into the mechanics model shown in **Fig. 6.2**, where the tool shank is considered to be clamped rigidly. We have therefore two types of boundary conditions for this problem. The first is the *displacement boundary condition*, which specifies that at any point on the surface of x = 0 of the tool, all the displacement components are rigidly constrained and thus are zero. Mathematically, such displacement boundary conditions can be expressed as

At $x = 0$, $0 \leq y \leq a$, $0 \leq z \leq b$,
$$u = 0, \; v = 0, \; w = 0. \tag{a}$$

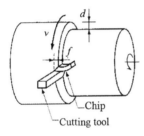 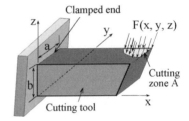

Figure 6.1 A turning operation.   **Figure 6.2** A mechanics model of the cutting tool.

Physically, the above displacement boundary conditions indicate that the displacements of the cutting tool due to deformation at its boundary surface of x = 0 must become consistent with the constraining by the tool post of the lathe.

The second type of boundary condition is the *stress boundary condition*. Except the surface at x = 0 that has been specified by displacements, all the other surfaces of the cutting tool are exposed to stress. However, only the tool-workpiece interaction zone, denoted as A in **Fig. 6.2**, is subjected to an externally applied stress F, while all the areas outside A are free from external stresses.[6.2] In Chapter 3, we have obtained the

---

[6.1] The distribution and direction of F and the size of the tool-workpiece interaction zone depend on the properties of the cutting tool material, workpiece material and lubricant. They also rely on the geometry of the cutting tool near the cutting edge and on the turning parameters, such as the depth of cut, d, feed rate, f, workpiece speed, v, and so on. Here we assume that F(x, y, z) is a known function.

[6.2] Air pressure is ignored.

expressions of stresses in any direction, Eq. (3-30). Now if we consider any of our tool surfaces with its direction cosines (l, m, n), then inside area A, we have

$$\sigma_{nx} = F_x, \ \sigma_{ny} = F_y, \ \sigma_{nx} = F_z,$$

where F_x, F_y and F_z are the resolved components of F in the x-, y- and z-directions, respectively. In all the other areas outside A,

$$\sigma_{nx} = 0, \ \sigma_{ny} = 0, \ \sigma_{nx} = 0.$$

Hence, according to Eq. (3-30), the stress boundary conditions of the cutting tool can be specified as follows.

Inside A(x, y, z)
$$F_x = l\sigma_{xx} + m\sigma_{yx} + n\sigma_{zx},$$
$$F_y = l\sigma_{xy} + m\sigma_{yy} + n\sigma_{yz}, \quad\quad\quad (b)$$
$$F_z = l\sigma_{xz} + m\sigma_{yz} + n\sigma_{zz}.$$

Outside A(x, y, z)

$$0 = l\sigma_{xx} + m\sigma_{yx} + n\sigma_{zx},$$
$$0 = l\sigma_{xy} + m\sigma_{yy} + n\sigma_{yz}, \quad\quad\quad (c)$$
$$0 = l\sigma_{xz} + m\sigma_{yz} + n\sigma_{zz}.$$

The stress boundary conditions, (b) and (c), indicate that stresses in the cutting tool must become consistent with the externally applied stresses at the tool surfaces.

With the boundary conditions defined above, the solution of the deformation of the cutting tool can now be obtained by solving the fifteen equations, *i.e.*, three equations of motion, Eq. (3-46), six geometric equations, Eq. (4-7), and six stress-strain relations, Eq. (5-11).

The above example demonstrates an approach of mechanics modelling and shows that there are two basic types of boundary conditions, that is, displacement and stress boundary conditions. In general, if the displacements at a surface point (x', y', z') of a solid are known as u', v' and w', which are constrained externally, the correct solution of displacements of the solid, u, v, and w, must satisfy the following displacement boundary conditions, *i.e.*,

at point (x', y', z'),
$$u = u', \quad v = v', \quad w = w'. \tag{6-1}$$

Similarly, if the stresses at a surface point (x", y", z") of the solid are known as $\sigma''_{nx}, \sigma''_{nx}$ and σ''_{nx}, which are applied externally, the correct solution of stresses of the solid, $\sigma_{xx}, \sigma_{yy}, \sigma_{zz}, \sigma_{xy}, \sigma_{yz}$ and σ_{zx}, must satisfy the following stress boundary conditions, i.e.,

at point (x", y", z"),
$$\begin{aligned}\sigma''_{nx} &= l\,\sigma_{xx} + m\,\sigma_{yx} + n\,\sigma_{zx}, \\ \sigma''_{ny} &= l\,\sigma_{xy} + m\,\sigma_{yy} + n\,\sigma_{yz}, \\ \sigma''_{nz} &= l\,\sigma_{xz} + m\,\sigma_{yz} + n\,\sigma_{zz},\end{aligned} \tag{6-2}$$

where l, m and n are the direction cosines of the external normal of the surface at point (x", y", z").

Mathematically, the requirement of boundary conditions in solving differential equations is obvious. Without boundary conditions, we obtain only a general solution. The special boundary conditions applied to determine the integral constants in the general solution give rise to the special solution for a specific problem, such as the above cutting tool deformation problem.

Example 6.1 A prismatic beam with square cross-section of side length a is clamped firmly into two rigid walls, as shown in **Fig. E6.1**. A heavy cubic object of side length a is placed on the beam. The density of the object is ρ and the friction between the object and beam surfaces is negligible. Describe the boundary conditions of the beam.

Solution: The weight of the cube is $\rho g a^3$, where g is the gravity acceleration. As the contact area between the object and beam is a^2, the contact stress over the area can be assumed to be

$$\sigma = \rho g a^3/a^2 = \rho g a,$$

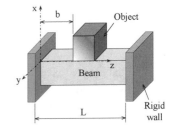

Figure E6.1

which is normal to the beam surface, because friction does not appear, and is acting towards the beam surface. (Note that the exact distribution of σ should be determined by the contact mechanics to be introduced in Chapter 7.) Here, σ is certainly an external stress on the beam and should be described as the boundary stress when considering the deformation of the beam.

It is clear that the beam has got two end surfaces on which displacements are restrained. Thus using Eqs. (6-1) and (6-2), the boundary conditions of the beam can be expressed as follows.

At the surfaces specified by $-a \leq x \leq 0$, $-a \leq y \leq 0$, $z = 0$ and
$$-a \leq x \leq 0, -a \leq y \leq 0, z = L,$$
$u = v = w = 0$. (displacement boundary conditions)

At the surfaces specified by $-a \leq x \leq 0$, $0 \leq z \leq L$, $y = 0$ and
$$-a \leq x \leq 0, 0 \leq z \leq L, y = -a,$$
$\sigma_{yy} = \sigma_{yx} = \sigma_{yz} = 0$. (stress boundary conditions)

At the surface defined by $-a \leq y \leq 0$, $0 \leq z \leq L$, $x = -a$,
$\sigma_{xx} = \sigma_{xy} = \sigma_{xz} = 0$. (stress boundary conditions)

At the surface defined by $x = 0$, $-a \leq y \leq 0$, $0 \leq z \leq b$ and
$$x = 0, -a \leq y \leq 0, a+b \leq z \leq L,$$
$\sigma_{xx} = \sigma_{xy} = \sigma_{xz} = 0$. (stress boundary conditions)

At the surface defined by $x = 0$, $-a \leq y \leq 0$, $b \leq z \leq a+b$
$\sigma_{xx} = -\rho g a$, $\sigma_{xy} = \sigma_{xz} = 0$. (stress boundary conditions)

Example 6.2 A circular tube under internal pressure p is welded on the rigid ends. The inner and outer radii of the tube are a and b, respectively. Describe its boundary conditions.

Solution: Due to the special geometry of the tube, it is more convenient to use a polar coordinate system, as illustrated in **Fig. E6.2**. Clearly the displacements at the two ends are known as zero and the inner and outer surfaces are stress boundaries. Again, using Eqs. (6-1) and (6-2), the boundary conditions of the tube can be written as

at $z = 0$, $a \leq r \leq b$, $0 \leq \theta \leq 2\pi$ and
$z = L$, $a \leq r \leq b$, $0 \leq \theta \leq 2\pi$,
$u = v = w = 0$;
at $r = a$, $0 \leq z \leq L$, $0 \leq \theta \leq 2\pi$,
$\sigma_{rr} = -p$, $\sigma_{r\theta} = \sigma_{rz} = 0$;
at $r = b$, $0 \leq z \leq L$, $0 \leq \theta \leq 2\pi$,
$\sigma_{rr} = \sigma_{r\theta} = \sigma_{rz} = 0$.

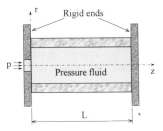

Figure E6.2

Example 6.3 Figure E6.3 shows the concrete dam cross-section of a reservoir that is in the xy-plane. The density of the concrete material is ρ_c and that of water is ρ_w. It is found that the stress components in the dam can be described by

$$\sigma_{xx} = ax + by,$$
$$\sigma_{yy} = cx + dy - \rho_c gy,$$
$$\sigma_{xy} = -dx - ay,$$
$$\sigma_{zz} = \nu(\sigma_{xx} + \sigma_{yy}),$$

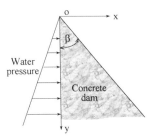

Figure E6.3

$$\sigma_{zx} = \sigma_{zy} = 0,$$

where a, b, c and d are undetermined constants. Find these constants by using the boundary conditions of the dam.

Solution: According to Eq. (6-2), the boundary conditions of the dam can be expressed below.
At $x = 0$, $0 \leq y$, $\sigma_{xx} = -\rho_w g y$, $\sigma_{xy} = \sigma_{xz} = 0$, where g is the gravity acceleration.
At $x = y \tan\beta$, $0 \leq y$, since $l = \cos\beta$, $m = -\sin\beta$ and $n = 0$ while $\sigma''_{nx} = \sigma''_{ny} = \sigma''_{nz} = 0$, Eq. (6-2) leads to

$$\cos\beta\, \sigma_{xx} - \sin\beta\, \sigma_{xy} = 0,$$
$$\sin\beta\, \sigma_{yy} - \cos\beta\, \sigma_{xy} = 0.$$

The substitution of the given stress expressions into the above boundary conditions then gives rise to
$$a = 0,\ b = -\rho_w g,\ c = \rho_c g/\tan\beta - 2\rho_w g/(\tan\beta)^3,\ d = \rho_w g/(\tan\beta)^2.$$

6.2.2 Mixed Boundary Conditions

When an element or structure possesses both stress and displacement boundary conditions, it becomes a problem with *mixed boundary conditions*. For example, the above cutting tool has mixed boundary conditions because it has got a displacement boundary condition specified on its end surface of $z = 0$ and stress boundary conditions on all the other surfaces. Similarly, the beam in Example 6.1 and the tube in Example 6.2 are all with mixed boundary conditions.

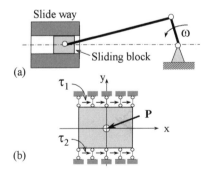

Figure 6.3 Modelling of a sliding block in a mechanical mechanism.

However, there are many other more complex cases where we need to specify different types of boundary conditions *in different directions* at the same surface point of a component. For example, when the block of a mechanical mechanism slides, as shown in **Fig. 6.3**, its upward and downward motions are completely constrained by the wall of the slide-way. Meanwhile, the upper and lower surfaces of the block are subjected to frictional stresses, τ_1 and τ_2, that are opposite to the sliding direction. Thus at any point on both the upper and lower surfaces, we must specify the displacement condition in the normal direction but the stress boundary condition in the tangential direction. Using Eqs. (6-1) and (6-2) and considering that $l = n = 0$ and $m = 1$ for both the surfaces, the boundary conditions become

at $y = b$, $-a \leq x \leq a$,
$$v = 0, \quad \sigma_{xy} = -\tau_1.$$
at $y = -b$, $-a \leq x \leq a$,
$$v = 0, \quad \sigma_{xy} = -\tau_2.$$

In these expressions, 2a is the width of the sliding block and 2b is its height. The above boundary conditions are also called mixed boundary conditions because at the same point we specified displacement in one direction and stress in the other. It must be noted that in any one direction at a point we usually cannot specify both stress and displacement conditions. The reader should think about the reasons.

Example 6.4 A cubic solid of side length L resting on a rigid foundation is subjected to a normal compressive stress, p(x, y, z), on its top surface, as shown in **Fig. E6.4**. The contact between the solid and foundation is frictionless. Describe the boundary conditions of the cubic solid.

Solution: Because the foundation is rigid, the cubic solid cannot penetrate into the foundation under p. Thus in the normal direction of the bottom surface of the solid, displacement w should vanish. In addition, because of the frictionless contact, there should be no shear stress on the bottom surface. All the other boundary conditions are obvious. Using Eq. (6-2), the boundary conditions of the solid can be written as

at $z = -L$, $-L \leq x \leq 0$, $0 \leq y \leq L$,
$$w = 0, \sigma_{zx} = \sigma_{zy} = 0; \text{ (mixed boundary conditions)}$$
at $z = 0$, $-L \leq x \leq 0$, $0 \leq y \leq L$,
$$\sigma_{zz} = -p(x, y, z), \sigma_{zx} = \sigma_{zy} = 0;$$
at $x = 0$ and $-L$, $0 \leq y \leq L$, $-L \leq z \leq 0$,
$$\sigma_{xx} = \sigma_{xy} = \sigma_{xz} = 0;$$
at $y = 0$ and L, $-L \leq x \leq 0$, $-L \leq z \leq 0$,
$$\sigma_{yy} = \sigma_{yx} = \sigma_{yz} = 0.$$

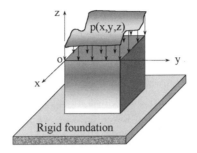

Figure E6.4

6.2.3 Understanding of Special Deformation Characteristics

Many structures and machine components possess special geometries and are subjected to special loading and constraining conditions. As a result, they deform with some special characteristics. A good understanding of the deformation characteristics will often greatly facilitate mechanics modelling, improve solution efficiency and accuracy and even help to avoid solution mistakes. This becomes particularly important when using a numerical solution method, such as the finite element method to be introduced in Chapter 10. In the following, we shall demonstrate the way to analyse deformation through some examples.

6.2.3.1 *Symmetry of Deformation*

When an underground tunnel is subjected to compression from its surrounding soil, as illustrated in **Fig. 6.4a**, the deformation of the tunnel cross-section must be symmetrical to the z-axis, because the loading, base supporting and the geometry of the tunnel are all symmetrical about the axis. In other words, the material points on the z-axis before deformation only move along the axis during deformation. Hence, at any point along the z-axis, the displacement in y-direction in the tunnel is zero. Mathematically, such a symmetric condition can be written as

$$\text{at } y = 0, \ v = 0. \tag{d}$$

Clearly, if we obtain a solution that violates the above symmetric condition, it must not be a correct solution. If a numerical method is used to analyse the tunnel deformation, only half of the tunnel shown in **Fig. 6.4b** needs to be considered to reduce the amount of calculation and improve the accuracy of the numerical result. In this way, of course, Eq. (d) should be used as the displacement boundary condition on ab and cd.

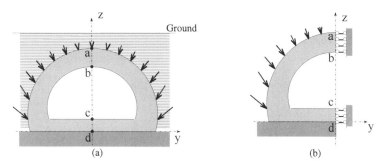

Figure 6.4 An underground tunnel subjected to compression.

When a rectangular plate with a central hole is subjected to a uniaxial tension in the x-direction, as shown in **Fig. 6.5a**, the deformation of the plate will be symmetrical to both the x- and y-axes, if the tensile force is symmetrically distributed about the x-axis. Thus at any point on the x-axis, $v = 0$ and $\sigma_{xy} = 0$, and at any point on the y-axis, $u = 0$ and $\sigma_{yx} = 0$. Again, if a solution does not satisfy these symmetry conditions of deformation, it cannot be a correct solution for the plate. When using a numerical method, we only need to consider a quarter of the plate by introducing the above symmetrical conditions as the boundary conditions, as shown in **Fig. 6.5b**. The complete description of the boundary conditions of the quarter plate is, using Eqs. (6-1) and (6-2),

$$\text{at } x = 0, c \le y \le b, u = 0, \sigma_{xy} = 0;$$
$$\text{at } x = a, c \le y \le b, \sigma_{xx} = \sigma, \sigma_{xy} = 0;$$
$$\text{at } y = 0, c \le x \le a, v = 0, \sigma_{yx} = 0;$$
$$\text{at } y = b, c \le x \le a, \sigma_{yy} = 0, \sigma_{yx} = 0.$$

The above are mixed boundary conditions.

Another type of symmetry commonly encountered is *axisymmetric deformation*. For instance, when a circular cylinder is subjected to an indentation by a hemispherical indenter along the central axis of the cylinder, z, as illustrated in **Fig. 6.6**, the deformation of the cylinder is axisymmetric and a polar coordinate system is more convenient. It is easy to imagine that there must be a frictional stress, q, and a normal stress, p, on the cylinder-indenter interface. However, due to the symmetry with respect to the z-axis, both p and q are functions only of the radial coordinate r, *i.e.*, p = p(r) and q = q(r). For the same reason, all the stress, strain and displacement components are independent of the circumferential coordinate θ. Thus straightforwardly, at any point on z-axis in the cylinder the radial displacement, u, and circumferential displacement, v, are zero. The bottom surface of the cylinder is fully clamped so that all the displacement components vanish there.

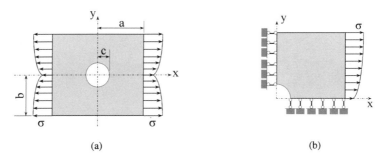

(a) (b)

Figure 6.5 A rectangular plate with a central hole subjected to a tension.

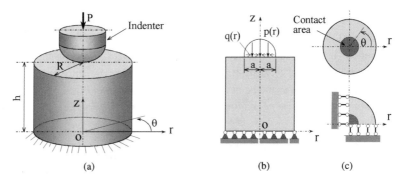

(a) (b) (c)

Figure 6.6 A circular cylinder under axisymmetric indentation.

The boundary conditions of the cylinder can therefore be described as follows.

At $z = 0$, $r \leq R$, $u = v = w = 0$.
At $z = h$,
when $a \leq r \leq R$, $\sigma_{zz} = \sigma_{zr} = \sigma_{z\theta} = 0$,
when $r \leq a$, $\sigma_{zz} = -p(r)$, $\sigma_{zr} = -q(r)$, $\sigma_{z\theta} = 0$.

At r = R, $0 \leq z \leq h$, $\sigma_{rr} = \sigma_{rz} = \sigma_{r\theta} = 0$.

The symmetrical conditions are

At r = 0, $0 \leq z \leq h$, u = v, $\sigma_{rr} = \sigma_{\theta\theta}$.

If we are going to carry out a numerical analysis, we only need to consider a quarter of the cylinder. In this case, the boundary conditions must reflect the nature of axisymmetric deformation in the cylinder. According to the above discussion and referring to **Fig. 6.6c**, the boundary conditions should be

at z = 0, $r \leq R$, $0 \leq \theta \leq 90°$, u = v = w = 0;
at z = h, $0 \leq \theta \leq 90°$,
　　when $a \leq r \leq R$, $\sigma_{zz} = \sigma_{zr} = \sigma_{z\theta} = 0$;
　　when $r \leq a$, $\sigma_{zz} = -p(r)$, $\sigma_{zr} = -q(r)$, $\sigma_{z\theta} = 0$;
at r = R, $0 \leq z \leq h$, $\sigma_{rr} = \sigma_{rz} = \sigma_{r\theta} = 0$;
at $\theta = 0$, $r \leq R$, $0 \leq z \leq h$, v = 0, $\sigma_{z\theta} = \sigma_{r\theta} = 0$;
at $\theta = 90°$, $r \leq R$, $0 \leq z \leq h$, u = 0, $\sigma_{z\theta} = \sigma_{r\theta} = 0$.

Owing to the axisymmetric nature of deformation, shear stresses $\sigma_{r\theta}$ and $\sigma_{z\theta}$, shear strains $\varepsilon_{r\theta}$ and $\varepsilon_{z\theta}$ and circumferential displacement v must be zero throughout the cylinder and all the differentiation about coordinate θ must vanish. The equations of motion, geometrical equations and stress-strain relations in this case can all be simplified to a large extent so that the solution of the problem becomes much easier. The equations of motion in a polar coordinate system, Eq. (3-50), can now be written as

$$\begin{cases} \dfrac{\partial \sigma_{rr}}{\partial r} + \dfrac{\partial \sigma_{rz}}{\partial z} + \dfrac{\sigma_{rr} - \sigma_{\theta\theta}}{r} = 0, \\ \dfrac{\partial \sigma_{rz}}{\partial r} + \dfrac{\partial \sigma_{zz}}{\partial z} + \dfrac{\sigma_{rz}}{r} = 0, \end{cases} \quad (6\text{-}3)$$

because the indentation is quasi-static and body forces can be ignored. Similarly, the geometrical equations, Eq. (4-8), reduce to

$$\begin{cases} \varepsilon_{rr} = \dfrac{\partial u}{\partial r}, \\ \varepsilon_{\theta\theta} = \dfrac{u}{r}, \\ \varepsilon_{zz} = \dfrac{\partial w}{\partial z}, \\ \varepsilon_{rz} = \dfrac{1}{2}\left(\dfrac{\partial u}{\partial z} + \dfrac{\partial w}{\partial r}\right) \end{cases} \qquad (6\text{-}4)$$

and the stress-strain relations, Eq. (5-11), become

$$\begin{cases} \sigma_{rr} = \dfrac{E}{(1+\nu)(1-2\nu)}\left[(1-\nu)\varepsilon_{rr} + \nu(\varepsilon_{\theta\theta} + \varepsilon_{zz})\right], \\ \sigma_{\theta\theta} = \dfrac{E}{(1+\nu)(1-2\nu)}\left[(1-\nu)\varepsilon_{\theta\theta} + \nu(\varepsilon_{zz} + \varepsilon_{rr})\right], \\ \sigma_{zz} = \dfrac{E}{(1+\nu)(1-2\nu)}\left[(1-\nu)\varepsilon_{zz} + \nu(\varepsilon_{rr} + \varepsilon_{\theta\theta})\right], \\ \sigma_{zr} = \dfrac{E}{1+\nu}\varepsilon_{zr}. \end{cases} \qquad (6\text{-}5)$$

Compared with the general basic equations, Eqs. (6-3) to (6-5) are much simpler and therefore are much easier to solve.

We must emphasise that the above discussions are based on isotropic materials. When the solid is anisotropic, the deformation symmetry in the above examples will no longer exist even if all the other conditions remain the same.

Let us take the bending of a clamped circular plate under uniform pressure as an example, as illustrated in **Fig. 6.7a**. If the plate material is a metal, which can be reasonably regarded as an isotropic material, the deformation is axisymmetric, and similar to the indentation problem above, only a quarter of the plate needs to be considered, see **Fig. 6.7b**. However, if it is a fibre-reinforced composite plate with fibres orientated as shown in **Fig. 6.7c**, the mechanics model of **Fig. 6.7b** becomes incorrect, because in this case the material property does not possess the axi-symmetry. Sometimes, even an experienced engineer may make such mistakes in mechanics modelling, see for example the case discussed by Helbawi, Zhang and Zarudi (1999).

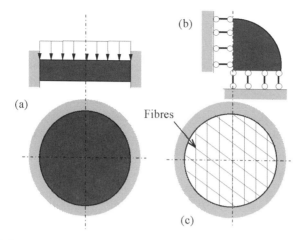

Figure 6.7 Axisymmetric and non-axisymmetric bending of clamped circular plates under uniform pressure.

Example 6.5 A cylindrical element is holding firmly on the table of a drilling machine, as illustrated in **Fig. E6.5**. The drill is drilling along the axisymmetric axis. In analysing the stress and deformation in the element during drilling, can it be considered as an axisymmetric problem?

Solution: During the drilling operation, the two cutting edges of the drill apply non-axisymmetric cutting forces to the element. Hence, the deformation of the element cannot be viewed as an axisymmetric problem, although the drilling operation is along the axisymmetric axis and the shape of the element as well as the displacement boundary conditions are all axisymmetric.

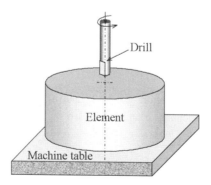

Figure E6.5

6.2.3.2 *Plane-Stress Deformation*

In addition to the symmetry of deformation discussed above, many structures and engineering elements possess another type of geometrical and loading characteristics. The gear shown in **Fig. 3.12** is a typical example, which has a uniform thickness much less than its plane dimensions (*e.g.*, one-twentieth) and is loaded within its surface plane at the peripheral boundary. The load is uniformly distributed through the gear thickness but its front and back surfaces are stress-free. The most general case of a thin plate with such loading and geometrical features has been shown in **Fig. 3.13**. Clearly, on the two stress-free surfaces of the plate,

$$\sigma_{zz} = 0, \sigma_{zx} = 0, \sigma_{zy} = 0. \tag{e}$$

Because the plate is thin and because the external stresses are uniformly distributed on the peripheral boundary through the plate thickness, it is reasonable to consider that the above three stress components vanish everywhere in the plate. Thus the non-zero stress components are σ_{xx}, σ_{yy} and σ_{xy} and they are independent of coordinate z, *i.e.*,

$$\sigma_{xx} = \sigma_{xx}(x, y), \sigma_{yy} = \sigma_{yy}(x, y), \sigma_{xy} = \sigma_{xy}(x, y). \tag{f}$$

Equation (f) shows that under the above special conditions, all the non-vanishing stresses are in the xy-plane. Such deformation is therefore called *plane-stress deformation*. Since stresses are functions of x and y, the use of the generalised Hooke's law directly concludes that all strain components are also independent of z. Equations (e), (f) and (5-12) thus give rise to

$$\varepsilon_{xx} = \varepsilon_{xx}(x, y), \varepsilon_{yy} = \varepsilon_{yy}(x, y), \varepsilon_{xy} = \varepsilon_{xy}(x, y). \tag{g}$$

$$\varepsilon_{zx} = 0, \varepsilon_{zy} = 0. \tag{h}$$

$$\varepsilon_{zz} = \varepsilon_{zz}(x, y) = -\nu(\sigma_{xx} + \sigma_{yy})/E. \tag{6-6}$$

Recalling the geometrical equation, Eq. (4-7), and using Eq. (h), we find that the in-plane displacement components u and v are also the functions of x and y only. The displacement in the z-direction, w, is not an independent variable because it is determined by Eq. (6-6), *i.e.*,

$$\frac{\partial w}{\partial z} = \varepsilon_{zz} = -\frac{\nu}{E}(\sigma_{xx} + \sigma_{yy}). \tag{6-7}$$

In summary, under plane-stress deformation, all the independent unknowns are functions of x and y only. The non-vanishing independent variables are three stress components, σ_{xx}, σ_{yy} and σ_{xy}, three strain components, ε_{xx}, ε_{yy} and ε_{xy}, and two displacements, u and v. ε_{zz} and w are dependent on others and can be determined by Eq. (6-6) and Eq. (6-7), respectively. Consequently, the equations of motion, Eq. (3-46), reduce to

$$\begin{cases} \dfrac{\partial \sigma_{xx}}{\partial x} + \dfrac{\partial \sigma_{xy}}{\partial y} + \rho f_x = \rho a_x, \\ \dfrac{\partial \sigma_{yx}}{\partial x} + \dfrac{\partial \sigma_{yy}}{\partial y} + \rho f_y = \rho a_y. \end{cases} \quad (6\text{-}8)$$

The geometrical equations, Eq. (4-7), become

$$\varepsilon_{xx} = \frac{\partial u}{\partial x}, \quad \varepsilon_{yy} = \frac{\partial v}{\partial y}, \quad \varepsilon_{xy} = \frac{1}{2}\left(\frac{\partial u}{\partial y} + \frac{\partial v}{\partial x}\right). \quad (6\text{-}9)$$

There only exists a single compatibility equation, that is,

$$\frac{\partial^2 \varepsilon_{xx}}{\partial y^2} + \frac{\partial^2 \varepsilon_{yy}}{\partial x^2} = 2\frac{\partial^2 \varepsilon_{xy}}{\partial x \partial y}. \quad (6\text{-}10)$$

And finally, the stress-strain relationships, Eq. (5-12), are simplified to

$$\begin{cases} E\varepsilon_{xx} = \sigma_{xx} - \nu\sigma_{yy}, \\ E\varepsilon_{yy} = \sigma_{yy} - \nu\sigma_{xx}, \\ E\varepsilon_{xy} = (1+\nu)\sigma_{xy}. \end{cases} \quad (6\text{-}11)$$

All the equations become much simpler under plane-stress deformation. Thus if an engineering problem can be modelled as a plane-stress problem, its solution will become easier.

6.2.3.3 *Plane-Strain Deformation*

Contrary to the thin geometry of a plate under plane-stress deformation, we often need to analyse the deformation of an element or a structure that is very long, say in the z-direction, where the cross-section in this longitudinal direction is uniform (unchanging) and where external loads distribute uniformly along the length. The long pressure vessel under uniform inner pressure shown in **Fig. 3.2a,** the dam of a reservoir in **Fig. 3.2c**, the long pin and ring after interference fit in **Fig. 1.6,** the underground tunnel in **Fig. 6.4** and both the plate and working roll in a four-high rolling mill in **Fig. 6.8** are all good examples having the above geometrical and loading characteristics.

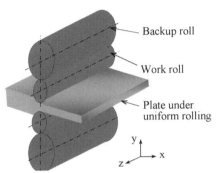

Figure 6.8 Plate rolling by a four-high rolling mill.

It is easy to understand that in all the above mentioned elements and structures, the stresses, strains and displacements in any two cross-sections parallel to the xy-plane must be identical and the displacement in the z-direction, w, vanishes.[6.3] This directly leads to the conclusion that all stresses, strains and displacements are functions of x and y only and that

$$\varepsilon_{zz} = 0, \ \varepsilon_{zx} = 0, \ \varepsilon_{zx} = 0 \tag{i}$$

due to the relationship between strain and displacements specified by Eq. (4-7). Clearly, the non-vanishing strains are ε_{xx}, ε_{yy} and ε_{xy} that are all within the xy-plane. Thus such a special deformation is called *plane-strain deformation*.

Under the plane-strain condition, it is easy to find that equations of motion, geometrical equations and compatibility equations become identical to those for

---

[6.3] In the neighbourhood of the ends of a long component, deformation is still three-dimensional and is complex. Here we have ignored the end effect. In practice, we often require that the length of a component is 20 times greater than its minimum cross-sectional dimension.

problems under plane-stress deformation, as listed in Eqs. (6-8), (6-9) and (6-10). The only difference is that the stress-strain relations now reduce to

$$\begin{cases} \left(\dfrac{E}{1-v^2}\right)\varepsilon_{xx} = \sigma_{xx} - \left(\dfrac{v}{1-v}\right)\sigma_{yy}, \\ \left(\dfrac{E}{1-v^2}\right)\varepsilon_{yy} = \sigma_{yy} - \left(\dfrac{v}{1-v}\right)\sigma_{xx}, \\ \left(\dfrac{E}{1-v^2}\right)\varepsilon_{xy} = \left(1 + \dfrac{v}{1-v}\right)\sigma_{xy}, \end{cases} \qquad (6\text{-}12)$$

and

$$\sigma_{zz} = v(\sigma_{xx} + \sigma_{yy}). \qquad (j)$$

Hence, if an engineering problem can be modelled as a plane-strain problem, its solution also becomes much easier.

6.2.3.4 *Similarity and Difference in Plane-Stress and Plane-Strain*

We have seen that some similarities and differences exist in the two plane deformation states. A careful comparison will bring about a very useful method for solution.

Similarity: In both cases, deformation occurs in a plane (we have used the xy-plane throughout the above discussion), all the independent stresses, strains and displacements are functions of coordinates x and y only, and all out-of-plane shear stresses and strains, i.e., ε_{zx}, ε_{zy}, σ_{zx} and σ_{zy}, vanish. Equations of motion, geometrical equations and the compatibility equation are identical. The stress-strain relations for plane-strain deformation, Eq. (6-12), are different from those for plane-stress deformation, Eq. (6-11). However, if we introduce the following notations,

$$E^* = \begin{cases} E, & \text{under plane-stress deformation} \\ \dfrac{E}{1-v^2}, & \text{under plane-strain deformation} \end{cases} \qquad (6\text{-}13)$$

$$v^* = \begin{cases} v, & \text{under plane-stress deformation} \\ \dfrac{v}{1-v}, & \text{under plane-strain deformation} \end{cases} \qquad (6\text{-}14)$$

Eqs. (6-11) and (6-12) become the same in form, i.e.,

$$\begin{cases} E^* \varepsilon_{xx} = \sigma_{xx} - v^* \sigma_{yy}, \\ E^* \varepsilon_{yy} = \sigma_{yy} - v^* \sigma_{xx}, \\ E^* \varepsilon_{xy} = (1 + v^*) \sigma_{xy}. \end{cases} \qquad (6\text{-}15)$$

This tells us that once we have got the solution of a plane-stress problem, we have also got the solution of the corresponding plane-strain problem provided that the elastic constants, E and v in the solution of the plane-stress problem are replaced by $\dfrac{E}{1-v^2}$ and $\dfrac{v}{1-v}$, respectively. Similarly, by replacing the E and v in the solution of a plane-strain problem by $\dfrac{E(1+2v)}{(1+v)^2}$ and $\dfrac{v}{1+v}$, respectively, we obtain the solution of the corresponding plane-stress problem. The above is a very useful method for finding solutions of plane deformation problems in engineering practice.

Difference: Although both plane-stress and plane-strain are plane deformation problems and possess the above similarities, they are different in nature. A plane-stress deformation concerns a very thin plate with stress-free surfaces normal to the z-axis and has the following properties:

$$\begin{cases} \sigma_{zz} = 0, \\ w \neq 0, \\ \varepsilon_{zz} = -\dfrac{v}{E}(\sigma_{xx} + \sigma_{yy}) \neq 0. \end{cases} \qquad (6\text{-}16)$$

However, a plane-strain deformation concerns a very long cylindrical component and possesses the properties of

$$\begin{cases} \sigma_{zz} = v(\sigma_{xx} + \sigma_{yy}) \neq 0, \\ w = 0, \\ \varepsilon_{zz} = 0. \end{cases} \qquad (6\text{-}17)$$

Plane-stress and plane-strain are two special states of deformation. When the dimensions of a component in three directions are comparable, the simplification of plane deformation does not apply and the component must be treated as a three-dimensional problem.

Example 6.6 A thin circular plate of radius R is subjected to a uniform pressure p_1 on its edge. The two plane surfaces are free, as illustrated in **Fig. E6.6**. The in-plane (xy-plane) displacements in the plate are found to be

$$u_{plate} = -p_1 \frac{1-v}{E} x, \quad v_{plate} = -p_1 \frac{1-v}{E} y.$$

Find the corresponding displacements in a long circular bar under a uniform pressure p_2. The length of the bar is much longer than its radius R.

Solution: The circular plate in **Fig. E6.6** is clearly under plane-stress deformation while the long bar is in plane-strain deformation. According to the relationship between the solutions for plane-stress and plane-strain problems, the in-plane displacements in the long bar can be obtained directly by replacing E and v in the given displacement expressions with $E/(1-v^2)$ and $v/(1-v)$, respectively. In this way, we easily get

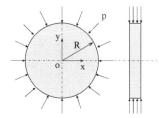

Figure E6.6

$$u_{bar} = -p_2 \frac{(1+v)(1-2v)}{E} x, \quad v_{bar} = -p_2 \frac{(1+v)(1-2v)}{E} y.$$

It is clear by comparing the above solutions that the displacements in the plate (plane-stress) are always greater that those in the bar (plane-strain) when they are made of the same material and subjected to the same external pressure ($p_1 = p_2$). For example, when the material is a steel of $v = 0.3$, the displacements in the plate are 1.3 times greater. If the material is incompressible, *i.e.*, $v = 0.5$, u_{bar} and v_{bar} will vanish but u_{plate} and v_{plate} will not.

6.2.4 Use of the Principle of Superposition

Most engineering problems are complex and normally difficult to solve directly. In some cases, however, we can make use of the principle of superposition to resolve an originally complicated problem by breaking it down into several simpler problems whose solutions are feasible. We can do so because when a material is in its elastic regime of small deformation all the equations are linear and the solution to such an elastic problem is unique and independent of loading history.[6.4] However, when elastic deformation is large or plasticity occurs, superposition can no longer be used. Let us see some examples.

Consider a thin plate with a central circular hole subjected to a biaxial tension, as shown in **Fig. 6.9**. The solution of the problem can be obtained by superposing the solution of case (a) that undergoes σ_1 in the x-direction and that of case (b) that is subjected to σ_2 in the y-direction.

If a component is under more complicated loading conditions, it can be resolved into more simple problems. For example, the solution of a bar subjected to torsion, bending, compression and rotation can be obtained by superposing the solutions of the bar under individual torsion, bending, compression and rotation.

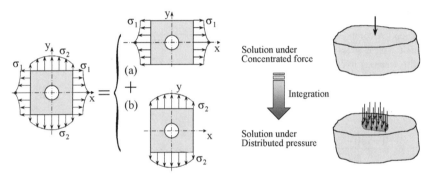

Figure 6.9 An application of the principle of Superposition.

Figure 6.10 A foundation under arbitrarily distributed normal pressure.

The above examples are with a finite number of superpositions. In the problem of **Fig. 6.9**, the solution is obtained by a single superposition of case (a) with case (b). However, the principle can be applied to an infinite number of superpositions, *i.e.*, through integration. For instance, if we know the stresses, strains and displacements of a

---

[6.4] It must be noted that the applicability of the superposition principle is also limited by the linearity of boundary conditions. Furthermore, if two sets of external loads have a coupled non-linear effect on the deformation, the superposition principle does not apply. In the buckling analysis of a bar, for instance, if the summation of two axial loads exceeds the critical buckling load of the bar while each of them is below the critical load, one cannot get a correct solution by superposing those with a single axial load. Detailed discussions can be found in many other textbooks such as those by Fung (1965) and Xie, Lin and Ding (1988).

foundation under a normal concentrated force, the solution for a foundation under an arbitrarily distributed pressure over a surface area can be obtained by integrating the concentrated force solution over the area, see **Fig. 6.10**. This is the way for us to obtain the solution for contact problems such as the contact in bearings. In Chapter 7, we shall discuss it further.

In short, a wise use of the principle of superposition can greatly facilitate the solution of complex engineering problems. In the following chapters, we shall see more examples of the application of the principle.

Example 6.7 The stress components in a plane-stress elastic plate under an in-plane bending, as shown in **Fig. E6.7a**, are known as $\sigma_{yy} = \sigma_{xy} = 0$ and $\sigma_{xx} = 6ay$. On the other hand, those in the same plate under an in-plane shearing, as illustrated in **Fig. E6.7b**, are $\sigma_{xx} = \sigma_{yy} = 0$ and $\sigma_{xy} = -a$. Find the stresses in the same elastic plate under the combined loading shown in **Fig. E6.7c**.

Solution: Since the plate is under small elastic deformation, the principle of superposition applies. Thus the stresses in the plate under combined loading in **Fig. E6.7c** are simply the summation of those of cases (a) and (b). Hence, the solution is

$$\sigma_{xx} = 6ay, \sigma_{yy} = 0, \sigma_{xy} = -a.$$

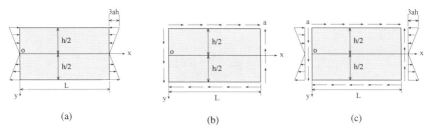

Figure E6.7

6.3 SOLUTION APPROACHES AND SKILLS

6.3.1 Approaches

When a proper mechanics model with correctly specified boundary conditions is available, the next step is to get the solution to three groups of unknowns, *i.e.*, displacements (u, v, w), strains (ε_{xx}, ε_{yy}, ε_{zz}, ε_{xy}, ε_{yz}, ε_{zx}) and stresses (σ_{xx}, σ_{yy}, σ_{zz}, σ_{xy}, σ_{yz}, σ_{zx}), by solving the equations. Although under some special conditions, such as plane deformation, the equations can be simplified to a great extent, it is generally impossible to solve the unknowns all together. We often take one group of the unknowns or some unknowns from different groups as the basic variables, solve them first, and then obtain the rest. In this way, we have four essential solution strategies: *displacement approach*, *stress approach*, *strain approach* and *mixed approach*. These approaches are important not only to analytical solutions but also to the development of numerical methods, such as the finite element method. [6.5]

In the displacement approach, u, v and w are taken as the basic unknown variables to solve first. To do this, the other two groups of unknowns, strains (ε_{xx}, ε_{yy}, ε_{zz}, ε_{xy}, ε_{yz}, ε_{zx}) and stresses (σ_{xx}, σ_{yy}, σ_{zz}, σ_{xy}, σ_{yz}, σ_{zx}), must be eliminated from the basic equations. The substitution of the geometrical equations, Eq. (4-7), into the stress-strain relations, Eq. (5-11), leads to the relations between stresses and displacements. When these relations are then substituted into the equations of motion, Eq. (3-46), we obtain three equations that involve displacements only,[6.6] *i.e.*,

$$\begin{cases} (\lambda+\mu)\dfrac{\partial I_1^\varepsilon}{\partial x}+\mu\nabla^2 u+\rho f_x = \rho a_x, \\ (\lambda+\mu)\dfrac{\partial I_1^\varepsilon}{\partial y}+\mu\nabla^2 v+\rho f_y = \rho a_y, \\ (\lambda+\mu)\dfrac{\partial I_1^\varepsilon}{\partial z}+\mu\nabla^2 w+\rho f_z = \rho a_z, \end{cases} \qquad (6\text{-}18)$$

where λ and μ are Lamé's constants given by Eq. (5-10), ∇^2 is Laplace operator defined by

$$\nabla^2 = \frac{\partial^2}{\partial x^2}+\frac{\partial^2}{\partial y^2}+\frac{\partial^2}{\partial z^2} \qquad (6\text{-}19)$$

[6.5] The displacement approach has been a key in the development of the finite element method. The stress and mixed approaches have also played an important role. Chapter 10 of this book explains the fundamentals of the finite element method using the displacement approach.

[6.6] Equation (6-18) is called the Navier equation to recognise the French scientist, Louis M. H. Navier (1785-1836), who first obtained this set of equations. It can be proved that displacements are biharmonic functions.

and I_1^ε is the first strain invariant given by Eq. (4-10a) or Eq. (5-17) that can alternatively be expressed by displacements as

$$I_1^\varepsilon = \frac{\partial u}{\partial x} + \frac{\partial v}{\partial y} + \frac{\partial w}{\partial z}$$

by using the strain-displacement relations, Eq. (4-7). When u, v and w are obtained by solving Eq. (6-18) in conjunction with the corresponding displacement boundary conditions of a problem, strains can be calculated from geometrical equations and then stresses from the generalised Hooke's law. Of course, the stresses calculated must also satisfy the stress boundary conditions of the problem.

Similarly, in the stress approach, we take stress components[6.7] as the basic unknowns, solve them first with corresponding stress boundary conditions of a problem and then calculate strains[6.8] and displacements. The displacements must satisfy the displacement conditions of the problem. The strain approach is also similar in principle but in this approach strains[6.9] are first solved.

In many cases, it may be easier to solve some stresses, some displacements and some strains but not all the variables of a single group. This is the so-called 'mixed approach'. Solution procedures here differ from case to case. This approach is effective when using the semi-inverse methods to be discussed later.

In principle, all the approaches above are equivalent. However, when using different approaches, the degree of difficulty in solving the same problem may be very different. This mainly depends on the number of unknown variables, the complexity of the equations and the nature of the boundary conditions. Usually, according to a given problem, we would choose a method with fewer unknowns and simpler forms of equations and boundary conditions.

6.3.2 Solution Skills

With any of the above solution approaches, we need to solve the resulting partial differential equations in terms of stresses (stress approach), displacements (displacement approach), strains (strain approach), or a combination of them (mixed approach) for

---

[6.7] It has been proved that stress components are also biharmonic functions.
[6.8] In this case, we must make sure that the strains obtained satisfy all the compatibility equations, Eq. (4-15).
[6.9] Mathematically, strains have also been proved to be biharmonic functions.

unknowns. Among various existing methods,[6.10] the *semi-inverse method* proposed by Saint-Venant[6.11] is the most effective in practice for obtaining analytical solutions. This method first assumes part of the solution based on the understanding of the deformation of a problem, such as the deformation symmetry discussed before, or obtains part of the solution through some other analysis, and then tries to determine the rest rationally so that all the basic equations and boundary conditions are satisfied. As part of the solution has been assumed or obtained, the solution may be proceeded with less difficulty.

The following examples will help us to gain a deeper understanding of the solution processes associated with the displacement approach, the stress approach and the mixed approach, respectively, when using the semi-inverse method. From there we can also experience the way of using the solutions for practical design.

6.3.2.1 *A Cylinder under Inner and Outer Pressure: Displacement Approach*

A circular hollow cylinder subjected to uniform pressure both internally and externally, as shown in **Fig. 6.11**, is a general mechanics model for a large number of engineering structures. For example, the pipes that transport oil from the bottom of the sea are subjected to an inner oil pressure and an outer water pressure and can be approximately treated as the above mechanics problem. A long pressure vessel is a special case of the model when the external pressure vanishes.

Now let us see how to obtain the displacements, strains and stresses in the cylinder by using the solution skills just outlined.

Since the loading and shape of the cylinder are both symmetrical about the z-axis, it is certainly more convenient to use a polar coordinate system in our analysis. When the cylinder is very long, it is under plane-strain deformation. The boundary conditions can be described as

---

[6.10] These include the inverse method, which requires high skills, the direct integration method, which is only applicable to very simple problems, the semi-inverse method that we are going to discuss in the following section in detail, and the numerical methods to be introduced briefly in Chapter 10 of this book.

[6.11] Barre de Saint-Venant (1797-1886), a great French scientist, made remarkable contributions to solid mechanics. His general solution of the problems of torsion and bending of prismatical bars, which first used the semi-inverse method, is of great importance to engineering.

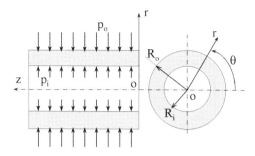

Figure 6.11 A cylinder under inner and outer pressure.

$$\text{at } r = R_i, \ \sigma_{rr} = -p_i, \ \sigma_{r\theta} = 0, \ \sigma_{rz} = 0; \quad (6\text{-}20a)$$

$$\text{at } r = R_o, \ \sigma_{rr} = -p_o, \ \sigma_{r\theta} = 0, \ \sigma_{rz} = 0. \quad (6\text{-}20b)$$

Because the deformation is axisymmetric and under plane-strain, our previous discussion on mechanics modelling indicates that the deformation is independent of coordinates z and θ and that circumferential and axial displacements, v and w, are zero, *i.e.*,

$$\begin{cases} u = u(r), \\ v = 0, \\ w = 0. \end{cases} \quad (6\text{-}21)$$

Using the above displacement expressions, the geometrical equations, Eq. (4-8), reduce to

$$\varepsilon_{rr} = \frac{du}{dr}, \ \varepsilon_{\theta\theta} = \frac{u}{r}, \ \varepsilon_{zz} = 0, \ \varepsilon_{z\theta} = 0, \ \varepsilon_{r\theta} = 0, \ \varepsilon_{rz} = 0. \quad (6\text{-}22)$$

Thus the substitution of such a strain-displacement relationship into the generalised Hooke's law, Eq. (5-11) or Eq. (6-5), gives rise to

$$\begin{cases} \sigma_{rr} = \dfrac{E}{(1+v)(1-2v)}\left[(1-v)\dfrac{du}{dr}+v\dfrac{u}{r}\right], \\ \sigma_{\theta\theta} = \dfrac{E}{(1+v)(1-2v)}\left[(1-v)\dfrac{u}{r}+v\dfrac{du}{dr}\right], \\ \sigma_{zz} = \dfrac{Ev}{(1+v)(1-2v)}\left[\dfrac{du}{dr}+\dfrac{u}{r}\right], \\ \sigma_{r\theta} = 0, \\ \sigma_{z\theta} = 0, \\ \sigma_{rz} = 0. \end{cases} \qquad (6\text{-}23)$$

On the other hand, as the cylinder is in static equilibrium, the equations of motion become equations of equilibrium. With this special axisymmetric and plane-strain stress state described by Eq. (6-23), the second and third equations of Eq. (3-50) are satisfied automatically and the first one leads to

$$\frac{d\sigma_{rr}}{dr}+\frac{\sigma_{rr}-\sigma_{\theta\theta}}{r}=0. \qquad (6\text{-}24)$$

The substitution of Eq. (6-23) into Eq. (6-24) immediately brings about

$$\frac{d^2u}{dr^2}+\frac{1}{r}\frac{du}{dr}-\frac{u}{r^2}=0. \qquad (6\text{-}25)$$

It is a linear and ordinary differential equation of the second order. Its solution[6.12] can be easily obtained as

$$u = c_1 r + c_2 \frac{1}{r}, \qquad (6\text{-}26)$$

where c_1 and c_2 are constants to be determined. Now, strains can be obtained directly by Eq. (6-22) and stresses be derived using Eq. (6-23), *i.e.*,

---

[6.12] Most textbooks on ordinary differential equations describe the solution procedures of this equation. A simple way is to let $u = \Sigma c_n r^n$, where c_n is an arbitrary constant. The substitution of u into Eq. (6-25) gives rise to $n^2 - 1 = 0$. Thus $n = \pm 1$. Hence, $u = c_{-1}/r + c_1 r$, which is the solution given in Eq. (6-26).

$$\begin{cases} \varepsilon_{rr} = c_1 - c_2 \dfrac{1}{r^2}, \\ \varepsilon_{\theta\theta} = c_1 + c_2 \dfrac{1}{r^2}, \\ \varepsilon_{zz} = 0, \ \varepsilon_{\theta z} = 0, \ \varepsilon_{zr} = 0, \end{cases} \qquad (6\text{-}27)$$

$$\begin{cases} \sigma_{rr} = A - \dfrac{B}{r^2}, \\ \sigma_{\theta\theta} = A + \dfrac{B}{r^2}, \end{cases} \qquad (6\text{-}28)$$

where

$$A = \frac{Ec_1}{(1+v)(1-2v)}, \quad B = \frac{Ec_2}{1+v}, \qquad (6\text{-}29)$$

which are still the constants to be determined.

In the above process of obtaining the solution of Eqs. (6-26) to (6-28), we have used equations of equilibrium, geometrical equations and stress-strain relations. Since we solved displacement first, strains obtained must have satisfied the equations of compatibility automatically. Hence, the solution satisfies all the basic equations.

The problem left now is to see whether the solution can satisfy the boundary conditions or not. To examine this, substituting the stresses, Eq. (6-28), into Eq. (6-20), we get two linear equations with two unknown constants, A and B, that is

$$A - \frac{B}{R_i^2} = -p_i,$$

$$A - \frac{B}{R_o^2} = -p_o,$$

which results in

$$A = \frac{p_i R_i^2 - p_o R_o^2}{R_o^2 - R_i^2}, \quad B = \frac{(p_i - p_o) R_i^2 R_o^2}{R_o^2 - R_i^2}. \qquad (6\text{-}30)$$

Hence, the complete solutions that satisfy both the basic equations and boundary conditions are

(a) displacements:

$$\begin{cases} u = \dfrac{(1+v)(1-2v)}{E}\left(\dfrac{p_i R_i^2 - p_o R_o^2}{R_o^2 - R_i^2}\right)r + \dfrac{1+v}{E}\left[\dfrac{(p_i - p_o)R_i^2 R_o^2}{R_o^2 - R_i^2}\right]\dfrac{1}{r}, \\ v = 0, \ w = 0; \end{cases} \quad (6\text{-}31)$$

(b) strains:

$$\begin{cases} \varepsilon_{rr} = \dfrac{(1+v)(1-2v)}{E}\left(\dfrac{p_i R_i^2 - p_o R_o^2}{R_o^2 - R_i^2}\right) - \dfrac{1+v}{E}\left[\dfrac{(p_i - p_o)R_i^2 R_o^2}{R_o^2 - R_i^2}\right]\dfrac{1}{r^2}, \\ \varepsilon_{\theta\theta} = \dfrac{(1+v)(1-2v)}{E}\left(\dfrac{p_i R_i^2 - p_o R_o^2}{R_o^2 - R_i^2}\right) + \dfrac{1+v}{E}\left[\dfrac{(p_i - p_o)R_i^2 R_o^2}{R_o^2 - R_i^2}\right]\dfrac{1}{r^2}, \\ \varepsilon_{zz} = 0, \ \varepsilon_{\theta z} = 0, \ \varepsilon_{zr} = 0, \ \varepsilon_{r\theta} = 0; \end{cases} \quad (6\text{-}32)$$

(c) stresses:

$$\begin{cases} \sigma_{rr} = \dfrac{p_i R_i^2 - p_o R_o^2}{R_o^2 - R_i^2} - \left[\dfrac{(p_i - p_o)R_i^2 R_o^2}{R_o^2 - R_i^2}\right]\dfrac{1}{r^2}, \\ \sigma_{\theta\theta} = \dfrac{p_i R_i^2 - p_o R_o^2}{R_o^2 - R_i^2} + \left[\dfrac{(p_i - p_o)R_i^2 R_o^2}{R_o^2 - R_i^2}\right]\dfrac{1}{r^2}, \\ \sigma_{zz} = 2v\left(\dfrac{p_i R_i^2 - p_o R_o^2}{R_o^2 - R_i^2}\right), \\ \sigma_{z\theta} = 0, \ \sigma_{r\theta} = 0, \ \sigma_{zr} = 0. \end{cases} \quad (6\text{-}33)$$

The whole solution process above represents the general way of using the displacement approach, which can be summarised more clearly in the flow chart shown in **Fig. 6.12**, where 'known' displacements indicate that the expressions of displacements may have been obtained by another method or some of them may have been determined during the modelling analysis. In the above example of cylinder deformation, we know, based on the analysis of symmetry and plane-deformation, that v and w are all zero and thus are 'known'. We therefore only need to solve the radial displacement u. However, we must check all the basic equations and boundary conditions to see if the 'known' displacements are the correct solution or not. This is the characteristic of the semi-inverse method.

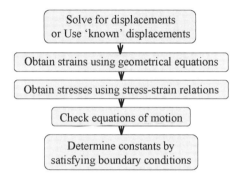

Figure 6.12 The solution procedure when using the semi-inverse method with displacement approach.

As shown previously, when the solution of a plane-strain problem is developed, the corresponding plane-stress solution can be obtained easily by replacing the relevant elastic constants in the solution. Now, if we replace E and v in the above long cylinder solution, Eq. (6-31) to Eq. (6-33), by $\dfrac{E(1+2v)}{(1+v)^2}$ and $\dfrac{v}{1+v}$, respectively, we shall get the solution for a thin circular ring with the same inner and outer radii under the same inner and outer pressure. By doing this and considering the difference in σ_{zz}, ε_{zz} and w as listed in Eqs. (6-16) and (6-17), we get

(a) displacements:

$$\begin{cases} u = \dfrac{1-v}{E}\left(\dfrac{p_i R_i^2 - p_o R_o^2}{R_o^2 - R_i^2}\right)r + \dfrac{1+v}{E}\left[\dfrac{(p_i - p_o)R_i^2 R_o^2}{R_o^2 - R_i^2}\right]\dfrac{1}{r}, \\ v = 0, \\ w = -\dfrac{2v}{E}\left(\dfrac{p_i R_i^2 - p_o R_o^2}{R_o^2 - R_i^2}\right)z + C; \end{cases} \qquad (6\text{-}34)$$

(b) strains:

$$\begin{cases} \varepsilon_{rr} = \dfrac{1-v}{E}\left(\dfrac{p_i R_i^2 - p_o R_o^2}{R_o^2 - R_i^2}\right) - \dfrac{1+v}{E}\left[\dfrac{(p_i - p_o)R_i^2 R_o^2}{R_o^2 - R_i^2}\right]\dfrac{1}{r^2}, \\ \varepsilon_{\theta\theta} = \dfrac{1-v}{E}\left(\dfrac{p_i R_i^2 - p_o R_o^2}{R_o^2 - R_i^2}\right) + \dfrac{1+v}{E}\left[\dfrac{(p_i - p_o)R_i^2 R_o^2}{R_o^2 - R_i^2}\right]\dfrac{1}{r^2}, \\ \varepsilon_{zz} = -\dfrac{v}{E}\left(\dfrac{p_i R_i^2 - p_o R_o^2}{R_o^2 - R_i^2}\right), \\ \varepsilon_{r\theta} = 0,\ \varepsilon_{\theta z} = 0,\ \varepsilon_{zr} = 0; \end{cases} \quad (6\text{-}35)$$

(c) stresses:

$$\begin{cases} \sigma_{rr} = \dfrac{p_i R_i^2 - p_o R_o^2}{R_o^2 - R_i^2} - \left[\dfrac{(p_i - p_o)R_i^2 R_o^2}{R_o^2 - R_i^2}\right]\dfrac{1}{r^2}, \\ \sigma_{\theta\theta} = \dfrac{p_i R_i^2 - p_o R_o^2}{R_o^2 - R_i^2} + \left[\dfrac{(p_i - p_o)R_i^2 R_o^2}{R_o^2 - R_i^2}\right]\dfrac{1}{r^2}, \\ \sigma_{zz} = 0,\ \sigma_{z\theta} = 0,\ \sigma_{r\theta} = 0,\ \sigma_{zr} = 0. \end{cases} \quad (6\text{-}36)$$

In the expression of w, which is the displacement in the z-direction, C is an integration constant and can be determined by displacement boundary conditions. For instance, if at $z = 0$, $w = 0$, then $C = 0$.

The above solutions imply a piece of very important information for our design and material selection. *The two major stresses in the cylinder, σ_{rr} and $\sigma_{\theta\theta}$, whether it is under plane-stress or plane-strain deformation, are independent of material constants.* This means that a cylinder made of any material, *e.g.*, mild steel, stainless steel, or plastics, will have the same values of stresses. Thus if stresses σ_{rr} and $\sigma_{\theta\theta}$ are the major concern in a design, we can select any material provided that it will not fail under the specified inner and outer pressure. However, the above solutions also show that the major displacement, u, does depend on material properties, E and v. Since u is proportional to 1/E, a cylinder made of a material with a larger E (*e.g.*, steel) will deform less than that with a smaller E (*e.g.*, polymer). Hence, if high stiffness is a concern in a design, we must select a material with a sufficiently large E. Poisson's ratio, v, does not contribute much to the variation of u because it only varies in the range of 0 to 0.5 for isotropic materials.

In addition, we also find that the above result, Eqs. (6-34) to (3-36), is the summation of the solution of the cylinder under an inner pressure p_i and the solution under an outer pressure p_o. This means that we can also use the principle of superposition to get the solution.

Let us now have a further discussion about the solution of the case with only an inner pressure because of its importance to engineering structures. Since p_o is zero, σ_{rr} and $\sigma_{\theta\theta}$ reduce to

$$\begin{cases} \sigma_{rr} = \dfrac{p_i R_i^2}{R_o^2 - R_i^2} - \left(\dfrac{p_i R_i^2 R_o^2}{R_o^2 - R_i^2}\right)\dfrac{1}{r^2}, \\ \sigma_{\theta\theta} = \dfrac{p_i R_i^2}{R_o^2 - R_i^2} + \left(\dfrac{p_i R_i^2 R_o^2}{R_o^2 - R_i^2}\right)\dfrac{1}{r^2}. \end{cases} \quad (6\text{-}37)$$

Because $r \le R_o$, the radial stress σ_{rr} is always negative and the circumferential stress $\sigma_{\theta\theta}$ is always positive. Thus $\sigma_{\theta\theta} \ge \sigma_{zz} \ge \sigma_{rr}$. As shear stresses are all zero, $\sigma_{\theta\theta}$, σ_{zz} and σ_{rr} are principal stresses and $\sigma_1 = \sigma_{\theta\theta}$, $\sigma_2 = \sigma_{zz}$ and $\sigma_3 = \sigma_{rr}$. It is important to note that when the outer radius of the cylinder becomes very large, or mathematically lets $R_o \to \infty$, the above stresses reduce to

$$\begin{cases} \sigma_{rr} = -p_i R_i^2 \dfrac{1}{r^2}, \\ \sigma_{\theta\theta} = p_i R_i^2 \dfrac{1}{r^2}. \end{cases}$$

Stresses at $r = R_i$ do not vary much due to the increase of R_o, which means that if we want to reduce the stresses near the inner surface of a cylinder in a design, increasing the thickness of the cylinder (*i.e.*, increasing R_o) is not cost-effective. It wastes the material without much improvement.

6.3.2.2 A Rotating Disk under Outer Pressure: Mixed Approach

Stresses and deformation in disks rotating at high speed are critical in the design of spinning devices and machine components, such as steam and gas turbines, disk saws and grinding wheels. We need to know the stresses and displacements at a given rotation speed so that proper materials can be selected or adequate manufacturing methods can be used to guarantee a reliable strength. In many cases the disks are not flat and are often thicker near the hub. However, with a disk of variable thickness, we have to use a numerical method for the solution. Since our objective here is to understand the mechanics and solution skills involved, we prefer to consider a flat disk rotating at a uniform angular velocity ω when it is subjected to an outer pressure p_o at the same time, as shown in **Fig. 6.13a**. The two flat surfaces are stress-free. When the radius of the disk is much larger than its thickness, it is a plane-stress problem. Since the disk is under uniform rotation, there is no circumferential acceleration. The only body force is the centrifugal force in the radial direction r, which is $\rho r\omega^2 (dr) r(d\theta)$ when we consider an infinitesimal element at r with a radial dimension dr and circumferential dimension $r(d\theta)$, as illustrated in **Fig. 6.13b**. Obviously, the disk is under an axisymmetric deformation about its rotating axis z. Therefore, the stresses and strains in the disk could be

$$\begin{cases} \sigma_{rr} = \sigma_{rr}(r), \ \sigma_{\theta\theta} = \sigma_{\theta\theta}(r), \\ \sigma_{zz} = 0, \ \sigma_{r\theta} = 0, \ \sigma_{rz} = 0, \ \sigma_{z\theta} = 0, \\ u = u(r), \\ v = 0, \\ w = w(r,z) = -\frac{v}{E} \int (\sigma_{rr} + \sigma_{\theta\theta}) \ dz. \end{cases} \quad (6\text{-}38)$$

If all the basic equations and boundary conditions can be satisfied, the above analysis will be correct.

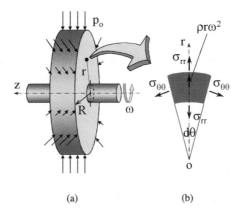

(a)   (b)

Figure 6.13 Forces on an infinitesimal element isolated from the rotating disk.

The boundary conditions of the disk can be described as

$$\text{at } r = R, \quad \sigma_{rr} = -p_0, \quad \sigma_{r\theta} = 0, \quad \sigma_{rz} = 0. \tag{6-39a}$$

In addition, at the centre of the disk along z-axis, we must have

$$\sigma_{rr} = \sigma_{\theta\theta}, \quad u = 0, \tag{6-39b}$$

because θ- and r-directions are identical at r = 0. If u ≠ 0 at the disk centre, a gap appears along z-axis.

Now we try to use the 'known solution', Eq. (6-38), to find the rest of the stresses, strains and displacements that must satisfy all the basic equations and boundary conditions. Using Eq. (6-38), the second and third equations of motion are satisfied directly and the first leads to

$$\frac{d}{dr}(r\sigma_{rr}) - \sigma_{\theta\theta} + \rho\omega^2 r^2 = 0. \tag{6-40}$$

The geometrical equations, Eq. (4-8), reduce to

$$\varepsilon_{rr} = \frac{du}{dr}, \quad \varepsilon_{\theta\theta} = \frac{u}{r}, \quad \varepsilon_{zz} = \frac{\partial w}{\partial z}, \quad \varepsilon_{z\theta} = 0, \quad \varepsilon_{r\theta} = 0, \quad \varepsilon_{rz} = 0. \tag{6-41}$$

Then according to the generalised Hooke's law, Eq. (5-12), and the first two equations of the above, we obtain

$$\frac{du}{dr} = \frac{1}{E}(\sigma_{rr} - \nu\sigma_{\theta\theta}) \tag{6-42}$$

and

$$\frac{u}{r} = \frac{1}{E}(\sigma_{\theta\theta} - \nu\sigma_{rr}). \tag{6-43}$$

Equations (6-40), (6-42) and (6-43) are the ones to solve for three unknowns, u, $\sigma_{\theta\theta}$ and σ_{rr}. (Note that the unknown w is not independent.) Let us first eliminate two of the unknowns, say, u and $\sigma_{\theta\theta}$ from these equations. To do this, substitute Eq. (6-43) into Eq. (6-42), which brings about

$$\frac{d}{dr}\left[r(\sigma_{\theta\theta} - \nu\sigma_{rr})\right] = \sigma_{rr} - \nu\sigma_{\theta\theta}. \tag{6-44}$$

On the other hand, Eq. (6-40) indicates that

$$\sigma_{\theta\theta} = \frac{d}{dr}(r\sigma_{rr}) + \rho\omega^2 r^2. \tag{6-45}$$

Thus Eq. (6-44) becomes an ordinary differential equation with σ_{rr} only, *i.e.*,

$$\frac{d^2}{dr^2}(r\sigma_{rr}) + \frac{1}{r}\frac{d}{dr}(r\sigma_{rr}) - \frac{1}{r^2}(r\sigma_{rr}) + (3+\nu)\rho\omega^2 r = 0. \tag{6-46}$$

Comparing this with Eq. (6-25), we can see that the above equation should have a similar general solution to Eq. (6-26), *i.e.*,

$$r\sigma_{rr} = Ar - B\frac{1}{r},$$

or

$$\sigma_{rr} = A - B\frac{1}{r^2}.$$

However, because of the last term in Eq. (6-46), it must have a particular solution that can be found as

$$-\frac{3+\nu}{8}\rho\omega^2 r^2.$$

Thus the complete solution of Eq. (4-46) is

$$\sigma_{rr} = A - B\frac{1}{r^2} - \frac{3+\nu}{8}\rho\omega^2 r^2 \tag{6-47}$$

with two integration constants, A and B, to be determined by the corresponding boundary conditions. Using Eqs. (6-40) and (6-43) again, we obtain

$$\sigma_{\theta\theta} = A + B\frac{1}{r^2} - \frac{1+3\nu}{8}\rho\omega^2 r^2, \qquad (6\text{-}48)$$

$$u = \frac{r}{E}\left[(1-\nu)A + (1+\nu)B\frac{1}{r^2} - \frac{1-\nu^2}{8}\rho\omega^2 r^2\right]. \qquad (6\text{-}49)$$

As specified in Eq. (6-39b), at the centre of the disk (r = 0), u must be zero. This means that B must be zero in the above expressions, because otherwise $u \to \infty$ when $r \to 0$. With B = 0, the other condition of $\sigma_{rr} = \sigma_{\theta\theta}$ is met automatically. The other constant, A, is determined by the boundary condition, Eq. (6-39a), which yields

$$A = \frac{3+\nu}{8}\rho\omega^2 R^2 - p_o.$$

We therefore finally obtain all the unknowns based on the result of our understanding of the disk deformation, Eq. (6-38), *i.e.*,

(a) stresses:

$$\begin{cases} \sigma_{rr} = -p_o + \dfrac{3+\nu}{8}\rho\omega^2\left(R^2 - r^2\right), \\ \sigma_{\theta\theta} = -p_o + \dfrac{3+\nu}{8}\rho\omega^2\left(R^2 - \dfrac{1+3\nu}{3+\nu}r^2\right), \\ \sigma_{zz} = 0,\ \sigma_{zr} = 0,\ \sigma_{z\theta} = 0,\ \sigma_{r\theta} = 0; \end{cases} \qquad (6\text{-}50)$$

(b) strains:

$$\begin{cases} \varepsilon_{rr} = -p_o\dfrac{1-\nu}{E} + \dfrac{1-\nu}{8E}\rho\omega^2\left[(3+\nu)R^2 - 3(1+\nu)r^2\right], \\ \varepsilon_{\theta\theta} = -p_o\dfrac{1-\nu}{E} + \dfrac{1-\nu}{8E}\rho\omega^2\left[(3+\nu)R^2 - (1+\nu)r^2\right], \\ \varepsilon_{zz} = \dfrac{\nu}{4E}\left\{8p_o - \rho\omega^2\left[(3+\nu)R^2 - 2(1+\nu)r^2\right]\right\}, \\ \varepsilon_{zr} = 0,\ \varepsilon_{z\theta} = 0,\ \varepsilon_{r\theta} = 0; \end{cases} \qquad (6\text{-}51)$$

(c) displacements:

$$\begin{cases} u = -p_o \dfrac{1-v}{E} r + \dfrac{1-v}{8E} \rho\omega^2 r\left[(3+v)R^2 - (1+v)r^2\right], \\ v = 0, \\ w = \dfrac{v}{4E}\left\{8p_o - \rho\omega^2\left[(3+v)R^2 - 2(1+v)r^2\right]\right\} z + C, \end{cases} \quad (6\text{-}52)$$

where C in the expression of w in Eq. (6-52) is an integration constant and depends on the displacement boundary conditions of the disk. For example, if $w = 0$ at $z = 0$, $C = 0$.

The above solution of the rotating disk satisfies equations of motion, geometrical equations, the generalised Hooke's law and all the boundary conditions. Although we used a mixed approach and solved for u, σ_{rr} and $\sigma_{\theta\theta}$ together, the strains were calculated based on the displacement u obtained. Thus the equations of compatibility of strains must have been satisfied automatically. Hence, the solution, Eqs. (6-50) to (6-52), is a correct solution for the rotating disk.

Similar to the solution to the cylinder subjected to inner and outer pressure discussed previously, the present solution of the rotating disk also consists of two parts, with one part proportional to the external pressure p_o and the other proportional to the centrifugal force $\rho\omega^2$. This means that the stresses, strains and displacements due to p_o are independent of those induced by $\rho\omega^2$. Thus once again the superposition principle applies. We can show easily that the above solution can be obtained by the summation of the solutions of two individual problems, a static disk under an external pressure p_o and a rotating disk with a uniform angular speed ω, as shown in **Fig. 6.14**. Then in each solution, the formulation will be less cumbersome. The reader may be interested in experiencing the individual solution process as an exercise.

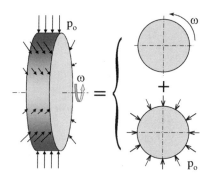

Figure 6.14 Application of superposition principle.

The centrifugal stresses, *i.e.*, the stresses caused by rotation, reach their largest value at the centre of the disk. According to Eq. (6-50),

$$\sigma_{rr}\big|_{r=0} = \sigma_{\theta\theta}\big|_{r=0} = \frac{3+\nu}{8}\rho(\omega R)^2.$$

This provides a useful piece of information for our design when the strength of a disk is concerned. For example, this solution directly predicts that the maximum allowable diameter of a rotating steel disk ($\rho = 7,860$ kg/m^3 and $\nu = 0.29$) with $\omega = 3,000$ rpm can be $2R = 1.585$ m, if the maximum normal stress allowable in the disk is 200 MPa.

The above solution process shows that when using the mixed approach to solve a problem there is no definite procedure to follow. However, the essential is that all the basic equations and boundary conditions of the problem must be satisfied.

Because of the importance of the solid and hollow cylinders to the design of apparatus and structures, we list their major stresses and displacements in **Table 6.1** under individual loading conditions. The solution of a problem with any combination of different loads can be easily obtained using the principle of superposition. The solution to a hollow rotating cylinder in the table has not been discussed in the above text. The reader should be able to get it without any difficulty by determining the constants A and B in Eqs. (6-47) to (6-49) again using the corresponding boundary conditions.

Now we can go back to solving the problem of the interference fit of a pin onto a collar, as discussed in section **1.2.1** and illustrated in **Fig. 1.6**, using the solution listed in **Table 6.1**. Let us assume that after the assembly the interface stress is in the radial direction with a magnitude of σ_{int}, which is obviously compressive. Thus we can treat the collar as a hollow cylinder subjected to an inner pressure σ_{int} and the pin as a solid cylinder subjected to an outer pressure σ_{int}. From **Table 6.1**, the radial displacement at any point on the surface of the pin is

$$u_p = -\sigma_{int}\frac{1-\nu_p}{E_p}R_m, \tag{k}$$

where subscript 'p' is used to denote 'pin'. The radial displacement at any point on the inner surface of the collar can also be obtained from the table as

$$u_c = \sigma_{int}\frac{R_i}{E_c}\frac{R_i^2}{R_o^2-R_i^2}\left[(1-\nu_c)+(1+\nu_c)\frac{R_o^2}{R_i^2}\right], \tag{l}$$

where subscript 'c' indicates 'collar'. In terms of displacements, the deformation compatibility at the fitting interface, represented by Eq. (1-1), can be re-written as

$$u_c - u_p = \delta. \tag{m}$$

Hence, the interface pressure σ_{int} can be obtained easily when Eqs. (k) and (l) are substituted into Eq. (m). This answers the question of the design problem in Chapter 1.

Table 6.1 Major stresses and displacements in solid and hollow circular disk under plane-stress deformation

	Load	σ_{rr}	$\sigma_{\theta\theta}$	u
Solid disk of radius R	Outer pressure p_o	$-p_o$	$-p_o$	$-p_o \dfrac{1-v}{E} r$
	Angular Speed ω	$\dfrac{3+v}{8}\rho\omega^2(R^2 - r^2)$	$\dfrac{3+v}{8}\rho\omega^2\left(R^2 - \dfrac{1+3v}{3+v}r^2\right)$	$\dfrac{1-v}{8E}\rho\omega^2 r\left[(3+v)R^2 - (1+v)r^2\right]$
Hollow disk of radii R_i and R_o	Inner pressure p_i	$p_i \dfrac{R_i^2}{R_o^2 - R_i^2}\left(1 - \dfrac{R_o^2}{r^2}\right)$	$p_i \dfrac{R_i^2}{R_o^2 - R_i^2}\left(1 + \dfrac{R_o^2}{r^2}\right)$	$p_i \dfrac{r}{E}\dfrac{R_i^2}{R_o^2 - R_i^2}\left[(1-v) + (1+v)\dfrac{R_o^2}{r^2}\right]$
	Outer pressure p_o	$-p_o \dfrac{R_o^2}{R_o^2 - R_i^2}\left(1 - \dfrac{R_i^2}{r^2}\right)$	$-p_o \dfrac{R_o^2}{R_o^2 - R_i^2}\left(1 + \dfrac{R_i^2}{r^2}\right)$	$-p_o \dfrac{r}{E}\dfrac{R_o^2}{R_o^2 - R_i^2}\left[(1-v) + (1+v)\dfrac{R_i^2}{r^2}\right]$
	Angular speed ω	$\dfrac{3+v}{8}\rho\omega^2\left(R_o^2 + R_i^2 - \dfrac{R_o^2 R_i^2}{r^2} - r^2\right)$	$\dfrac{3+v}{8}\rho\omega^2\left(R_o^2 + R_i^2 + \dfrac{R_o^2 R_i^2}{r^2} - \dfrac{1+3v}{3+v}r^2\right)$	$\dfrac{(1-v)(3+v)}{8}\dfrac{r}{E}\rho\omega^2\left(R_o^2 + R_i^2 + \dfrac{1+v}{1-v}\dfrac{R_o^2 R_i^2}{r^2} - \dfrac{1+v}{3+v}r^2\right)$

6.3.2.3 A Uniform Bar under Gravity: Stress Approach

To experience the stress approach to solving elastic deformation problems, let us consider a prismatic bar of length L hanging vertically under gravity, as shown in **Fig. 6.15**. The cross-section of the bar can be of any shape, but for convenience of description, let us assume that it is a square of side a. The z-axis coincides with the central axis of the bar and the origin of the Cartesian system is at the hanging end.

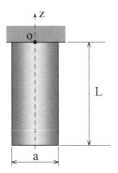

Figure 6.15 A bar under gravity.

The boundary conditions of the bar can be described as

at $z = 0$,
$$u = 0, \; v = 0, \; w = 0. \tag{6-53}$$

On all the other surfaces, both normal and shear stresses are zero. For instance, at the end with $z = -L$,

$$\sigma_{zz} = \sigma_{zy} = \sigma_{zx} = 0. \tag{6-54}$$

Since the bar is only subjected to its own weight in the negative z-direction, *i.e.*, $f_z = -g$, where g is the gravitational acceleration, it may be reasonable to assume that all stress components in the bar vanish except σ_{zz}. Thus as a start of the semi-inverse method, we assume that

$$\sigma_{zz} \neq 0, \; \sigma_{xx} = \sigma_{yy} = \sigma_{xy} = \sigma_{xz} = \sigma_{yz} = 0. \tag{6-55}$$

Now the questions to answer are: 'What is σ_{zz}?' and 'Are the assumed stresses of Eq. (6-55) correct?' These must be examined by satisfying the basic equations and boundary conditions. Nevertheless, the above 'assumed known solution' has met all the stress boundary conditions on the free surfaces of the bar except $\sigma_{zz} = 0$ at $z = -L$.

Because of Eq. (6-55), the first and second equilibrium equations in Eq. (3-47) are satisfied automatically. The third one brings about

$$\frac{\partial \sigma_{zz}}{\partial z} - \rho g = 0, \tag{6-56}$$

where ρ is the density of the bar material. A direct integration of the above gives rise to

$$\sigma_{zz} = \rho g z + C(x, y), \tag{6-57}$$

where $C(x, y)$ is an arbitrary function of x and y due to partial integration with respect to z. To satisfy the boundary condition $\sigma_{zz} = 0$ at $z = -L$ as specified by Eq. (6-54), we must have

$$C(x, y) = \rho g L.$$

Hence,

$$\sigma_{zz} = \rho g (z + L).$$

Therefore, our stress solution

$$\begin{cases} \sigma_{zz} = \rho g(z + L), \\ \sigma_{xx} = 0, \ \sigma_{yy} = 0, \ \sigma_{xy} = 0, \ \sigma_{xz} = 0, \ \sigma_{yz} = 0 \end{cases} \tag{6-58}$$

meets the requirements of both stress boundary conditions and equilibrium equations.

We can now use Eq. (6-58) to obtain strains and displacements. The generalised Hooke's law, Eq. (5-12), leads to

$$\begin{cases} \varepsilon_{xx} = \varepsilon_{yy} = -\frac{\nu}{E}\rho g(z + L), \ \varepsilon_{zz} = \frac{1}{E}\rho g(z + L), \\ \varepsilon_{xy} = 0, \ \varepsilon_{xz} = 0, \ \varepsilon_{yz} = 0. \end{cases} \tag{6-59}$$

As the above strains are calculated from the stress field, Eq. (6-58), which may not be a correct solution, we must examine whether the strains are compatible. Fortunately, they satisfy all six compatibility equations given by Eq. (4-15).

To get displacements, we can use the geometrical equations, Eq. (4-7). Due to the above strain field, the strain-displacement relations become

$$\begin{cases} \dfrac{\partial u}{\partial x} = -\dfrac{v}{E}\rho g(z+L), \\ \dfrac{\partial v}{\partial y} = -\dfrac{v}{E}\rho g(z+L), \\ \dfrac{\partial w}{\partial z} = \dfrac{1}{E}\rho g(z+L), \\ \dfrac{\partial u}{\partial y} + \dfrac{\partial v}{\partial x} = 0, \\ \dfrac{\partial v}{\partial z} + \dfrac{\partial w}{\partial y} = 0, \\ \dfrac{\partial u}{\partial z} + \dfrac{\partial w}{\partial x} = 0. \end{cases} \quad (6\text{-}60)$$

The integration of the first three equations above yields

$$\begin{cases} u = -\dfrac{v}{E}\rho g(z+L)x + f_1(y,z), \\ v = -\dfrac{v}{E}\rho g(z+L)y + f_2(x,z), \\ w = \dfrac{1}{E}\rho g\left(\dfrac{1}{2}z^2 + Lz\right) + f_3(x,y), \end{cases} \quad (6\text{-}61)$$

where f_1, f_2 and f_3 are undetermined functions due to partial integration. By substituting Eq. (6-61) into the last three equations of Eq. (6-60), we obtain

$$\begin{cases} \dfrac{\partial f_1(y,z)}{\partial y} + \dfrac{\partial f_2(x,z)}{\partial x} = 0, \\ \dfrac{\partial f_3(x,y)}{\partial y} + \dfrac{\partial f_2(x,z)}{\partial z} = \dfrac{v}{E}\rho gy, \\ \dfrac{\partial f_3(x,y)}{\partial x} + \dfrac{\partial f_1(y,z)}{\partial z} = \dfrac{v}{E}\rho gx. \end{cases} \quad (6\text{-}62)$$

The above set of partial differential equations is inhomogeneous and is not so easy to solve. However, if we introduce a transform, *i.e.*, let

$$f_3(x,y) = F_3(x,y) + \frac{1}{2}\frac{\nu\rho g}{E}\left(x^2 + y^2\right), \tag{6-63}$$

these equations can be simplified to a set of homogeneous ones,

$$\begin{cases} \dfrac{\partial f_1(y,z)}{\partial y} + \dfrac{\partial f_2(x,z)}{\partial x} = 0, \\ \dfrac{\partial F_3(x,y)}{\partial y} + \dfrac{\partial f_2(x,z)}{\partial z} = 0, \\ \dfrac{\partial F_3(x,y)}{\partial x} + \dfrac{\partial f_1(y,z)}{\partial z} = 0. \end{cases} \tag{6-64}$$

Equation (6-64) can be integrated easily and then the integration constants are normally determined by displacement boundary conditions. (*e.g.*, Timoshenko and Goodier, 1951). Here, we prefer to use a more physical argument to determine these functions. We know that f_1, f_2 and f_3 are part of the displacements of the bar under gravitation. But Eq. (6-64) indicates that f_1, f_2 and F_3 are the displacements arising from the case of $\rho g = 0$, when compared with Eq. (6-62), because it is a set of homogeneous equations. The bar does not displace if the only external force, *i.e.*, the gravitation, vanishes unless there is a rigid body motion. This is certainly not true because the bar is fixed at the end of $z = 0$ and no rigid body displacement can actually happen. Hence, we must have $f_1 = f_2 = F_3 = 0$. Using Eq. (6-63) this conclusion directly leads to

$$\begin{cases} f_1(y,z) = 0, \\ f_2(x,z) = 0, \\ f_3(x,y) = \dfrac{1}{2}\dfrac{\nu\rho g}{E}\left(x^2 + y^2\right). \end{cases}$$

Therefore, the displacements in the bar are

$$\begin{cases} u = -\dfrac{v}{E}\rho g(z+L)x, \\ v = -\dfrac{v}{E}\rho g(z+L)y, \\ w = \dfrac{\rho g}{2E}\left[z^2 + 2Lz + v\left(x^2 + y^2\right)\right] \end{cases} \quad (6\text{-}65)$$

according to Eq. (6-61).

The above displacements satisfy the displacement boundary condition of Eq. (6-53) only at the origin o(0, 0, 0). Hence, although the solution obtained, including stresses, Eq. (6-58), strains, Eq. (6-59), and displacements, Eq. (6-65), satisfies the basic equations and all the other boundary conditions, it is not a correct solution strictly. Fortunately, Saint-Venant's principle to be introduced in the next section will give us an encouraging answer. We will see that this solution is correct in the region of a distance away from the hanging end (z = 0).

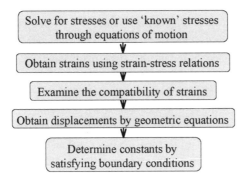

Figure 6.16 Solution procedures when using the stress approach.

Similar to the case in section **6.3.2.1** using the displacement approach, the present solution process in the stress approach can also be summarised by the flow chart shown in **Fig. 6.16**. It is important to emphasise that when using the stress approach the compatibility of strains must be examined as explained in Chapter 4.

According to the displacements given in Eq. (6-65), the shape of the bar after deformation can be obtained, as illustrated in **Fig. 6.17**. It is clear that its originally flat ends are no longer flat after deformation and its peripheral surfaces that are parallel to the z-axis before deformation become inclined after deformation.

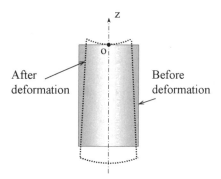

Figure 6.17 The shape of the bar before and after deformation.

6.4 SAINT-VENANT'S PRINCIPLE

In discussing the pure bending of a beam in the nineteenth century, Saint-Venant pointed out that the linear stress distribution over the beam cross-section found by the elementary bending theory, as shown in **Fig. 6.18a**, conforms a rigorous solution only when the external forces applied at the ends of the beam are distributed over the end cross-sections in exactly the same manner as the internal stress distribution, *i.e.*, linear distribution. But the solution obtained will be accurate enough for any other distribution of the forces at the ends, provided that the resultant force and couple of the applied forces always remain unchanged. He concluded that if a system of self-equilibrating forces is distributed on a small portion of the surface of a prism, a substantial deformation will be produced only in the vicinity of these forces. For instance, the two equal and opposite forces that act on a rubber bar, as illustrated in **Fig. 6.18b**, produce only a local deformation at the end, but the remainder is practically unaffected. This finding, which was later on called *Saint-Venant's principle*, has been widely used in engineering practice since it brings about significant simplification in solving stress and deformation problems.

As we have experienced in the previous section, even a simple three-dimensional problem of elasticity may present formidable complications in solution because of the difficulty of satisfying boundary conditions precisely. However, using Saint-Venant's principle we can seek an approximate solution, that is, a solution that satisfies some of the boundary conditions in an approximate manner. Thus the solution of the bar under gravitation given by Eq. (6-65) is a good approximate solution in the zone away from the top end, say in the zone of $|z| > a$, although it only partly satisfies the boundary conditions of Eq. (6-53) if they are examined over the whole end surface in a point wise manner. This is because the deformation at $z = 0$ implied by solution (6-65), as shown in **Fig. 6.17**, only replaces the end loading statically equivalently[6.13] and according to Saint-Venant's principle it does not affect the solution in the zone away from the end.

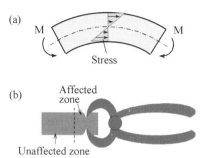

Figure 6.18 (a) A uniform beam under pure bending. (b) a rubber bar subjected to a set of self-equilibrium forces in the local zone near an end.

A general engineering interpretation of Saint-Venant's principle may be described as follows: 'If the forces acting *on a small portion of the surface* of an elastic body are replaced by another *statically equivalent* system of forces acting *on the same portion of the surface*, such redistribution of loading produces substantial changes in the stresses locally but has a negligible effect on the stresses at distances which are large in comparison with the linear dimensions of the surface on which the forces are changed.'

There are two critical aspects in the above description, which are the key to the correct application of Saint-Venant's principle. First, the portion of the surface subjected to force change must be much smaller than the characteristic dimension of the component for solution so that the size of the zone affected by the change is small and the approximate solution is applicable to the unaffected zone. For example, in the solution to the bar under gravity, the length of the bar, L, must be much larger than its cross-sectional dimension, a, so that the boundary condition change at its top end will only have an affected zone of about | z | = a. If L and a are of comparable dimensions, the whole bar will be affected by the boundary condition change and the solution will become meaningless. Secondly, the force replacement must be statically equivalent. This means that the replacement must not change either the resultant force or the resultant couple.

[6.13] We have known that the bar will deform in the manner illustrated by **Fig. 6.17** according to solution (6-64). At the top end of the bar in this case, material moves upwards and inwards. However, to satisfy the hanging condition of **Fig. 6.15** described by Eq. (6-53), the rigid constraint prevents such deformation. This is equivalent to applying a set of self-equilibrium shear stress to the end surface of the bar and distributing the total hanging force, which is equal to the weight of the bar, to the whole end surface.

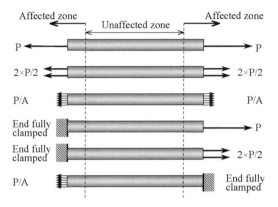

Figure 6.19 A uniform circular bar under tension, where A is the cross-sectional area of the bar.

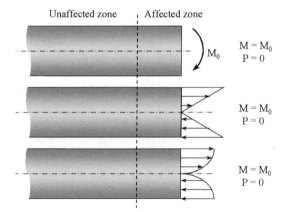

Figure 6.20 A beam under symmetrical bending, where M is the resultant bending moment and P is the resultant axial force.

Figures 6.19 and **6.20** illustrate more examples of equivalent boundary conditions under Saint-Venant's principle. As long as the replacement of the boundary conditions is statically equivalent, a solution obtained under one set of boundary conditions will be applicable to the unaffected zone of the problem under any other set of boundary conditions.

Saint-Venant's principle can also be used to facilitate mechanics modelling and experimental investigation. For example, in designing a long solid steel shaft to transmit power from the engine to the generator shown in **Fig. 6.21a**, at the stage of considering the twisting stiffness of the shaft, we do not need to worry about the effect of connection details at the two ends of the shaft according to Saint-Venant's principle, *i.e.*, the connection of the shaft to the engine and the generator by keyways, couplings or other methods does not affect the stress and deformation in the shaft away from the two ends. Thus we can use a simple mechanics model, a long rotating solid circular bar under a pair of torques, Q, at the two ends, to generate a useful solution to the design, as illustrated in **Fig. 6.21b**. Moreover, it is also a common practice, for instance, that in a uniaxial tensile test the way of holding a specimen has no effect on the stress and deformation in the middle region of the specimen when its dimensions follow the testing standard.[6.14]

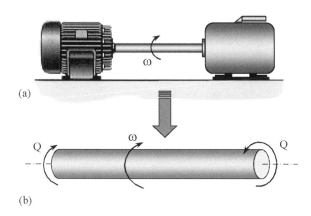

Figure 6.21 An application of Saint-Venant's principle in the mechanics modelling of a transmission shaft.

Example 6.8 A solid circular shaft of a machine, as shown in **Fig. E6.8**, is subjected to tension by a concentrated force P, torsion by a torque Q and bending by a bending moment M in the xz-plane. The radius of the shaft is R and the length is L (L is much larger than R). Describe the boundary conditions.

Solution: Although we know the resultant forces, P, M and Q on the two ends of the shaft, we do not know the exact stress distributions that make these resultant forces. However, since the bar is long (L >> R), Saint-Venant's principle allows us to use statically equivalent conditions by ignoring the detailed stress distributions. Hence, the boundary conditions can be described as follows:

[6.14] A testing standard, such as the Australian, European and US standards, specifies a sufficient length of the specimen to avoid the end effect on the testing result. It is an application of Saint-Venant's principle.

At $z = 0$ and $z = -L$, $0 \leq r \leq R$, $0 \leq \theta \leq 2\pi$,

$$\int_0^{2\pi}\int_0^R \sigma_{zz} r\, dr\, d\theta = P, \quad (a)$$

$$\int_0^{2\pi}\int_0^R \sigma_{zz} r^2 \cos\theta\, dr\, d\theta = M, \quad (b)$$

$$\int_0^{2\pi}\int_0^R \sigma_{zz} r^2 \sin\theta\, dr\, d\theta = 0, \quad (c)$$

$$\int_0^{2\pi}\int_0^R \sigma_{z\theta} r^2\, dr\, d\theta = -Q, \quad (d)$$

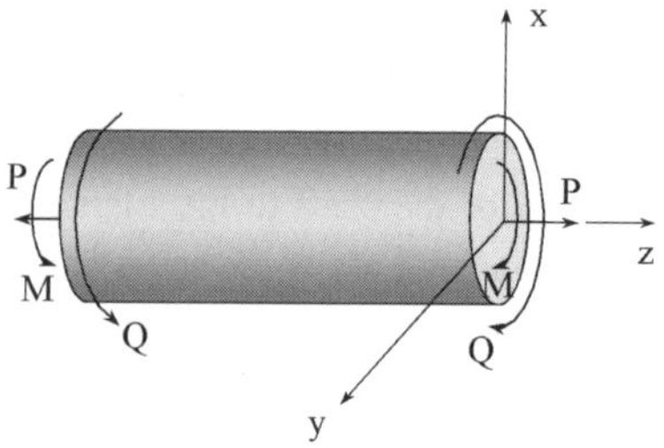

$$\sigma_{zr} = 0. \quad (e)$$

At $r = R$, $-L \leq z \leq 0$, $0 \leq \theta \leq 2\pi$,

$$\sigma_{rr} = \sigma_{r\theta} = \sigma_{rz} = 0. \quad (f)$$

Figure E6.8

In the above, condition (a) means that the resultant axial force of σ_{zz} must be equal to the external load P applied. It is unknown what distribution of the external surface stress in the z-direction causes P. Thus the condition does not specify, in the point-wise manner, the consistency between the internal stress σ_{zz} and the externally applied stress, but instead, uses a statically equivalent condition of axial resultant force in the direction. According to Saint-Venant's principle, it will not introduce approximation in the zone far away from the ends of the shaft. Similarly, conditions (b) to (d) above specify the equivalence of the resultant bending moment in the xz-plane, the resultant bending moment in the yz-plane and the resultant torque in the xy-plane. Condition (e) specifies that the internal shear stress in the r-direction must become zero at any point of the end surface of the shaft and condition (f) defines the stress-free peripheral surface.

References

DeGarmo, EP, Black, JT and Kohser, RA (1997), *Materials and Processes in Manufacturing*, 8th edition, Prentice-Hall, Inc., Englewood Cliffs, NJ, pp.50.

Fung, YC (1965), *Foundations of Solid Mechanics*, Prentice-Hall, Inc., Englewood Cliffs, NJ.

Groover, MP (1996), *Fundamentals of Modern Manufacturing*, Prentice-Hall, Inc., Englewood Cliffs, NJ, pp.597.

Helbawi, H, Zhang, L and Zarudi, I (1999), Applicability of the bonded-interface technique for subsurface damage evaluation, in: *Abrasive Technology: Current Development and Applications I*, edited by J Wang, W Scott and L Zhang, World Scientific, Singapore, pp.445-452. Also to appear in: *International Journal of Mechanical Science*.

Hibbeler, RG (1997), *Mechanics of Materials*, 3rd edition, Prentice-Hall, Inc., Englewood Cliffs, NJ.

Monoruki, N (1996), Study on precise micromachining, *Micromachine*, June, No.15, 3-4.

Timoshenko, SP and Goodier, JN (1951), *Theory of Elasticity*, McGraw-Hill, Singapore.

Xie, Y-Q, Lin, Z-X and Ding, H-J (1988), *Elasticity* (in Chinese), Zhejiang University Press, Zhejiang.

Zhang, L and Tanaka, H (1998), Atomic scale deformation in silicon induced by two-body and three-body contact sliding, *Tribology International*, **31**, 425-433.

Important Concepts

Mechanics modelling	Boundary condition
Stress boundary condition	Displacement boundary condition
Symmetry of deformation	Axisymmetric deformation
Plane-stress	Plane-strain
Stress approach	Displacement approach
Mixed approach	Affected zone
Semi-inverse method	Saint-Venant's principle

Questions

6.1 Why is a correct mechanics modelling a central step in solving an engineering problem?

6.2 Why can an understanding of deformation characteristics of a problem facilitate the solution of the problem?

6.3 What do boundary conditions mean physically and mathematically?

6.4 What is the similarity and difference between two plane deformation states?

6.5 When the solution of a plane-stress problem has been obtained, what is the most convenient way to obtain the solution for a corresponding plane-strain problem?

6.6 What are the major solution approaches?

6.7 Why do we need to examine the strain compatibility in the stress approach to solution but not in the displacement approach? When using a mixed approach, should we examine the strain compatibility?

6.8 How do we judge if a solution obtained is correct or not?

6.9 Why is Saint-Venant's principle so important to the solution of deformation problems in engineering? What are the critical points in the application of this principle? Give some examples in design and manufacturing practice that use the principle.

6.10 The length of a beam is comparable to its cross-sectional dimensions and the bending moment is the resultant of the normal stress over the whole end area. Can Saint-Venant's principle be used in this case for an approximate solution?

Problems

6.1 A rigidly clamped beam with a square cross-section of side length a is subjected to a set of external loads, as illustrated in **Fig. P6.1**, where P is a concentrated shear force on the end surface along the negative direction of the x-axis and M is a bending moment in the xz-plane. Describe the boundary conditions of the beam.

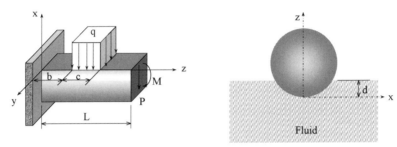

Figure P6.1   **Figure P6.2**

6.2 A ball of radius R and density ρ_b is floating on a fluid of density ρ_f, as illustrated in **Fig. P6.2**. ρ_b is smaller than ρ_f. Describe the boundary conditions of the ball. (*Hint*: Find the depth d first by the overall equilibrium condition of the ball.)

6.3 **Figure P6.3** illustrates the cross-section of the dam of a reservoir. The left side of the dam is subjected to water pressure and the right side and top surface are open to air. The bottom of the dam is embedded rigidly into the rock stratum. The top surface width is w_1, that of the bottom is w_2 and the height is h. Describe the boundary conditions of the dam.

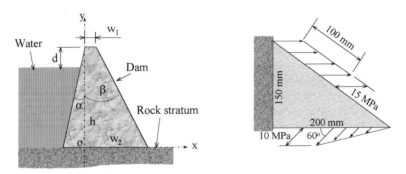

Figure P6.3   **Figure P6.4**

6.4 A plane-strain plate is clamped rigidly by the wall, as illustrated in **Fig. P6.4**. Describe the boundary conditions.

6.5 Describe two examples in engineering practice where the boundary conditions on a certain region of an elastic element are such that two components of surface stresses and one component of displacements are prescribed.

6.6 A plane-stress plate of variable width is subjected to a concentrated tensile force P (**Fig. P6.5**). Find the stresses σ_{yx} and σ_{xx} on an infinitesimal element at a point on the curved surface expressed by σ_{yy}. (*Hint*: Make use of Eq. (6-2).)

6.7 To design an airport it is critical to understand the stress and deformation in the runways during a landing of an aircraft. Assume that the contact stress between one of the aircraft wheels is F(x, y) with its normal component $F_x(x, y)$ and tangential component $F_y(x, y)$. The contact area is an ellipse whose lengths of major and minor axes are 2a and 2b, respectively, as illustrated in **Fig. P6.6**. Define the boundary conditions of the runway during the contact with a single wheel. (*Hint*: Treat the runway foundation as a half space, *i.e.*, it is infinite in the xy-plane and in the positive z-direction.)

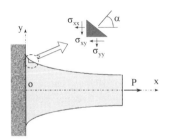

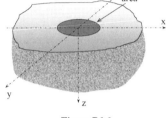

Figure P6.5   **Figure P6.6**

6.8 A bar of constant mass density ρ hangs under its own weight and is supported by the uniform stress $\sigma_0 = \rho g L$ as shown in **Fig. P6.7**. Assume that the stress components σ_{xx}, σ_{yy}, σ_{xy}, σ_{xz}, and σ_{zy} all vanish.

(a) Describe the boundary conditions of the bar.

(b) Based on the above assumption, reduce the fifteen governing equations to seven in terms of σ_{zz}, ε_{xx}, ε_{yy}, ε_{zz}, u, v and w.

(c) Integrate the equilibrium equation to show that $\sigma_{zz} = \rho g z$ and that the prescribed boundary conditions are satisfied by this solution.

(d) Find ε_{xx}, ε_{yy}, ε_{zz} from the generalised Hooke's law.

(e) If the translation and rotation components are zero at the point (x, y, z) = (0, 0, L), determine the displacement components u, v and w.

(f) Compare the case with the example shown in section **6.3.2.3**. Discuss the reason that the present solution is an exact one to the problem of **Fig. P6.7** but that of the example shown in section **6.3.2.3** is only an approximate solution to the problem of **Fig. 6.15**.

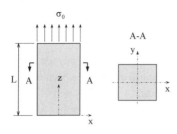
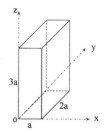

Figure P6.7   **Figure P6.8**

6.9 Show that the stresses $\sigma_{xx} = kxy$, $\sigma_{yy} = kx$, $\sigma_{zz} = \nu kx(1+y)$, $\sigma_{xy} = -ky^2/2$ and $\sigma_{xz,} = \sigma_{yz} = 0$, where k is a constant, represent the solution to a plane-strain problem without body forces if the translation and rotation at the origin are zero. Determine the displacements and the restraining stress in the z-direction. What surface stress distribution must be applied to the prism shown in **Fig. P6.8** so that the given stress components are the solution for this body?

6.10 Show that the stresses $\sigma_{xx} = ky$, $\sigma_{yy} = \sigma_{zz} = \sigma_{xy} = \sigma_{xz,} = \sigma_{yz} = 0$, where k is a constant, are a solution to an elastic body subjected to deformation. Determine the external loading that must be applied to the bar of **Fig. P6.9** so that the given stress components are the solution of the bar. If the applied external load is a couple (bending moment), find the relationship between the stress component σ_{xx} and the couple by making use of Saint-Venant's principle.

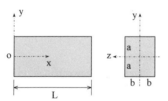

Figure P6.9   **Figure P6.10**

6.11 It is known that
$\sigma_{xx} = pyx^3 - 2axy + by$, $\sigma_{yy} = pxy^3 - 2px^3y$, $\sigma_{xy} = 3px^2y^2/2 + ay^2 + px^4/2 + c$,
$\sigma_{zz} = \sigma_{xz,} = \sigma_{yz} = 0$,
where p, a, b and c are constants.
(a) Show that this stress distribution represents a plane-stress elasticity solution for a thin plate.

(b) If the given stresses act in the thin plate as shown in **Fig. P6.10**, determine constants a, b and c such that there is no shear stress on the edges $x = \pm h/2$ and no normal stress on the edge $x = -h/2$.

(c) Determine the surface forces on each boundary of the plate.

6.12 An irregularly shaped body is subjected to a constant pressure p on its whole surface. Determine the stress, strain and displacement components in the body. Assume that the displacements and rotations at the origin (placed at a point inside the body) are zero. Compare the displacements obtained with the solution to the problem of Example 6.6. (*Hint*: Based on the surface stress, it is reasonable to assume that all normal stresses are equal to a constant and shear stresses vanish. Then use the semi-inverse method to get the solution by following the stress approach.)

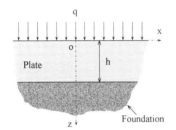

Figure P6.11

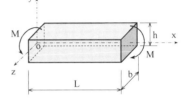

Figure P6.12

6.13 A very large plate is resting on a rigid foundation, as shown in **Fig. P6.11**. The friction between the plate and foundation is very large so that no surface point of the plate can move horizontally. The top surface of the plate is subjected to a uniform pressure p. If the density of the plate is ρ and its thickness is h which is much smaller than the plane dimension of the plate, use the semi-inverse method to find displacements, strains and stresses by (a) displacement approach and (b) stress approach, respectively. (*Hint*: (i) The plate can be considered as infinite in the xy-plane because its thickness is much smaller than its plane dimension. (ii) Based on the loading and supporting conditions, it is reasonable to assume that $u = v = 0$, $w = w(z)$ and that the deformation is axisymmetric about z-axis.)

6.14 (a) Describe the boundary conditions of a uniform beam subjected to pure bending in the xy-plane, see **Fig. P6.12**. (b) Use the displacement approach and the semi-inverse method to find the displacements, strains and stresses in the beam (Body forces are neglected and the beam is in static equilibrium.) To begin with, use the following 'known' displacement field:

$$u = \frac{M}{EI} x y + \omega_y z - \omega_z y + u_0,$$

$$v = -\frac{M}{2EI}\left(x^2 + v y^2 - v z^2\right) + \omega_z x - \omega_x z + v_0,$$

$$w = -\frac{v M}{EI} y z + \omega_x y - \omega_y x + w_0,$$

where E is Young's modulus, v is Poisson's ratio, M is bending moment in the xy-plane, I is the second moment of cross-sectional area of the beam, and u_0, v_0, w_0, ω_x, ω_y and ω_z are undetermined constants.

6.15 Through an approximate analysis one obtained the stress distribution in the plate shown in **Fig. P6.13** as given in strength of material is $\sigma_{xx} = \sigma_0 a/x$ and $\sigma_{yy} = \sigma_{xy} = \sigma_{zz} = \sigma_{xz,} = \sigma_{yz} = 0$. Show that
(a) these stresses do not satisfy the basic equations,
(b) the boundary conditions are not satisfied.

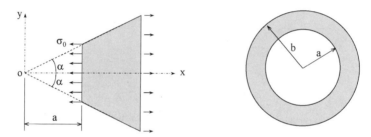

Figure P6.13 **Figure P6.14**

6.16 Find the stress and displacement distribution for an infinitely long hollow cylinder (**Fig. P6.14**) under the following boundary conditions:
(a) At $r = a$, $\sigma_{rr} = \sigma_a$; at $r = b$, $u = u_b$.
(b) At $r = a$, $u = u_a$; at $r = b$, $u = u_b$.
Repeat the problem for the case where the cylinder is very thin.

Chapter 7

APPLICATIONS

This chapter discusses more engineering examples using the solution approaches introduced in Chapter 6. It is expected that the discussion can further enhance our understanding of the process of mechanics modelling, solutions and the analysis of results obtained. Although more complicated problems can only be solved by numerical methods, the fundamental understanding achieved here lays the foundation of successful numerical solutions.

Improvement of structure and machine designs often depends on mechanics analysis. For example, it is not excessive to say that many major developments of the structures of transportation vehicles, such as high-speed trains, aeroplanes and spacecrafts are due to the wise applications of solid mechanics. The above figure shows a super-high-speed train at test (Courtesy of Railway Technical Research Institute, Japan).

7.1 TORSION OF A CIRCULAR SHAFT

Let us consider the deformation of the transmission shaft discussed in the last chapter, as shown in **Fig. 6.20**. Assume that the shaft subjected to a constant torque Q is rotating uniformly with an angular velocity ω, see **Fig. 7.1a**.

7.1.1 Modelling and Solution

According to the principle of superposition, the problem can be resolved into two mechanics models, *i.e.*, a long circular solid bar rotates uniformly with an angular velocity ω, as shown in **Fig. 7.1b**, and a long circular solid bar deforms under the torque Q at its right end and is fully clamped at its left end, as shown in **Fig. 7.1c**. The model in **Fig. 7.1b** is obvious and that in **Fig. 7.1c** is due to Saint-Venant's principle because on the clamped end the total torque is still Q, the total axial load is still zero and thus the change of boundary conditions at the left end is statically equivalent.

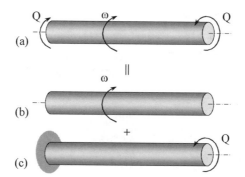

Figure 7.1 Mechanics modelling for the transmission shaft.

We obtained the solution to a rotating circular disk under plane-stress deformation in the last chapter, *i.e.*, Eqs. (6-50) to (6-52). The rotating bar in **Fig. 7.1b** is under plane-strain deformation because it is long. Hence, its solution can be directly obtained by replacing E and ν in Eqs. (6-50) to (6-52) by $\dfrac{E}{1-\nu^2}$ and $\dfrac{\nu}{1-\nu}$, respectively, letting $p_0 = 0$, $w = \varepsilon_{zz} = 0$ and $\sigma_{zz} = \nu(\sigma_{rr} + \sigma_{\theta\theta})$.

Consider the solution to the mechanics model of **Fig. 7.1c**. Assume that the radius of the bar is R and the length is L. For convenience, we use a polar coordinate system with its origin at the left end centre and z-axis coinciding with the central axis of the bar, as shown in **Fig. 7.2**. The boundary conditions can then be described as

$$\text{at } z = 0, \ 0 \leq r \leq R, \ u = v = w = 0; \tag{7-1a}$$

$$\text{at } 0 \leq z \leq L, \ r = R, \ \sigma_{rr} = \sigma_{r\theta} = \sigma_{rz} = 0; \tag{7-1b}$$

$$\text{at } z = L, \ 0 \leq r \leq R, \ \sigma_{zz} = \sigma_{zr} = 0, \tag{7-1c}$$

$$\int_0^{2\pi} \int_0^R r^2 \sigma_{z\theta} \, dr d\theta = Q. \tag{7-1d}$$

Condition (7-1d) is due to the application of Saint-Venant's principle. This is because the difference in distribution of shear stress $\sigma_{z\theta}$ at the end of $z = L$ only alters the stress and deformation in the neighbourhood of the end. According to Saint-Venant's principle, any statically equivalent $\sigma_{z\theta}$ at the end of the bar will give rise to the same solution in the zone far away from $z = L$. We understand that the above argument is reasonable only when L is much larger than R.

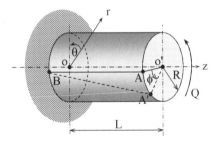

Figure 7.2 The deformation of a long circular solid bar subjected to a pure torsion.

Now consider the possible deformation of the bar. Since the only non-zero stress component over all the stress boundaries of the bar is $\sigma_{z\theta}$ and the end surface at $z = 0$ is fully fixed with $u = v = w = 0$, it is reasonable to assume that the radial displacement, u, and axial displacement, w, are all zero throughout the bar. If we do a simple twisting experiment, we can observe that a straight line O'A on the right end ($z = L$) before deformation will remain a straight line, O'A', after deformation. The included angle, ∠A'O'A, can be denoted by ϕ. Similarly, we can also observe that the straight line, AB, on the peripheral surface before deformation becomes a straight line A'B after deformation. Thus at any point on the surface of the right end, the circumferential displacement, v, is a function of the rotational angle ϕ and the radial coordinate r of the point, *i.e.*, $v|_{z=L} = r\phi$. On the other hand, the circumferential displacement at a point on the peripheral surface should be a linear function of z according to the above observation, *i.e.*, $v|_{r=R} = z/L$, which satisfies the displacement boundary condition (7-1a), *i.e.*, v = 0 at z = 0. Hence, it will be reasonable to assume that at any point throughout the bar, $v = \phi rz/L$. The above consideration gives us the following displacements in the bar under pure torsion:

$$\begin{cases} u = 0, \\ v = \phi \dfrac{zr}{L}, \\ w = 0. \end{cases} \quad (7-2)$$

Now we must check all the basic equations and boundary conditions to see whether the above is the correct displacement solution or not. Substituting Eq. (7-2) into the geometrical equation, Eq. (4-8), gives rise to strain components, *i.e.*,

$$\begin{cases} \varepsilon_{z\theta} = \dfrac{\phi}{2L} r, \\ \varepsilon_{rr} = \varepsilon_{\theta\theta} = \varepsilon_{zz} = \varepsilon_{r\theta} = \varepsilon_{zr} = 0, \end{cases} \quad (7-3)$$

with $\varepsilon_{z\theta}$ being the only non-vanishing strain. Since we obtain strains by differentiating displacements, the compatibility equations of strains must have been satisfied automatically. Using the generalised Hooke's law, Eq. (5-11), we get

$$\begin{cases} \sigma_{z\theta} = \dfrac{\phi G}{L} r, \\ \sigma_{rr} = \sigma_{\theta\theta} = \sigma_{zz} = \sigma_{r\theta} = \sigma_{zr} = 0, \end{cases} \quad (7-4)$$

where G is the shear modulus of the bar material as defined by Eq. (5-15). In obtaining the above solutions, we have used all the basic equations and displacement boundary conditions. On the other hand, we can see that the stress solution, Eq. (7-4), satisfies conditions (7-1b) and (7-1c) automatically. However, to meet condition (7-1d), we must have

$$\int_0^{2\pi}\int_0^R r^3 \frac{\phi G}{L} dr d\theta = Q, \tag{7-5}$$

which results in

$$\phi = \frac{2LQ}{\pi GR^4}. \tag{7-6}$$

Hence, when we use the φ determined above, the stresses in Eq. (7-4) satisfy all the stress boundary conditions. The displacements, Eq. (7-2), strains, Eq. (7-3), and stresses, Eq. (7-4), are therefore the correct solutions to the problem. The complete solution of the bar of **Fig. 7.1a** considering its rotation can be easily obtained by superposition.

It is clear that we have used the semi-inverse method following the displacement approach in the above solution. For this problem, we can also use the stress approach. The stress boundary conditions indicate that the non-vanishing stress in the bar is $\sigma_{z\theta}$ and that it is a function of r only because under pure torsion we can imagine that any cross-section in the bar will have the same stress state. Thus we can start with

$$\begin{cases} \sigma_{z\theta} = \sigma_{z\theta}(r), \\ \sigma_{rr} = \sigma_{\theta\theta} = \sigma_{zz} = \sigma_{r\theta} = \sigma_{zr} = 0, \end{cases}$$

to get strains and displacements. However, compared with the displacement approach used above, the derivation involved in the stress approach will be relatively lengthy because we need to solve equations of equilibrium to obtain $\sigma_{z\theta}$ and then solve the geometrical equations to get displacements. It is similar to the solution of a bar under gravity in Chapter 6. This reminds us that when using different approaches we will encounter different difficulties. The reader is recommended to try to solve the above problem again using the stress approach, experience the process and see if the same solution can be obtained.

7.1.2 Analysis of the Solution

In designing a transmission shaft or a similar component, an important factor to consider is whether it will fail under a certain load and if so what will be the failure mechanism. To achieve a deeper understanding, let us analyse the principal stresses in the bar. For simplicity, consider only the case with pure torsion. The reader may wish to study the combined case with both torsion and rotation.

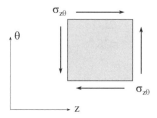

Figure 7.3 The stress state at any point in the bar under pure torsion.

Since the non-vanishing stress is $\sigma_{z\theta}$, at any point in the bar, as shown in **Fig. 7.3**, the principal stresses can be calculated as

$$\left. \begin{array}{c} \sigma_1 \\ \sigma_3 \end{array} \right\} = \pm\sqrt{(\sigma_{z\theta})^2} = \pm\sigma_{z\theta},$$

according to Eq. (3-21). The intermediate principal stress is clearly $\sigma_2 = \sigma_{rr} = 0$. The direction of the maximum and minimum principal stresses with respect to z-axis can be determined by Eq. (3-20), *i.e.*,

$$\tan(2\theta) = \frac{2\sigma_{z\theta}}{\sigma_{zz} - \sigma_{\theta\theta}} = \infty.$$

Thus θ_1, the angle between the first principal direction and z-axis, is 45°. Hence, if the bar material is brittle, such as cast iron or ceramics, it will fail under the maximum tensile stress. The failure angle must be in the direction of 45° with respect to the z-axis. However, if the material is ductile, such as mild steel, the failure will be controlled by the maximum shear stress.[7.1] As discussed in Chapter 3, the maximum shear stress determined by Eq. (3-44) is in the direction of 45° with respect to the maximum principal stress. Thus in this case, the failure will be along the surface perpendicular to the z-axis. It has been shown that the above theoretical predictions are exactly those observed in practice, as shown in **Fig. 7.4**.

[7.1] The details of the failure theory will be discussed in Chapter 9, which will show the reasons for ductile and brittle failure modes.

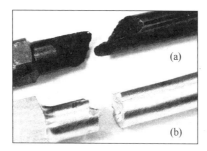

Figure 7.4 The failure modes of bars under pure torsion. (a) grey cast iron (a brittle material). (b) aluminium alloy 2024-T351 (a ductile material) (Dowling, 1993).

7.2 BEAM BENDING

7.2.1 Modelling and Solution

For convenience, let us consider the pure bending of a uniform prismatic beam with a rectangular cross-section, as illustrated in **Fig. 7.5**. The origin of the Cartesian coordinate system is at the centre of the left end of the beam with the z-axis coinciding with the beam's central axis. The bending is in the yz-plane. Thus the stress boundary conditions of the beam can be written as

at $x = \pm b/2$, $-h/2 \leq y \leq h/2$, $0 \leq z \leq L$,

$$\sigma_{xx} = \sigma_{xy} = \sigma_{xz} = 0; \tag{7-7a}$$

at $y = \pm h/2$, $-b/2 \leq x \leq b/2$, $0 \leq z \leq L$,

$$\sigma_{yy} = \sigma_{yx} = \sigma_{yz} = 0; \tag{7-7b}$$

at $z = 0$ or L, $-b/2 \leq x \leq b/2$, $-h/2 \leq y \leq h/2$,

$$\sigma_{zy} = \sigma_{zx} = 0, \tag{7-7c}$$

$$\int_{-h/2}^{h/2} \int_{-b/2}^{b/2} y\sigma_{zz} \, dxdy = M, \tag{7-7d}$$

$$\int_{-h/2}^{h/2}\int_{-b/2}^{b/2} x\sigma_{zz}\,dxdy = 0, \qquad (7\text{-}7e)$$

$$\int_{-h/2}^{h/2}\int_{-b/2}^{b/2} \sigma_{zz}\,dxdy = 0. \qquad (7\text{-}7f)$$

Similar to the case in torsion discussed above, conditions (7-7d) to (7-7f) mean that we have used Saint-Venant's principle. This also implies that the length of the beam L must be much larger than its cross-sectional dimensions, b and h.

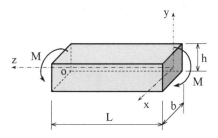

Figure 7.5 A rectangular beam under pure bending.

Since we are interested in deformation induced stresses, strains and displacements, possible rigid body motion must be eliminated. Again according to Saint-Venant's principle, if we fix the left end of the beam, the deformation in the zone far away from the end will not be affected. We therefore have the following displacement boundary conditions:

at the origin, *i.e.*, at $x = y = z = 0$,

$$u = v = w = 0; \qquad (7\text{-}7g)$$

at $z = 0$, $-b/2 \le x \le b/2$, $-h/2 \le y \le h/2$,

$$w = 0. \qquad (7\text{-}7h)$$

Condition (7-7h) prevents the beam from its rigid rotation about the x- and y-axes.

Because the bending is in the symmetrical plane, *i.e.*, the yz-plane, the beam deformation must obey the following symmetric condition:

$$u = 0 \qquad (7\text{-}7\text{i})$$

at $x = 0$, $-h/2 \le y \le h/2$, $0 \le z \le L$.

The elementary theory of beam bending[7.2] shows that under pure bending, shear strains, shear stresses and transverse normal stresses in a beam are zero and that a plane cross-section of the beam before bending remains a plane during bending so that the axial strain, ε_{zz}, is a linear function of coordinate y. Mathematically, these conclusions can be written as

$$\begin{cases} \varepsilon_{zz} = cy, \ \varepsilon_{xy} = \varepsilon_{xz} = \varepsilon_{yz} = 0, \\ \sigma_{zz} = cEy, \ \sigma_{xx} = \sigma_{yy} = \sigma_{xy} = \sigma_{xz} = \sigma_{yz} = 0, \end{cases} \qquad (7\text{-}8)$$

where c is a constant. Equation (7-8) can be used for our solution together with the semi-inverse method.

The stresses in Eq. (7-8) satisfy the equations of equilibrium. The generalised Hooke's law brings about

$$\varepsilon_{xx} = \varepsilon_{yy} = -\nu cy \qquad (7\text{-}9)$$

In the present case the compatibility equations, Eq. (4-15), must be examined because the strains have been calculated from the assumed stresses.[7.3] Fortunately, all the compatibility equations are satisfied. Thus the strain field is compatible and there must exist a unique displacement solution.

The stress components meet the requirement of all the stress boundary conditions except condition (7-7d). To satisfy this condition, substitute σ_{zz} of Eq. (7-8) into Eq. (7-7d), which leads to

$$c = \frac{M}{IE}, \qquad (7\text{-}10)$$

[7.2] The elementary theory of beam bending can be found in most textbooks on Mechanics of Materials, such as that by Timoshenko and Gere (1972). Detailed discussions about the limitation of the theory can be found in the monograph by Yu and Zhang (1996).

[7.3] We are using the stress approach to solve the problem.

where $I = \dfrac{1}{12}bh^3$ is the moment of area of the beam cross-section. Hence,

$$\sigma_{zz} = \dfrac{M}{I}y. \tag{7-11}$$

Equation (7-11) is exactly the stress obtained by the elementary theory of beam bending. Thus in terms of stresses, the elementary theory gives an exact solution to elasticity equations.

To find displacements, we need to integrate the geometrical equations, Eq. (4-7), after the substitution of the strain components of Eqs. (7-8) and (7-9), *i.e.*, we need to integrate the following equations

$$\begin{cases} \dfrac{\partial u}{\partial x} = -vcy, \\ \dfrac{\partial v}{\partial y} = -vcy, \\ \dfrac{\partial w}{\partial z} = cy, \\ \dfrac{\partial u}{\partial y} + \dfrac{\partial v}{\partial x} = 0, \\ \dfrac{\partial w}{\partial y} + \dfrac{\partial v}{\partial z} = 0, \\ \dfrac{\partial u}{\partial z} + \dfrac{\partial w}{\partial x} = 0. \end{cases} \tag{7-12}$$

The partial integration of Eqs. (7-12a) to (7-12c) gives rise to

$$\begin{cases} u = -vcxy + f_1(y,z), \\ v = -\dfrac{1}{2}vcy^2 + f_2(x,z), \\ w = cyz + f_3(x,y), \end{cases}$$

where f_1, f_2 and f_3 are functions due to partial integration and should be determined by both the displacement boundary conditions, Eqs. (7-7g) to (7-7i), and the vanishing shear strain conditions, Eqs. (7-12d) to (7-12f). Obviously, to satisfy Eq. (7-7g), $f_3(x, y)$ must vanish. Similarly, to satisfy Eq. (7-7h), $f_1(y, z) = 0$. We can see that the boundary condition (7-7i) and Eq. (7-12f) can now be satisfied automatically. The substitution of the above u, v and w into Eqs. (7-12d) and (7-12e) leads to

$$\begin{cases} \dfrac{\partial f_2(x,z)}{\partial x} = vcx, \\ \dfrac{\partial f_2(x,z)}{\partial z} = -cz. \end{cases}$$

They bring about

$$f_2(x,z) = \frac{1}{2}vcx^2 + g_1(z),$$

and

$$f_2(x,z) = -\frac{1}{2}cz^2 + g_2(x),$$

respectively. Since function f_2 must be unique, we get

$$g_1(z) = -\frac{1}{2}cz^2 + d_1, \quad g_2(x) = \frac{1}{2}vcx^2 + d_2,$$

where d_1 and d_2 are constants. Hence,

$$f_2(x,z) = \frac{1}{2}c\left(vx^2 - z^2\right) + d,$$

where d is a constant. Consequently,

$$v = \frac{1}{2}c\left(v\left[x^2 - y^2\right] - z^2\right) + d.$$

Furthermore, since $v = 0$ at $x = y = z = 0$, we must have $d = 0$. The v obtained above obviously satisfies the symmetry condition $v(x, y, z) = v(-x, y, z)$ and is consistent with the plane cross-section assumption because $v(x, -y, z) = v(x, -y, z)$. Thus the displacements in the beam are finally determined as

$$\begin{cases} u = -vcxy, \\ v = \dfrac{1}{2}c\left(v\left[x^2 - y^2\right] - z^2\right), \\ w = cyz. \end{cases} \quad (7\text{-}13)$$

The above solutions are under the displacement boundary conditions (7-7g) to (7-7h). According to Saint-Venant's principle, any statically equivalent change of the boundary conditions will not affect the solution in the zone far away from the area. The reader can try to set different displacement boundary conditions at the left end of the beam, find the solutions of displacements and compare them with Eq. (7-13) to see in which zone they are very different and from where they become the same. It will also be a helpful exercise to solve the bending of beams under different loading and constraining conditions, e.g., a cantilever subjected to a concentrated tip load, by using the solutions from elementary bending theory as a start of the semi-inverse method.

7.2.2 Discussion

In the solution, the details of the beam cross-section are not used in determining the integration constants. Thus the above solution is valid for symmetrical bending of solid beams with any cross-section, provided that the moment of area of the beam cross-section, I in Eq. (7-10), is replaced accordingly.

Now let us consider the deformed centre-line of the beam. The solution (7-13) shows that $u = w = 0$ along the line because $x = y = 0$. The vertical displacement, normally called *deflection* of the beam in mechanics of materials, becomes

$$v = -\frac{M}{2IE}z^2. \quad (7\text{-}14)$$

This is exactly the solution to deflection of a cantilever beam subjected to a tip bending moment M in the elementary bending theory. However, the elementary theory cannot offer any information about the shape change of a beam cross-section during bending. As we can see from Eq. (7-13), at a specific cross-section, say at $z = z_0$, we have

$$\begin{cases} u = -v c x y, \\ v = \frac{1}{2} c \left(v \left[x^2 - y^2 \right] - z_0^2 \right). \end{cases} \qquad (7\text{-}15)$$

The relationships offer an alternative method for measuring Poisson's ratio v of a material by taking a simple beam bending without the use of the strain gauge facilities.

7.3 THERMAL STRESS ANALYSIS

As discussed in Chapter 5, when a component or a structure is subjected to a temperature change, thermal deformation will take place and may cause thermal stresses. In solving such a thermal problem with a known field of temperature change, all the basic equations are the same as before provided that we now use the stress-strain relationships, Eq. (5-19), that include the thermal strains generated by the temperature change. Obviously, the solution of a thermal deformation problem involves two parts. The first is to obtain the temperature field so that we can have the temperature change at any point in the body. The second is then to find the stresses, strains and displacements induced by the temperature change and any other external loads applied.

Deformation and temperature change is always coupled. This is understandable because deformation causes additional motion of atoms in a solid and certainly results in temperature change. Thus theoretically, deformation and temperature change should be treated as a whole. The solution to such coupled problems is difficult. Fortunately, the deformation process in many engineering components or structures is slow and the effect of deformation (particularly at the small elastic deformation stage) on temperature is negligible at the size scale of our interest. In this case, we can reasonably treat temperature variation and deformation independently. The principle of superposition also applies. The resultant stresses, strains and displacements in a body are the summation of those generated by thermal deformation and other external forces.

There are usually two ways to find the temperature change in a solid, experimental and theoretical. Experimentally, we can use various techniques to measure the temperature field throughout a body, such as the infrared and thermocouple techniques. Theoretically, we can make use of the theory of thermodynamics to find the temperature field. More detailed discussions can be found in relevant textbooks.[7.4]

[7.4] See for example, Carslaw and Jaeger (1959), Fung (1965), Little (1973), Xie, Lin and Ding (1988), Kurpisz and Nowak (1995), Gokcen and Reddy (1996) and Hetnarski (1996).

In the following, we shall only discuss some very simple cases in steady temperature fields.[7.5]

7.3.1 A General Method for Thermal Deformation Analysis

Let us use the displacement approach to solve thermal deformation problems. For simplicity, consider a solid in static equilibrium without body forces. We need to obtain a set of equations in terms of the three displacement components, u, v, and w, similar to those of Eq. (6-18). To do this, we can substitute the geometrical equations into the stress-strain relationships with thermal straining, Eq. (5-18), and get the expressions of stresses in terms of displacements. We then substitute these into the equations of equilibrium, Eq. (3-48), which gives rise to

$$\begin{cases} (\lambda + \mu) \dfrac{\partial I_1^\varepsilon}{\partial x} + \mu \nabla^2 u - \alpha \left(\dfrac{E}{1-2v} \right) \dfrac{\partial (\Delta T)}{\partial x} = 0, \\ (\lambda + \mu) \dfrac{\partial I_1^\varepsilon}{\partial y} + \mu \nabla^2 v - \alpha \left(\dfrac{E}{1-2v} \right) \dfrac{\partial (\Delta T)}{\partial y} = 0, \\ (\lambda + \mu) \dfrac{\partial I_1^\varepsilon}{\partial z} + \mu \nabla^2 w - \alpha \left(\dfrac{E}{1-2v} \right) \dfrac{\partial (\Delta T)}{\partial z} = 0. \end{cases} \quad (7\text{-}16)$$

Comparing the above with Eq. (6-18) we can see that for thermal deformation problems the basic equations to solve are equivalent to a normal static problem with body force densities of

$$f_x = -\alpha \left(\dfrac{E}{1-2v} \right) \dfrac{\partial (\Delta T)}{\partial x}, \quad f_x = -\alpha \left(\dfrac{E}{1-2v} \right) \dfrac{\partial (\Delta T)}{\partial y}, \quad f_x = -\alpha \left(\dfrac{E}{1-2v} \right) \dfrac{\partial (\Delta T)}{\partial z}. \quad (7\text{-}17)$$

Now let us see what should be the boundary conditions when thermal strains are involved. When displacements are specified as u', v' and w' at any point on the boundary surfaces of a solid, we have the following displacement boundary conditions:

$$u = u', \ v = v', \ w = w' \text{ at the surfaces with specified displacements.} \quad (7\text{-}18)$$

[7.5] A steady temperature field is independent of time. Thus temperature change ΔT is only a function of coordinates, *i.e.*, $\Delta T = \Delta T\,(x, y, z)$.

They are the same as the displacement boundary conditions, Eq. (6-1), for a standard non-thermal problem. On the stress boundaries,[7.6] by doing the same derivation as in obtaining Eq. (7-16), we get

$$\begin{cases} \alpha\left(\dfrac{E}{1-2v}\right)(\Delta T)l = l\,\sigma_{xx} + m\,\sigma_{yx} + n\,\sigma_{zx}\,, \\ \alpha\left(\dfrac{E}{1-2v}\right)(\Delta T)m = l\,\sigma_{xy} + m\,\sigma_{yy} + n\,\sigma_{yz}\,, \\ \alpha\left(\dfrac{E}{1-2v}\right)(\Delta T)n = l\,\sigma_{xz} + m\,\sigma_{yz} + n\,\sigma_{zz}\,. \end{cases} \quad (7\text{-}19)$$

Comparing the above with Eq. (6-2), we see clearly that thermal deformation is equivalent to a non-thermal problem with surface stresses of

$$\sigma''_{nx} = \alpha\left(\dfrac{E}{1-2v}\right)(\Delta T)l\,,\ \sigma''_{ny} = \alpha\left(\dfrac{E}{1-2v}\right)(\Delta T)m\,,\ \sigma''_{nz} = \alpha\left(\dfrac{E}{1-2v}\right)(\Delta T)n\,. \quad (7\text{-}20)$$

The above discussion concludes that *the solution of a thermal static deformation problem without body forces and surface stresses but with a steady field of temperature change, ΔT, is equivalent to that of a static problem subjected to a set of body forces described by Eq. (7-17) and a set of surface stresses specified by Eq. (7-20)*. Hence, when ΔT is known, we can easily obtain stresses, strains and displacements induced by thermal deformation using the solution methods and skills developed before.

The above conclusion also greatly facilitates experimentation in terms of thermal stress measurement. For instance, in doing an experiment with scale models, it is rather difficult to match the similarity of the model to the original problem in both thermodynamics and mechanics. However, the above conclusion makes it possible to use body forces and surface stresses to replace thermal heating. This transfers a thermomechanics problem to a pure mechanics one and thus enables one to do many experimental studies that are otherwise difficult or impossible.

Similar to the method used in Chapter 6, by carrying out a simplification into plane-stress and plane-strain deformation from the above three-dimensional equations, we can

---

[7.6] Since we can use the principle of superposition to solve a problem with combined thermal deformation and external loading on the surfaces, it is always convenient to assume, when solving the thermal deformation part, that there are no external stresses applied on the boundary surfaces.

find that when the solution to a plane-stress thermal problem is achieved, the solution to a corresponding plane-strain thermal problem can be obtained directly by replacing E, ν and α by $\dfrac{E}{1-\nu^2}$, $\dfrac{\nu}{1-\nu}$ and $(1+\nu)\alpha$, respectively. In the same way, when the solution to a plane-strain problem is available, the corresponding solution to a plane-stress problem can be directly derived by replacing E, ν and α by $\dfrac{E(1+2\nu)}{(1+\nu)^2}$, $\dfrac{\nu}{1+\nu}$ and $\dfrac{1+\nu}{1+2\nu}\alpha$, respectively.

Now let us go through an engineering example to experience the solution process of thermal problems and understand the mechanics associated with thermal deformation.

7.3.2 Thermal Fit of a Hollow Disk onto a Shaft

Similar to the interference fit discussed in sections **1.2.1** and **6.3.2.2**, which is a mechanical process, the thermal fit of a hollow disk onto a shaft is also a commonly used joining process in practice. For simplicity, consider the assembly between a circular hollow disk and a circular shaft, as illustrated in **Fig. 7.6**. Initially, the inner radius of the disk, R_i, is smaller than that of the shaft, R_s. By heating the disk, R_i will increase and when it becomes larger than R_s, the shaft can be inserted into the disk easily. Upon cooling, the disk shrinks and its inner radius decreases thus fastening itself onto the shaft. To have an optimal result, we must know the heating temperature that will be sufficient to make such a thermal fitting. A thermal deformation analysis of the disk is therefore necessary to find out the relationship of the variation of R_i with respect to temperature change and then, similar to the previous mechanical fitting problem, solve a contact problem with deformation compatibility at the disk-shaft interface while the disk shrinks during cooling.

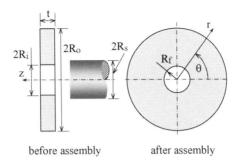

before assembly   after assembly

Figure 7.6 The disk and shaft before and after thermal assembly.

Let us assume that the temperature change at the inner surface is ΔT_i and at the outer surface is ΔT_o. When the disk is thin, *i.e.*, when R_i is much larger than its thickness t, the disk is under plane-stress deformation. Because of the special shape of the disk, it is convenient to use the polar coordinate system. The boundary conditions of the disk can therefore be described as

at $r = R_i$ and R_o,
$$\sigma_{rr} = \sigma_{r\theta} = 0; \qquad (7\text{-}21a)$$
at $r = R_i$,
$$\Delta T = \Delta T_i; \qquad (7\text{-}21b)$$
at $r = R_o$,
$$\Delta T = \Delta T_o. \qquad (7\text{-}21c)$$

If the field of temperature change, ΔT, is axisymmetrical, that is, if $\Delta T = \Delta T(r)$, the deformation of the disk is also axisymmetrical. Thus the circumferential displacement, v, in the disk must always be zero and the radial displacement, u, is the function of coordinate r only, *i.e.*,

$$u = u(r), \, v = 0. \qquad (7\text{-}22)$$

As discussed previously, we need to find ΔT first and then solve an equivalent static problem.

According to the theory of heat conduction (Carslaw and Jaeger, 1959), when ΔT is steady and axisymmetrical, it must satisfy the following equation

$$\frac{1}{r}\frac{d}{dr}\left[r\frac{d(\Delta T)}{dr}\right] = 0.$$

By integrating the equation twice directly, we get

$$\Delta T = A \ln r + B,$$

where A and B are integration constants and can be determined by the boundary conditions specified by Eqs. (7-21b) and (7-21c). This gives rise to

$$\Delta T = (\Delta T_i)\frac{\ln\left(\frac{R_o}{r}\right)}{\ln\left(\frac{R_o}{R_i}\right)} + (\Delta T_o)\frac{\ln\left(\frac{R_i}{r}\right)}{\ln\left(\frac{R_i}{R_o}\right)}. \tag{7-23}$$

If the disk is under uniform heating, *i.e.*, if $\Delta T_i = \Delta T_o = \Delta T^*$, the above solution of temperature change reduces to

$$\Delta T = \Delta T^*. \tag{7-24}$$

With the field of temperature change obtained above, now we need to use the basic equation, Eq. (7-16), to get a general solution and determine the integration constants using the boundary conditions specified by Eq. (7-19). Considering that the deformation of the disk is in plane-stress and axisymmetric and using the semi-known displacements of Eq. (7-22), Eq. (7-16) reduces to a single equation, that is,

$$\frac{d^2 u}{dr^2} + \frac{1}{r}\frac{du}{dr} - \frac{u}{r^2} = (1+v)\,\alpha\frac{d(\Delta T)}{dr}.$$

This can be re-written as

$$\frac{d}{dr}\left(\frac{1}{r}\frac{d}{dr}(ru)\right) = (1+v)\,\alpha\frac{d(\Delta T)}{dr}.$$

By integrating the above equation twice, we obtain

$$u = (1+v)\frac{\alpha}{r}\int_{R_i}^{r}(\Delta T)r dr + Ar + B\frac{1}{r}, \qquad (7\text{-}25)$$

where A and B are integration constants. Hence, we have the following strains and stresses

$$\begin{cases} \varepsilon_{rr} = (1+v)\alpha\left[\Delta T - \frac{1}{r^2}\int_{R_i}^{r}(\Delta T)r dr\right] + A - B\frac{1}{r^2}, \\ \varepsilon_{\theta\theta} = (1+v)\frac{\alpha}{r^2}\int_{R_i}^{r}(\Delta T)r dr + A + B\frac{1}{r^2}, \end{cases} \qquad (7\text{-}26)$$

$$\begin{cases} \sigma_{rr} = -\frac{E\alpha}{r^2}\int_{R_i}^{r}(\Delta T)r dr + \frac{E\alpha}{1-v}A - \frac{E\alpha}{1+v}\frac{B}{r^2}, \\ \sigma_{\theta\theta} = \frac{E\alpha}{r^2}\int_{R_i}^{r}(\Delta T)r dr - E\alpha(\Delta T) + \frac{E\alpha}{1-v}A - \frac{E\alpha}{1+v}\frac{B}{r^2}. \end{cases} \qquad (7\text{-}27)$$

The constants A and B can therefore be determined by boundary conditions, Eq. (7-21a), as

$$A = \frac{1-v}{R_o^2 - R_i^2}\int_{R_i}^{R_o}(\Delta T)r dr, \quad B = \frac{(1+v)R_i^2}{R_o^2 - R_i^2}\int_{R_i}^{R_o}(\Delta T)r dr. \qquad (7\text{-}28)$$

Consequently, when we substitute the temperature change, Eq. (7-23) or Eq. (7-24) into the above solution, we get the specific displacement, strains and stresses for the disk under thermal deformation.

To perform the thermal fitting, we must have

$$R_i + u\big|_{r=R_i} > R_s. \qquad (7\text{-}29)$$

Since $u|_{r=R_i}$ is given by Eq. (7-25), condition (7-29) gives rise to the required temperature change ΔT directly.

The calculation of the disk-shaft interface stress is exactly the same as that of the collar-pin mechanical fit discussed before. The reader should be able to solve the problem without any difficulty.

7.4 STRESS AND DEFORMATION DUE TO CONTACT

Contact is a commonly encountered problem in engineering practice and in many cases the stress and deformation caused by contact are of significant importance to the reliability and service life of a component. For example, the optimal design of indentation (see section **6.2.3.1**, **Fig. 6.6**), which is an important test to characterise materials properties, and plate rolling (see section **6.2.3.3**, **Fig. 6.8**), which is one of the main manufacturing processes for making plates and foils in industry, both rely on an in-depth understanding of the contact stress and deformation involved (*e.g.*, Fleck *et al.*, 1992; Zhang, 1995). **Fig. 7.7** shows some further examples of contact. The contact stresses in gear teeth, tyres, road surfaces, train wheels, rails and bearings govern their wear rates and failure.

When the size of the contact area is much smaller than the characteristic dimension of a component, the component can be considered to be semi-infinite bounded by a surface, either flat or curved, on which contact occurs. This is because the contact only introduces significant stress and deformation locally around the contact region according to Saint-Venant's principle. If the surface is flat or has a very small curvature, the component can be idealised by a half-space as illustrated in **Fig. 7.8a**.

The contact traction on the surface of a body with the above dimensional characteristics can be modelled by a half-space subjected to a general surface stress over the contact area, such as the stress F(x, y) over the area A(x, y) illustrated in **Fig. 7.8a**. Obviously, F(x, y) can always be resolved into its normal component, p(x, y), and tangential component, q(x, y). Hence, if we can find the solutions to the two basic problems,[7.7] *i.e.*, a half-space subjected to a normal concentrated force P (Boussinesq's problem) and that under a tangential concentrated force Q (Cerruti's problem), as shown in **Figs. 7.8b** and **7.8c**, we can get the general solution for the case of **Fig. 7.8a** using the

---

[7.7] A half-space subjected to a normal concentrated force was first solved in 1885 by a French scientist, J. V. Boussinesq (1842-1929), who was the most distinguished student of Saint-Venant. The deformation of a half-space under a tangential concentrated force was solved by V. Cerruti in 1882. These are thereafter called Boussinesq's and Cerruti's problems, respectively, in the literature.

principle of superposition. This is because an infinitesimal normal component of p(x, y), dp = pdA, over an infinitesimal surface area, dA, can be considered as a concentrated force on the half-space surface and thus the stresses and displacements caused by such dp is given by the solution of the case of **Fig. 7.8b** provided that all P there are replaced by pdA and that a corresponding coordinate transformation is performed because P applies at the origin whereas the application point of pdA varies in the contact area A. The resultant stresses and displacements due to p(x, y) over the whole area A(x, y) can then be obtained by superposing the individual contribution of dp over the area, that is by an integration over A. Similarly, the stresses and displacements due to q(x, y) can also be obtained by integrating the solution of the case of **Fig. 7.8c**. The final solution of stresses and displacements caused by F(x, y) is finally the summation of the above two parts, those due to p(x, y) and those due to q(x, y).

Hence, to solve a contact problem, the first step is to find the solutions to the cases of **Figs. 7.8b** and **7.8c**. We shall follow a more direct approach.

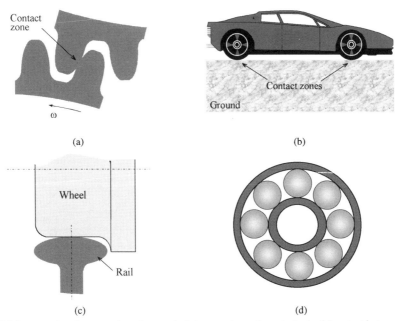

Figure 7.7 Some engineering examples where contact stresses play an important role. (a) contact between gear teeth, (b) contact between car tyre and ground surface, (c) contact between a wheel and a rail, and (d) contact in a ball bearing.

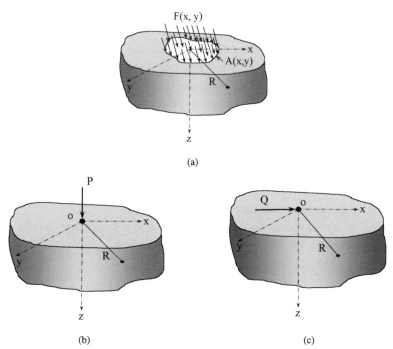

Figure 7.8 A half-space loaded by (a) a general surface stress, F(x, y), over a surface area A(x, y), (b) a concentrated normal force P, and (c) a concentrated tangential force Q.

Table 7.1 Some harmonic functions

General cases in Cartesian coordinate system	x	y	z	
	xy	xz	yz	
	xyz			
	$x/(R+z)$	$y/(R+z)$	$x/(R-z)$	$y/(R-z)$
	$x/[R(R+z)]$	$y/[R(R+z)]$	$1/R$	$\ln(R+z)$
	x/R^3	y/R^3	z/R^3	
Axisymmetric cases in polar coordinate system	z			
	$\ln r$			
	$2z^2 - r^2$			
	$2z^3 - 3zr^2$			
	$8z^4 - 24z^2r^2 + 3r^4$			
	$8z^5 - 40z^3r^2 + 15zr^4$			
In the above, ξ is an arbitrary constant and $R^2 = x^2 + y^2 + z^2$.				

The general solution of Eq. (6-18) without acceleration and body forces was developed by Papkovich in 1932 and Neuber in 1934 independently in terms of harmonic functions,[7.8] that is

$$\mathbf{d} = \mathbf{\psi} - \frac{1}{4(1-v)}\operatorname{grad}(\psi_0 + \mathbf{R} \bullet \mathbf{\psi}), \qquad (7\text{-}30)$$

where grad$(\cdots)$ means the gradient of $(\cdots)$[7.9], $\mathbf{d} = (u, v, w)$ is the displacement vector, $\mathbf{R} = (x, y, z)$ is the position vector of a point and ψ_0 and $\mathbf{\psi} = (\psi_x, \psi_y, \psi_z)$ are all harmonic functions that satisfy Laplace equation, *i.e.*, $\nabla^2 \psi_0 = 0$ and $\nabla^2 \mathbf{\psi} = \mathbf{0}$, where $\nabla^2 = \frac{\partial^2}{\partial x^2} + \frac{\partial^2}{\partial y^2} + \frac{\partial^2}{\partial z^2}$. The expansion of the Papkovich-Neuber solution, Eq. (7-30), is

$$\begin{cases} u = \psi_x - \dfrac{1}{4(1-v)}\dfrac{\partial}{\partial x}(\psi_0 + x\psi_x + y\psi_y + z\psi_z), \\ v = \psi_y - \dfrac{1}{4(1-v)}\dfrac{\partial}{\partial y}(\psi_0 + x\psi_x + y\psi_y + z\psi_z), \\ w = \psi_z - \dfrac{1}{4(1-v)}\dfrac{\partial}{\partial z}(\psi_0 + x\psi_x + y\psi_y + z\psi_z). \end{cases} \qquad (7\text{-}31)$$

Harmonic functions have been investigated in mathematics extensively. **Table 7.1** lists some of such functions. Thus ψ_0, ψ_x, ψ_y and ψ_z can be selected for the solution of a particular problem with given boundary conditions. There are many harmonic functions and the selection of specific functions for a particular deformation problem is indeed skill-dependent. What we are going to discuss in Chapter 8 about the selection of stress functions will facilitate our understanding of the selection rule. In the following, we temporarily assume that we know how to select the required harmonic functions ψ_0, ψ_x, ψ_y and ψ_z in Eq. (7-31), and we shall give the proper form of displacements directly.

[7.8] It is therefore called the Papkovich-Neuber solution. A more general treatise on the subject can be found in other books, *e.g.*, that by Love (1952).

[7.9] The gradient of a function $g(x, y, z)$ is $\operatorname{grad}(g) = \left(\dfrac{\partial g}{\partial x}, \dfrac{\partial g}{\partial y}, \dfrac{\partial g}{\partial z}\right)$.

7.4.1 A Half-Space under a Normal Concentrated Load

Consider the case of **Fig. 7.8b**. The boundary conditions of the problem are

at $z = 0$,

$$\sigma_{zz} = 0, \; \sigma_{zx} = 0, \; \sigma_{zy} = 0, \tag{7-32a}$$

except at the origin O where the concentrated force P acts. Since the deformation is axisymmetric about the z-axis, displacements u and v must vanish along z-axis, *i.e.*,

when $z \geq 0$, $x = y = 0$,

$$u = v = 0. \tag{7-32b}$$

In addition, because the body is a half-space, all the displacement components must approach zero when $R \to \infty$, that is,

when $R \to \infty$,

$$u = v = w = 0. \tag{7-32c}$$

On the other hand, the half-space is in equilibrium in all the directions. Since P is along the z-axis, the equilibrium in the x- and y-directions must be satisfied automatically. However, the equilibrium in the z-direction requires that

$$\int_{-\infty}^{\infty} \int_{-\infty}^{\infty} \sigma_{zz} dx dy + P = 0. \tag{7-32d}$$

By taking the harmonic functions

$$\psi_x = \psi_y = 0, \; \psi_z = C_1 \frac{1}{R}, \; \psi_0 = C_2 \ln(R+z)$$

in the general solution, Eq. (7-31), where C_1 and C_2 are constants, we get

$$\begin{cases} u = A\dfrac{zx}{R^3} + B\dfrac{x}{R(R+z)}, \\ v = A\dfrac{zy}{R^3} + B\dfrac{y}{R(R+z)}, \\ w = A\left[\dfrac{z^2}{R^3} + \dfrac{\lambda+3\mu}{(\lambda+\mu)R}\right] + B\dfrac{1}{R}, \end{cases} \qquad (7\text{-}33)$$

where A and B are constants to be determined. The corresponding stress components are therefore

$$\begin{cases} \sigma_{xx} = A\left(-\dfrac{2\mu z}{R^3}\right)\left[3\left(\dfrac{x}{R}\right)^2 - \dfrac{\mu}{\lambda+\mu}\right] + B(2\mu)\left[\dfrac{y^2+z^2}{R^3(R+z)} - \dfrac{x^2}{R^2(R+z)^2}\right], \\ \sigma_{yy} = A\left(-\dfrac{2\mu z}{R^3}\right)\left[3\left(\dfrac{y}{R}\right)^2 - \dfrac{\mu}{\lambda+\mu}\right] + B(2\mu)\left[\dfrac{x^2+z^2}{R^3(R+z)} - \dfrac{y^2}{R^2(R+z)^2}\right], \\ \sigma_{zz} = A\left(-\dfrac{2\mu z}{R^3}\right)\left[3\left(\dfrac{z}{R}\right)^2 + \dfrac{\mu}{\lambda+\mu}\right] - B(2\mu)\dfrac{z}{R^3}, \\ \sigma_{xy} = A\left(-\dfrac{6\mu xyz}{R^5}\right) - B\left[2\mu\dfrac{xy(z+2R)}{R^3(R+z)^2}\right], \\ \sigma_{yz} = A\left(-\dfrac{2\mu y}{R^3}\right)\left[3\left(\dfrac{z}{R}\right)^2 + \dfrac{\mu}{\lambda+\mu}\right] - B(2\mu)\dfrac{y}{R^3}, \\ \sigma_{zx} = A\left(-\dfrac{2\mu x}{R^3}\right)\left[3\left(\dfrac{z}{R}\right)^2 + \dfrac{\mu}{\lambda+\mu}\right] - B(2\mu)\dfrac{x}{R^3}. \end{cases}$$
$$(7\text{-}34)$$

Because the displacements, Eq. (7-33), and stresses, Eq. (7-34), are calculated from the general solution, Eq. (7-31), they have satisfied all the basic equations. To obtain the particular solution to the present problem, therefore, we only need to see whether all the boundary, symmetric and equilibrium conditions, *i.e.*, Eqs. (7-32a) to (7-32d), can be satisfied or not. The substitution of σ_{zz}, σ_{zx} and σ_{zy} of Eq. (7-34) into Eqs. (7-32a) and (7-32d) leads to

$$\frac{4\pi\mu(\lambda+2\mu)}{\lambda+\mu}A + 4\pi\mu B = P$$

and

$$\frac{\mu}{\lambda+\mu}A + B = 0.$$

Therefore,

$$A = \frac{P}{4\pi\mu}, \quad B = -\frac{P}{4\pi(\lambda+\mu)}. \tag{a}$$

Hence, by substituting (a) into Eqs. (7-33) and (7-34), we get the solution of Boussinesq's problem, which is,

(a) stresses:

$$\begin{cases} \sigma_{xx} = -\frac{P}{2\pi}\frac{z}{R^3}\left[3\left(\frac{x}{R}\right)^2 - \frac{\mu}{\lambda+\mu}\right] - \frac{P\mu}{2\pi(\lambda+\mu)}\left[\frac{y^2+z^2}{R^3(R+z)} - \frac{x^2}{R^2(R+z)^2}\right], \\ \sigma_{yy} = -\frac{P}{2\pi}\frac{z}{R^3}\left[3\left(\frac{y}{R}\right)^2 - \frac{\mu}{\lambda+\mu}\right] - \frac{P\mu}{2\pi(\lambda+\mu)}\left[\frac{x^2+z^2}{R^3(R+z)} - \frac{y^2}{R^2(R+z)^2}\right], \\ \sigma_{zz} = -\frac{3P}{2\pi}\frac{z^3}{R^5}, \\ \sigma_{xy} = -\frac{3xyz}{2\pi R^5} + \frac{P\mu}{2\pi(\lambda+\mu)}\frac{xy(z+2R)}{R^3(R+z)^2}, \\ \sigma_{yz} = -\frac{3P}{2\pi}\frac{yz^2}{R^5}, \\ \sigma_{zx} = -\frac{3P}{2\pi}\frac{xz^2}{R^5}. \end{cases} \tag{7-35}$$

(b) displacements:

$$\begin{cases} u = \dfrac{P}{4\pi\mu} \dfrac{zx}{R^3} - \dfrac{P}{4\pi(\lambda+\mu)} \dfrac{x}{R(R+z)}, \\ v = \dfrac{P}{4\pi\mu} \dfrac{zy}{R^3} - \dfrac{P}{4\pi(\lambda+\mu)} \dfrac{y}{R(R+z)}, \\ w = \dfrac{P}{4\pi\mu} \dfrac{z^2}{R^3} + \dfrac{P(\lambda+2\mu)}{4\pi\mu(\lambda+\mu)} \dfrac{1}{R}. \end{cases} \quad (7\text{-}36)$$

The displacements (7-36) obviously satisfy conditions (7-32b) and (7-32c) when $z > 0$. However, when $R \to 0$, both the displacements and stresses become singular. It is true because we have used an idealised concentrated force P and thus the singularity is expected. In engineering practice, forces always act through a finite area, as discussed in section **3.1**, although its dimension can be very small. (In nanometre scale devices of nanotechnological applications, an external force may act on an area with an atomic dimension.)

The above stress solution implies an interesting yet important phenomenon. We can see that on any plane in the half-space parallel to the boundary surface (xy-plane), σ_{zz}, σ_{zx} and σ_{zy} are independent of the elastic constants of material. In other words, these stresses will be the same for any half-space bodies, regardless of their materials, *e.g.*, steel, plastics, or ceramics, provided that they are subjected to an identical concentrated load P. This makes our experimental studies much easier because we can use any isotropic material that is convenient to handle. The material property independence of a solution is the theoretical basis of photoelasticity,[7.10] an important technique for stress measurement in experimental mechanics.

7.4.2 A Half-Space under a Tangential Concentrated Load

This is the case of **Fig. 7.8c**. The boundary conditions are

at $z = 0$,

$$\sigma_{zz} = 0, \ \sigma_{zx} = 0, \ \sigma_{zy} = 0, \quad (7\text{-}37a)$$

[7.10] See the books by Kuske and Robertson (1974) and Cloud (1995) for details of the technique.

except at the origin O where Q acts. On the other hand, the half-space is in equilibrium in all the directions. Since Q is along the x-axis, the equilibrium in the z- and y-directions must be satisfied automatically. However, the equilibrium in the x-direction requires that

$$\int_{-\infty}^{\infty}\int_{-\infty}^{\infty} \sigma_{zx}\,dxdy + Q = 0. \tag{7-37b}$$

Using the Papkovich-Neuber solution again and letting the harmonic functions be

$$\psi_x = C_1 \frac{1}{R},\ \psi_y = 0,\ \psi_z = C_2 \frac{x}{R(R+z)},\ \psi_0 = C_3 \frac{x}{R+z},$$

where C_1, C_2 and C_3 are constants, the displacements can be expressed as

$$\begin{cases} u = \dfrac{C-A}{R}\left[\dfrac{R}{R+z} - \dfrac{x^2}{(R+z)^2}\right] + \dfrac{(3-4v)B - C}{R} + \dfrac{(B+C)x^2}{R^3}, \\[4pt] v = \dfrac{(B+C)xy}{R^3} - \dfrac{(C-A)xy}{(R+z)^2 R}, \\[4pt] w = (B+C)\dfrac{xz}{R^3} + \left[A + (3-4v)C\right]\dfrac{x}{R(R+z)}, \end{cases} \tag{7-38}$$

where A, B and C are constants to be determined. Substituting Eq. (7-38) into conditions (7-37a, b), we get three equations containing A, B and C, which then give rise to

$$A = -\frac{(1-2v)^2 Q}{8(1-v)\pi\mu},\ B = \frac{Q}{8(1-v)\pi\mu},\ C = \frac{(1-2v)Q}{8(1-v)\pi\mu}. \tag{7-39}$$

By substituting the constants into Eq. (7-38) and using geometric equations and the generalised Hooke's law, the complete solution of Cerruti's problem is obtained as follows:

(a) stresses:

$$\begin{cases} \sigma_{xx} = \dfrac{Qx}{2\pi R^3}\left[-\dfrac{3x^2}{R^2}+\dfrac{1-2v}{(R+z)^2}\left(R^2-y^2-\dfrac{2Ry^2}{R+z}\right)\right], \\[2mm] \sigma_{yy} = \dfrac{Qx}{2\pi R^3}\left[-\dfrac{3y^2}{R^2}+\dfrac{1-2v}{(R+z)^2}\left(3R^2-x^2-\dfrac{2Rx^2}{R+z}\right)\right], \\[2mm] \sigma_{zz} = -\dfrac{3Qxz^2}{2\pi R^5}, \\[2mm] \sigma_{xy} = \dfrac{Qy}{2\pi R^3}\left[-\dfrac{3x^2}{R^2}+\dfrac{1-2v}{(R+z)^2}\left(-R^2+x^2+\dfrac{2Rx^2}{R+z}\right)\right], \\[2mm] \sigma_{yz} = -\dfrac{3Qxyz}{2\pi R^5}, \\[2mm] \sigma_{zx} = -\dfrac{3Qx^2 z}{2\pi R^5}. \end{cases} \qquad (7\text{-}40)$$

(b) displacements:

$$\begin{cases} u = \dfrac{Q}{4\pi\mu R}\left\{1+\dfrac{x^2}{R^2}+(1-2v)\left[\dfrac{R}{R+z}-\dfrac{x^2}{(R+z)^2}\right]\right\}, \\[2mm] v = \dfrac{Q}{4\pi\mu R}\left[\dfrac{xy}{R^2}-(1-2v)\dfrac{xy}{(R+z)^2}\right], \\[2mm] w = \dfrac{Q}{4\pi\mu R}\left[\dfrac{xz}{R^2}+(1-2v)\dfrac{x}{R+z}\right]. \end{cases} \qquad (7\text{-}41)$$

The above solution possesses similar properties to that of Boussinesq's problem. For example, on any plane with z = constant, stresses σ_{zz}, σ_{zx} and σ_{zy} are independent of material properties.

7.4.3 A Half-Space under a Local Uniform Pressure

When an area, A(x, y, z), on a half-space surface is subjected to a distributed stress, the solution can be obtained directly from those of Boussinesq's and Cerruti's problems by superposition. To experience the process, let us try to find the vertical displacement, w, at a surface point, M, of a half-space subjected to a uniform pressure in a circular area of radius a, as shown in **Fig. 7.9a**. The distance between point M and the centre of the area, which is the origin O, is r. We follow the method used by Timoshenko and Goodier (1970).

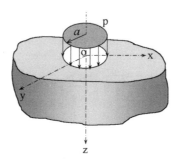

(a) uniform loading with centre at origin O

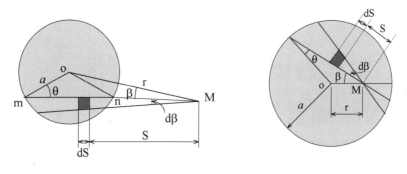

(b) M outside the loading area (c) M inside the loading area

Figure 7.9 A half-space subjected to a uniform pressure in a circular area.

The force in an infinitesimal area $dA = Sd\beta dS$, which is indicated by the shaded area in **Figs.7.9b** and **7.9c**, is pdA. Since the area is small, pdA can be treated as a concentrated force. Thus the vertical displacement dw caused by pdA at M is given by the third equation of Eq. (7-35) when letting $z = 0$ and $R = S$, that is,

$$dw = \frac{(1-v^2)p}{\pi E} d\beta dS.$$

The total displacement at M caused by p over the whole area is therefore

$$w = \frac{(1-v^2)p}{\pi E} \iint d\beta dS. \quad (b)$$

Note that when M is outside the loading area (**Fig. 7.9b**), the length of segment mn is $2\sqrt{a^2 - r^2 \sin^2 \beta}$ and that

$$d\beta = \frac{a \cos\theta \, d\theta}{r \cos\beta} = \frac{a \cos\theta \, d\theta}{r\sqrt{1 - \frac{a^2}{r^2}\sin^2\theta}}$$

because $a \sin\theta = r \sin\beta$. In addition, when β changes from 0 to its maximum value, θ changes from 0 to $\pi/2$. Hence,

$$w = \frac{4(1-v^2)p}{\pi E} \int_0^{\frac{\pi}{2}} \frac{a \cos\theta}{r\sqrt{1-\frac{a^2}{r^2}\sin^2\theta}} d\theta$$

$$= \frac{4(1-v^2)pr}{\pi E} \left[\int_0^{\frac{\pi}{2}} \sqrt{1-\frac{a^2}{r^2}\sin^2\theta} \, d\theta - \left(1-\frac{a^2}{r^2}\right) \int_0^{\frac{\pi}{2}} \frac{1}{r\sqrt{1-\frac{a^2}{r^2}\sin^2\theta}} d\theta \right]. \quad (7\text{-}42)$$

The above integrals are known as complete elliptic integrals whose values can be obtained from handbooks of mathematics (*e.g.*, Jahnke et al., 1960) when a/r is given. For instance, when M is just at the boundary of the loading zone, $r = a$, and thus Eq. (7-42) gives rise to

$$w = \frac{4(1-v^2)pa}{\pi E}. \quad (7\text{-}43)$$

On the other hand, if point M is inside the loading area, as shown in **Fig. 7.9c**, the length of segment mn is $2a\cos\theta$. In this case, Eq. (b) can be re-written as

$$w = \frac{4(1-v^2)pa}{\pi E} \int_0^{\frac{\pi}{2}} \sqrt{1 - \frac{r^2}{a^2}\sin^2\beta}\, d\beta. \tag{7-44}$$

The integral is also a complete elliptic integral. At the centre of the loading area, *i.e.*, when r = 0, we have

$$w = \frac{2(1-v^2)pa}{E}. \tag{7-45}$$

It is the maximum deflection on the surface of the half-space.

In a similar way, stresses due to uniform pressure can also be obtained (Timoshenko and Goodier, 1970).

7.4.4 Elastic Bodies in Contact

We are not unfamiliar with the solution of stresses and deformation in two elastic bodies in contact. Both the mechanical and thermal interference fittings of collars onto shafts, which we discussed previously in sections **6.3.2.2** and **7.3.2**, are examples of two bodies in contact. A common characteristic in these problems is that deformation in the two bodies must be compatible on the contact interface, *i.e.*, material in one body can never go beyond the contact interface and penetrate into the other.[7.11] In the collar-shaft fitting problems, the surfaces in contact were cylindrical and the deformation was circumferentially uniform so that the compatibility of deformation at the interface was easier to describe.

Generally, when two bodies are brought into contact, as shown in **Fig. 7.10**, we can use the solutions of Boussinesq's and Cerruti's problems to predict the shape of the contact area, the variation of its size with external loads and the stresses in the two bodies. The principle of analysis is still the deformation compatibility on the contact interface but will involve more complications because the shapes of the two bodies are not as simple as in the case of cylindrical fitting. Some commonly encountered cases of elastic contact are summarised in **Fig. 7.11**, which can be regarded as the mechanics models for various engineering components or devices, such as the contact between a wheel and a rail, a wheel and a road surface, a ball/roll and a ring of a ball/roll bearing, a shaft and a journal bearing, a ball-tipped detector of a robot and a target surface, a tip and a workpiece in an atomic force microscope and so on. Because of the importance of the contact cases in **Fig. 7.11**, Appendix B at the end of the book lists some useful formulae without considering friction. These formulae can be obtained by the superposition of Boussinesq's solutions. When friction becomes important, Cerruti's solution must also be used in the superposition.

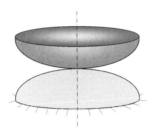

Figure 7.10 Two elastic bodies in contact.

Contact mechanics as a discipline has been the focus of engineers over decades, because it is so important to most engineering fields. In the above, we only studied some fundamentals. The reader can consult the books in the area, such as that by Johnson (1985), for further details and applications, and can try to use the principle of superposition demonstrated in the above sections to obtain the solutions for any special contact problems.

[7.11] The diffusion of materials during contact such as gold into silicon, which are common in advanced manufacturing process, is not considered in this text.

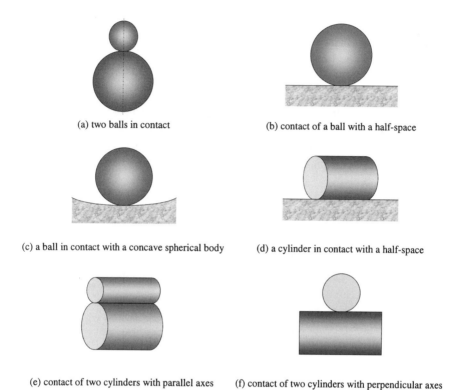

Figure 7.11 Mechanics models for some commonly encountered contact cases in engineering.

(a) two balls in contact
(b) contact of a ball with a half-space
(c) a ball in contact with a concave spherical body
(d) a cylinder in contact with a half-space
(e) contact of two cylinders with parallel axes
(f) contact of two cylinders with perpendicular axes

References

Ashley, S (1998), Bringing launch costs down to earth, *Mechanical Engineering*, **120** (10), 62-68.

Carslaw, HS and Jaeger, JC (1959), *Conduction of Heat in Solids*, 2nd edition, Oxford University Press, Oxford.

Cloud, GL (1995), *Optical Methods of Engineering Analysis*, Cambridge University Press, Cambridge.

Dowling, NE (1993), *Mechanical Behaviour of Materials*, Prentice-Hall, NJ, pp.193.

Fleck, NA, Johnson, KL, Mear, M and Zhang, L (1992), Cold rolling of thin foil, *Journal of Engineering Manufacture, IMechE*, **B206**, 119-131.

Fung, YC (1965), *Foundations of Solid Mechanics*, Prentice-Hall, Inc., Englewood Cliffs, NJ.

Gokcen, NA and Reddy, RG (1996), *Thermodynamics*, 2nd edition, Plenum Press, New York.

Hetnarski, RB (1996), *Thermal Stresses IV*, Elsevier, Amsterdam.
Jahnke, E, Emde, F and Losch, F (1960), *Tables of Higher Functions*, McGraw-Hill, New York.
Johnson, KL (1985), *Contact Mechanics*, Cambridge University Press, Cambridge.
Kurpisz, K and Nowak, AJ (1995), *Inverse Thermal Problems*, Computational Mechanics Publications, Southampton.
Kuske, A and Robertson, G (1974), *Photoelastic Stress Analysis*, Wiley, London.
Little, R (1973), *Elasticity*, Prentice-Hall, Inc., Englewood Cliffs, NJ.
Love, AEH (1952), *A Treatise on the Mathematical Theory of Elasticity*, 4th edition, Cambridge University Press, Cambridge.
Timoshenko, SP and Gere, J (1972), *Mechanics of Materials*, Van Nostrand, Reinhold.
Timoshenko, SP and Goodier, JN (1970), *Theory of Elasticity*, McGraw-Hill Book Company, Singapore.
Xie, YQ, Lin, ZX and Ding HJ (1988), *Elasticity* (in Chinese), Zhejiang University Press, Hangzhou.
Yu, T and Zhang, L (1996), *Plastic Bending: Theory and Applications*, World Scientific, Singapore.
Zhang, L (1995), On the mechanism of cold rolling thin foil, *International Journal of Machine Tools and Manufacture*, **35**, 363-372.

Important Concepts

Deformation compatibility in contact
Papkovich-Neuber solution

General solution to thermal deformation
Superposition through integration

Questions

7.1 What are the common ways to obtain a 'known' solution for the semi-inverse method?
7.2 Why do circular shafts made of ductile and brittle materials have different failure modes?
7.3 How can we design an experimental test to measure Poisson's ratio by using Eq. (7-15)?
7.4 What are the two central steps to solve a thermal deformation problem?
7.5 Why can we measure thermal stresses in a component that is subjected to a temperature change not by heating the component but by applying corresponding body and surface forces?
7.6 When we obtain the solution of a plane-stress thermal deformation problem, can we get the solution of the corresponding plane-strain problem without solving the basic equations again? How?

7.7   What are Boussinesq's and Cerruti's problems? Why are they fundamental to contact mechanics?

Problems

Unless otherwise stated, the material properties required in solving the following problems should be taken from **Table 5.2**.

7.1   A mechanical press shown in **Fig. P7.1** is to be so designed that the maximum axial and shear stresses in the power screw made of high-carbon steel (member A), which can be considered as a solid cylinder, do not exceed 50 MPa and 20 MPa, respectively. The torque Q for steadily rotating the cylinder to proceed with a press operation can be calculated by $Q = 0.3RF$, where R is the radius of the cylinder and F is the axial compressive force. (a) Model the cylinder for stress analysis. (b) Find the minimum diameter of the cylinder to ensure a safe operation with a safety factor of 2.0.

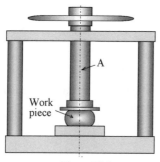

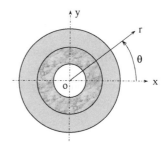

Figure P7.1   Figure P7.2

7.2   A tube as shown in **Fig. P7.2**, having nominal dimensions of inner diameter 25 mm and outer diameter 50 mm, over which a second tube having nominal dimensions of 50 mm and outer diameter of 75 mm, is to be shrink-fitted. The tube material is low-carbon steel. It is desired to fit these two members together to cause a stress of $\sigma_{\theta\theta} = 20$ MPa at the inner surface of the outer member. (a) Find the required original dimensions of the members. (b) Determine the resulting stress distributions.

7.3   A brake for a rotating system consists of a circular disk and a brake band as shown in **Fig. P7.3**. (a) Work out the mechanics model for analysing the stress and deformation of the circular disk when a braking force P is applied to the band. (b) Use the principle of superposition to resolve the above mechanics model to simpler problems so that solution can be carried out more easily.

7.4 Detailed stress solutions are important to the design of a deep-drawing system to avoid flange wrinkling in the metal sheet. **Fig. P7.4** shows the schematic of an axisymmetric deep-drawing process using an initially flat circular sheet. (a) Provide the mechanics model for stress analysis of the flange part. The diameter of the die opening is 2a and the sheet thickness is t (t is much smaller than a). The outer diameter of the flange is 2b. (*Hint*: When the flange is modelled as an annular plate, an equivalent set of stresses must be applied at the inner edge of the plate to replace the function of the material that originally exists in r < a.) (b) Use the principle of superposition to resolve the problem into simpler models. (c) If the friction between the metal sheet and the flange holder and that between the sheet and the die are ignored, find the stresses σ_{rr}, $\sigma_{\theta\theta}$ and σ_{zz} in the flange. (*Hint*: The plate is under a uniform tension at its inner surface and a compression due to the flange holder. Make use of the solution of a pressure vessel under uniform inner pressure as presented in **Table 6.1**.)

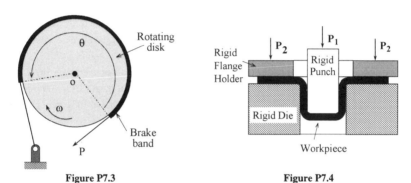

Figure P7.3   **Figure P7.4**

7.5 The operation shown in **Fig. 3.2d** (page 24) is a surface grinding process. (a) Provide the mechanics model for stress and deformation analysis of the grinding wheel during a grinding operation. (b) Assume that the wheel can be considered as an isotropic disk and its inner surface is rigidly mounted on the grinding spindle that is rotating with a uniform angular velocity ω. The wheel has an inner diameter 2a, outer diameter 2b and thickness t, where t is much smaller than a. Find the maximum angular velocity allowable before the wheel is loaded onto a workpiece if the maximum circumferential stress of the wheel material is σ_{max}.

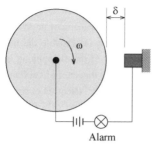
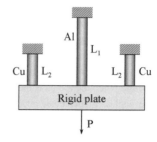

Figure P7.5   Figure P7.6

7.6 To monitor the angular velocity of a spindle, an alarm system is designed, as illustrated in **Fig. P7.5**, which consists of a thin disk mounted rigidly on the spindle and an electrical circuit with one pole on the spindle and the other on a conductor block. When the gap between the disk edge and the block, δ, vanishes due to the deformation of the disk, the alarm will be on. Find the relationship of δ to the angular velocity ω. (*Hint*: Consider it as a solid disk.)

7.7 A rigid plate subjected to a central load P = 90 kN is symmetrically suspended by an aluminium bar of L_1 = 600 mm and two copper bars of length L_2 = 300 mm, as shown in **Fig. P7.6**. The cross-sectional area of the aluminium bar is 1032 mm^2 and that of the copper bar is 516 mm^2. Calculate the stresses in the bars when the environmental temperature increases from 20 °C to 80°C.

7.8 There are two thick-walled tubes to be thermally fitted into the assembly illustrated in **Fig. P7.2**. One has an inner diameter 2a and outer diameter 2b and the other has an inner diameter 2(b − δ) and outer diameter 2c, where a < b < c and δ is the interference. If the thermal fitting is to be without any resistance, the larger tube must be heated so that its inner diameter becomes greater than 2b. (a) What should be the temperature rise for such an operation when the heating is uniform over the whole tube? (b) Determine the interaction stress between the outer surface of the smaller tube and the inner surface of the larger tube after cooling. (c) The tube material is low-carbon steel and the allowable maximum interaction stress is 35 MPa. When 2a = 100 mm, 2b = 200 mm and 2c = 300 mm, find the maximum allowable tolerance δ.

7.9 The temperature distribution in a long metal cylinder of radius R due to electric current can be described by $T = A(R^2 − r^2)$, where A is a known constant and r is the radial coordinate of the polar coordinate system whose origin is at the centre of the cylinder. All the surfaces of the cylinder are free. Find the thermal stresses σ_{rr}, $\sigma_{\theta\theta}$ and σ_{zz} in the cylinder. (*Answer*: $\sigma_{rr} = -E\alpha A(R^2 − r^2)/[4(1 − \nu)]$, $\sigma_{\theta\theta} = E\alpha A(3r^2 − R^2)/[4(1 − \nu)]$ and $\sigma_{zz} = E\alpha A(2r^2 − R^2)/[2(1 − \nu)]$.) (*Hint*: The deformation is axisymmetric and because of this the radial displacement along the axisymmetric axis of the cylinder vanishes, *i.e.*, u $|_{r=0} = 0$.)

7.10 In plate rolling using a four-high rolling mill, as illustrated in **Fig. 6.8** (page 130), the backup rolls are used to support the smaller work rolls to prevent them from bending so that loading on the plate becomes uniform in the direction of the rolls' axis. On a backup roll, rolling force, P, is applied through its two ends. (a) Provide the mechanics model for analysing the stress and deformation of a backup roll. (*Hint*: Assume a contact area and stress between the work and backup rolls.) (b) If one is only interested in the deformation of the backup roll in the vicinity of the contact zone and the radius of the backup roll can be considered to be much larger than that of the work roll, provide a simplified mechanics model by applying Saint-Venant's principle.

7.11 Find the stress and displacement components in a half-space that is subjected to a concentrated normal force P and a concentrated tangential force Q at a surface point O. (A combined loading of the cases is shown in **Fig. 7.8b** and **Fig. 7.8c**.)

7.12 Find the stress and displacement components in a half-space subjected to a concentrated normal line load, as shown in **Fig. P7.7**. (*Hint*: Superpose the solution to a half-space under a normal concentrated load q, Eqs.(7-35) and (7-36). In this case, integration is necessary.)

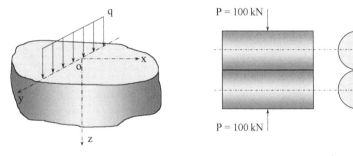

Figure P7.7   Figure P7.8

7.13 In the simplified model obtained in Problem 7.10(b), if the distribution of the contact stress is assumed to be uniform over the whole contact area, determine the stress and displacement components in the backup roll. (*Hint*: Use the principle of superposition and the solution to Problem 7.12.)

The solutions to the following problems need the formulae listed in Appendix B. Friction is ignored in all cases.

7.14 Two identical high-carbon steel cylinders are pressed together by load P of 100 kN, as illustrated in **Fig. P7.8**. If the maximum stress on the contact surface is not allowed to exceed 1000 MPa, determine the minimum diameter d of the cylinder. The length of the cylinders is 200 mm.

7.15 Find the maximum contact stress at the contact surface between a ball and the outer ring in a ball bearing. The ball diameter is 38 mm, the ball race radius of the ring is 25 mm while the inner diameter of the outer ring is 200 mm. The load on each ball is 20 kN. Both the ball and ring are made of high-carbon steel.

7.16 The contact between a wheel of a train and a rail is shown in **Fig. P7.9**. Both the wheel and rail are made of a steel with Young's modulus E = 210 GPa and Poisson's ratio ν = 0.3. The radius of the wheel is R_1 = 500 mm and that of the railway surface is R_2 = 300 mm. When the total load on the wheel is 5 kN, find the maximum contact stress on the wheel-rail interface.

7.17 The structure illustrated in **Fig. P7.10** is a common design of a type of bridge support. Both the rolls and platform are made of medium-carbon steel. When the maximum allowable contact stress between a roll and the platform is 1000 MPa, what is the possible maximum load P of such a support? The roll diameter is 100 mm and its length is 300 mm.

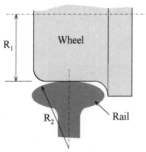
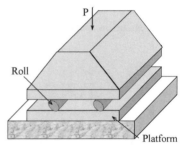

Figure P7.9   **Figure P7.10**

7.18 The Brinell hardness test is a commonly used non-destructive test for measuring material properties, which uses a very hard ball as an indenter, as shown in **Fig. 6.6a** (page 124). In a test on a stainless steel with P = 10 N, determine the maximum contact stress on the interface. The ball of diameter 10 mm is considered to be rigid.

Chapter 8

STRESS FUNCTION METHOD

Under certain conditions, stresses can be related to a scalar function and thus the solution process can be simplified and become more mechanical. This chapter discusses the stress function method for solving plane deformation problems, including the appropriate selection of the stress functions for a specified problem. Some important concepts in engineering design, such as stress concentration, will also be addressed.

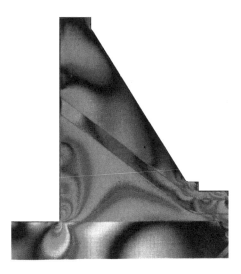

The above photograph shows the fringes in the model of a reservoir dam when subjected to water pressure on its left side surfaces. The fringes correspond to stresses in the dam according to photoelasticity, one of the important techniques for measuring stresses. Dense fringes stand for higher stresses. Some stress concentration zones in the dam can be observed clearly.

8.1 INTRODUCTION

Through the exercises in Chapters 6 and 7, we have seen that when an appropriate mechanics model of an engineering problem is ready the next important step is to solve the basic equations to find a solution that must satisfy all the boundary conditions. The task has been mathematical. However, our methods so far have dealt with the vector quantities directly, such as forces and displacements, and have been uneasy. In a way, if we can find the relationship between scalar and vector components, it will make our solution much more straightforward. We know that the gradient of a scalar function is a vector, as explained in footnote 7.9. This indicates that stresses, strains and displacements may be related to the derivatives of a scalar function. In this chapter, we shall introduce a simple scalar method for plane-deformation problems.

8.2 AIRY STRESS FUNCTION

Consider a solid in static equilibrium subjected to plane deformation in the xy-plane and use the stress approach with the semi-inverse method. If we introduce a scalar function, $\phi(x, y)$, and define its relationship with stress components as

$$\begin{cases} \sigma_{xx} = \dfrac{\partial^2 \phi}{\partial y^2}, \\ \sigma_{yy} = \dfrac{\partial^2 \phi}{\partial x^2}, \\ \sigma_{xy} = -\dfrac{\partial^2 \phi}{\partial x \partial y}, \end{cases} \qquad (8\text{-}1)$$

the equilibrium equations, Eq. (6-8), are satisfied automatically when body forces are negligible (acceleration terms vanish for static problems). By substituting the above stress expressions into the generalised Hooke's law, we can obtain strain components in terms of the function ϕ defined above. However, as we emphasised in Chapter 7 using the stress approach, the strains thus obtained must satisfy the compatibility equation, Eq. (6-10). This requires that

$$\nabla^2 \nabla^2 \phi = 0, \qquad (8\text{-}2)$$

where

$$\nabla^2 = \frac{\partial^2}{\partial x^2} + \frac{\partial^2}{\partial y^2} \qquad (8\text{-}3)$$

is the Laplace operator. Equation (8-2) is a biharmonic equation. Thus by introducing the relationship between the scalar function and stress components, the solution of a solid under plane deformation reduces to that of finding a single biharmonic function ϕ whose derivatives, *i.e.*, the stresses defined by Eq. (8-1), satisfy certain boundary conditions on the solid surfaces.[8.1] In this way, the solution process becomes much more straightforward and definite when compared with the vector methods discussed before. The above relationship between ϕ and stresses was first introduced in 1862 by G. B. Airy (1801-1892), a British scientist of Cambridge University, and therefore ϕ is called the *Airy stress function*, although Airy did not realise that ϕ must satisfy the compatibility equation (8-2).

It is important to note that the compatibility equation in terms of the stress function ϕ is independent of material constants, *i.e.*, Young's modulus E and Poisson's ratio ν. Hence the stresses in plane-stress or plane-strain deformation problems are independent of material properties, if the body is simply-connected and with stress boundary conditions. The conclusion forms the theoretical basis of photoelasticity (Kuske and Robertson, 1974) for the experimental stress measurement of components under plane deformation using some special plastic materials.

By taking a simple coordinate transform, we can rewrite the above relations in a polar coordinate system. In this case, corresponding to Eq. (8-1), stresses are determined by the Airy stress function by

$$\begin{cases} \sigma_{rr} = \frac{1}{r}\frac{\partial \phi}{\partial r} + \frac{1}{r^2}\frac{\partial^2 \phi}{\partial \theta^2}, \\ \sigma_{\theta\theta} = \frac{\partial^2 \phi}{\partial r^2}, \\ \sigma_{r\theta} = -\frac{\partial}{\partial r}\left(\frac{1}{r}\frac{\partial \phi}{\partial \theta}\right), \end{cases} \qquad (8\text{-}4)$$

[8.1] This is because a biharmonic function has satisfied equilibrium equations, stress-strain relations and compatibility equations. Hence, when we use a biharmonic function as a stress function, there is no need for us to check again the satisfaction of these equations.

and the Laplace operator of Eq. (8-3) becomes

$$\nabla^2 = \frac{\partial^2}{\partial r^2} + \frac{1}{r}\frac{\partial}{\partial r} + \frac{1}{r^2}\frac{\partial^2}{\partial \theta^2}. \tag{8-5}$$

Biharmonic functions have been studied extensively in physics and mathematics. (Some of them are listed in **Tables 8.1** and **8.2**.) Thus theoretically any biharmonic functions can be used as a stress function ϕ. However, the problem is that we normally do not know which should be chosen for a solid under given boundary stresses. For example, if we need to find the stresses in a thin rectangular plate subjected to pure bending in the xy-plane, as shown in **Fig. 8.1**, how do we select a proper ϕ from **Table 8.1** that can satisfy the stress boundary conditions of the plate?

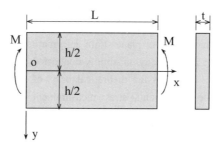

Figure 8.1 A thin rectangular plate under pure bending. The resultant bending moment is M. The front and back surfaces of the plate are stress-free. The plate is therefore under plane-stress deformation.

The most common approach[8.2] to stress function selection is via the semi-inverse method that we have been using for solving most of our previous problems. In this method, however, we have to know which biharmonic function produces what kind of boundary conditions and then use the principle of superposition to choose a number of them to satisfy the specified boundary conditions of a given problem. Such a process is certainly inefficient and relies greatly on one's solution experience. Nevertheless, the advantage is that for some simple problems we can see the physical indication of the

---

[8.2] If the boundary conditions of a problem are unmixed, *i.e.*, stress or displacement boundary conditions pertain, it is possible to solve for a stress function directly mathematically. However, this process often leaves us a difficult contour integral. Thus in this introductory text, we shall not discuss the details.

stress functions used. **Tables 8.3** and **8.4** list the stresses and displacements corresponding to some stress functions. In the following examples, we are going to demonstrate how to use them to solve some simple but interesting engineering problems. Through the process, we shall also experience the method of stress function selection. In section **8.5.3**, we shall demonstrate briefly how to obtain stress functions by directly solving the compatibility equation, Eq. (8-2), for a specific problem.

Table 8.1 Some biharmonic functions in the Cartesian coordinate system

1	-	-	-
x	y	-	-
x^2	xy	y^2	-
x^3	x^2y	xy^2	y^3
x^3y	xy^3	-	-
$\sinh(\zeta x)\sin(\zeta y)$	$\sinh(\zeta x)\cos(\zeta y)$	$\cosh(\zeta x)\sin(\zeta y)$	$\cosh(\zeta x)\cos(\zeta y)$
$\sin(\zeta x)\sinh(\zeta y)$	$\sin(\zeta x)\cosh(\zeta y)$	$\cos(\zeta x)\sinh(\zeta y)$	$\cos(\zeta x)\cosh(\zeta y)$
$x\sinh(\zeta x)\sin(\zeta y)$	$x\sinh(\zeta x)\cos(\zeta y)$	$x\cosh(\zeta x)\sin(\zeta y)$	$x\cosh(\zeta x)\cos(\zeta y)$
$y\sin(\zeta x)\sinh(\zeta y)$	$y\sin(\zeta x)\cosh(\zeta y)$	$y\cos(\zeta x)\sinh(\zeta y)$	$y\cos(\zeta x)\cosh(\zeta y)$
$x\sin(\zeta x)\sinh(\zeta y)$	$x\sin(\zeta x)\cosh(\zeta y)$	$x\cos(\zeta x)\sinh(\zeta y)$	$x\cos(\zeta x)\cosh(\zeta y)$
$y\sinh(\zeta x)\sin(\zeta y)$	$y\sinh(\zeta x)\cos(\zeta y)$	$y\cosh(\zeta x)\sin(\zeta y)$	$y\cosh(\zeta x)\cos(\zeta y)$

In the above, ζ is an arbitrary positive constant and
$$\sinh(\zeta x) = (e^{\zeta x} - e^{-\zeta x})/2, \quad \cosh(\zeta x) = (e^{\zeta x} + e^{-\zeta x})/2$$

Table 8.2 Some biharmonic functions in the polar coordinate system

1	r^2	$\ln r$	$r^2 \ln r$
θ	θr^2	$\theta \ln r$	$\theta r^2 \ln r$
$r \ln r \sin\theta$	$r \ln r \cos\theta$	$r\theta \ln r \sin\theta$	$r\theta \ln r \cos\theta$
$r\theta \sin\theta$	$r\theta \cos\theta$	-	-
$r^\zeta \sin(\zeta\theta)$	$r^\zeta \cos(\zeta\theta)$	$r^{2+\zeta} \sin(\zeta\theta)$	$r^{2+\zeta} \cos(\zeta\theta)$
$r^{-\zeta} \sin(\zeta\theta)$	$r^{-\zeta} \cos(\zeta\theta)$	$r^{2-\zeta} \sin(\zeta\theta)$	$r^{2-\zeta} \cos(\zeta\theta)$
$\sin(\zeta \ln r)\sinh(\zeta\theta)$	$\sin(\zeta \ln r)\cosh(\zeta\theta)$	$\cos(\zeta \ln r)\sinh(\zeta\theta)$	$\cos(\zeta \ln r)\cosh(\zeta\theta)$
$r^2 \sin(\zeta \ln r)\sinh(\zeta\theta)$	$r^2 \sin(\zeta \ln r)\cosh(\zeta\theta)$	$r^2 \cos(\zeta \ln r)\sinh(\zeta\theta)$	$r^2 \cos(\zeta \ln r)\cosh(\zeta\theta)$
$r\sin(\zeta \ln r)\sinh(\zeta\theta)\sin\theta$	$r\sin(\zeta \ln r)\cosh(\zeta\theta)\sin\theta$	$r\cos(\zeta \ln r)\sinh(\zeta\theta)\sin\theta$	$r\cos(\zeta \ln r)\cosh(\zeta\theta)\sin\theta$
$r\sin(\zeta \ln r)\sinh(\zeta\theta)\cos\theta$	$r\sin(\zeta \ln r)\cosh(\zeta\theta)\cos\theta$	$r\cos(\zeta \ln r)\sinh(\zeta\theta)\cos\theta$	$r\cos(\zeta \ln r)\cosh(\zeta\theta)\cos\theta$

In the above, ζ is an arbitrary positive constant.

Table 8.3 Examples of stresses and boundary conditions corresponding to individual stress functions in the Cartesian coordinate system

No	Stress Function ϕ	Corresponding stresses	Corresponding boundary conditions
1	$ax + by + c$	$\sigma_{xx} = 0$ $\sigma_{yy} = 0$ $\sigma_{xy} = 0$	
2	ax^2	$\sigma_{xx} = 0$ $\sigma_{xy} = 0$ $\sigma_{yy} = 2a$	
3	ay^2	$\sigma_{yy} = 0$ $\sigma_{xy} = 0$ $\sigma_{xx} = 2a$	
4	axy	$\sigma_{xx} = 0$ $\sigma_{yy} = 0$ $\sigma_{xy} = -a$	

5	ax^3	$\sigma_{xx} = 0$ $\sigma_{xy} = 0$ $\sigma_{yy} = 6ax$	
6	ax^2y	$\sigma_{xx} = 0$ $\sigma_{yy} = 2ay$ $\sigma_{xy} = -2ax$	
7	axy^2	$\sigma_{xx} = 2ax$ $\sigma_{yy} = 0$ $\sigma_{xy} = -2ay$	
8	ay^3	$\sigma_{yy} = 0$ $\sigma_{xy} = 0$ $\sigma_{xx} = 6ay$	
9	$a(x^4 - y^4)$	$\sigma_{xx} = -12ay^2$ $\sigma_{yy} = 12ax^2$ $\sigma_{xy} = 0$	

10	axy^3	$\sigma_{xx} = 6axy$ $\sigma_{yy} = 0$ $\sigma_{xy} = -3ay^2$	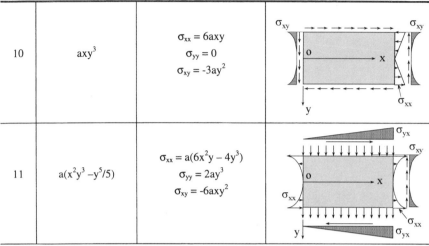
11	$a(x^2y^3 - y^5/5)$	$\sigma_{xx} = a(6x^2y - 4y^3)$ $\sigma_{yy} = 2ay^3$ $\sigma_{xy} = -6axy^2$	

In the above, a, b and c are arbitrary constants.

Table 8.4 Some examples of stresses and displacements corresponding to individual stress functions in the polar coordinate system

No	Stress function ϕ	σ_{rr}	$\sigma_{\theta\theta}$	$\sigma_{r\theta}$	$2\mu u$	$2\mu v$
1	1	0	0	0	0	0
2	r^2	2	2	0	$(\omega-1) r$	0
3	$r^2 \ln r$	$2 \ln r + 1$	$2 \ln r + 3$	0	$(\omega-1) r \ln r - r$	$(\omega+1) r \theta$
4	$\ln r$	r^{-2}	$-r^{-2}$	0	$-r^{-1}$	0
5	θ	0	0	r^{-2}	0	$-r^{-1}$
6	$r^3 \cos\theta$	$2r \cos\theta$	$6r \cos\theta$	$2r \sin\theta$	$(\omega-2) r^2 \cos\theta$	$(\omega+2) r^2 \sin\theta$
7	$r\theta \sin\theta$	$2r^{-1} \cos\theta$	0	0	$\frac{1}{2}[(\omega-1) \theta \sin\theta - \cos\theta + (\omega+1) \ln r \cos\theta]$	$\frac{1}{2}[(\omega-1) \theta \cos\theta - \sin\theta - (\omega+1) \ln r \sin\theta]$
8	$r \ln r \cos\theta$	$r^{-1} \cos\theta$	$r^{-1} \cos\theta$	$r^{-1} \sin\theta$	$\frac{1}{2}[(\omega+1) \theta \sin\theta - \cos\theta + (\omega-1) \ln r \cos\theta]$	$\frac{1}{2}[(\omega+1) \theta \cos\theta - \sin\theta - (\omega-1) \ln r \sin\theta]$
9	$r^{-1} \cos\theta$	$-2 r^{-3} \cos\theta$	$2 r^{-3} \cos\theta$	$-2 r^{-3} \sin\theta$	$r^{-2} \cos\theta$	$r^{-2} \sin\theta$
10	$r^3 \sin\theta$	$2r \sin\theta$	$6r \sin\theta$	$-2r \cos\theta$	$(\omega-2) r^2 \sin\theta$	$-(\omega+2) r^{-2} \cos\theta$
11	$r \theta \cos\theta$	$-2 r^{-1} \sin\theta$	0	0	$\frac{1}{2}[(\omega-1) \theta \cos\theta + \sin\theta - (\omega+1) \ln r \sin\theta]$	$\frac{1}{2}[-(\omega-1) \theta \sin\theta - \cos\theta - (\omega+1) \ln r \cos\theta]$

12	$r \ln r \sin\theta$	$r^{-1} \sin\theta$	$r^{-1} \sin\theta$	$-r^{-1} \cos\theta$	$\frac{1}{2}[-(\omega+1)\theta\cos\theta - \sin\theta + (\omega-1)\ln r \sin\theta]$	$\frac{1}{2}[(\omega+1)\theta\sin\theta + \cos\theta + (\omega-1)\ln r \cos\theta]$
13	$r^{-1} \sin\theta$	$-2 r^{-3} \sin\theta$	$2 r^{-3} \sin\theta$	$2 r^{-3} \cos\theta$	$r^{-2} \sin\theta$	$-r^{-2} \cos\theta$
14	$r^{2+\zeta} \cos(\zeta\theta)$	$-(\zeta+1)(\zeta-2) r^\zeta \cos(\zeta\theta)$	$(\zeta+1)(\zeta+2) r^\zeta \cos(\zeta\theta)$	$\zeta(\zeta+1) r^\zeta \sin(\zeta\theta)$	$(\omega-\zeta-1) r^{\zeta+1} \cos(\zeta\theta)$	$(\omega+\zeta+1) r^{\zeta+1} \sin(\zeta\theta)$
15	$r^{2-\zeta} \cos(\zeta\theta)$	$-(\zeta+2)(\zeta-1) r^{-\zeta} \cos(\zeta\theta)$	$(\zeta-2)(\zeta-1) r^{-\zeta} \cos(\zeta\theta)$	$-\zeta(\zeta-1) r^{-\zeta} \sin(\zeta\theta)$	$(\omega+\zeta-1) r^{-\zeta+1} \cos(\zeta\theta)$	$-(\omega-\zeta+1) r^{-\zeta+1} \sin(\zeta\theta)$
16	$r^\zeta \cos(\zeta\theta)$	$-\zeta(\zeta-1) r^{\zeta-2} \cos(\zeta\theta)$	$\zeta(\zeta-1) r^{\zeta-2} \cos(\zeta\theta)$	$\zeta(\zeta-1) r^{\zeta-2} \sin(\zeta\theta)$	$-\zeta r^{\zeta-1} \cos(\zeta\theta)$	$\zeta r^{\zeta-1} \sin(\zeta\theta)$
17	$r^{-\zeta} \cos(\zeta\theta)$	$-\zeta(\zeta+1) r^{-\zeta-2} \cos(\zeta\theta)$	$\zeta(\zeta+1) r^{-\zeta-2} \cos(\zeta\theta)$	$-\zeta(\zeta+1) r^{-\zeta-2} \sin(\zeta\theta)$	$\zeta r^{-\zeta-1} \cos(\zeta\theta)$	$\zeta r^{-\zeta-1} \sin(\zeta\theta)$
18	$r^{2+\zeta} \sin(\zeta\theta)$	$-(\zeta+1)(\zeta-2) r^\zeta \sin(\zeta\theta)$	$(\zeta+1)(\zeta+2) r^\zeta \sin(\zeta\theta)$	$-\zeta(\zeta+1) r^\zeta \cos(\zeta\theta)$	$(\omega-\zeta-1) r^{\zeta+1} \sin(\zeta\theta)$	$-(\omega+\zeta+1) r^{\zeta+1} \cos(\zeta\theta)$
19	$r^{2-\zeta} \sin(\zeta\theta)$	$-(\zeta+2)(\zeta-1) r^{-\zeta} \sin(\zeta\theta)$	$(\zeta-2)(\zeta-1) r^{-\zeta} \sin(\zeta\theta)$	$\zeta(\zeta-1) r^{-\zeta} \cos(\zeta\theta)$	$(\omega+\zeta-1) r^{-\zeta+1} \sin(\zeta\theta)$	$(\omega-\zeta+1) r^{-\zeta+1} \cos(\zeta\theta)$
20	$r^\zeta \sin(\zeta\theta)$	$-\zeta(\zeta-1) r^{\zeta-2} \sin(\zeta\theta)$	$\zeta(\zeta-1) r^{\zeta-2} \sin(\zeta\theta)$	$-\zeta(\zeta-1) r^{\zeta-2} \cos(\zeta\theta)$	$-\zeta r^{\zeta-1} \sin(\zeta\theta)$	$-\zeta r^{\zeta-1} \cos(\zeta\theta)$
21	$r^{-\zeta} \sin(\zeta\theta)$	$-\zeta(\zeta+1) r^{-\zeta-2} \sin(\zeta\theta)$	$\zeta(\zeta+1) r^{-\zeta-2} \sin(\zeta\theta)$	$\zeta(\zeta+1) r^{-\zeta-2} \cos(\zeta\theta)$	$\zeta r^{-\zeta-1} \sin(\zeta\theta)$	$-\zeta r^{-\zeta-1} \cos(\zeta\theta)$

In the above expressions, ζ is an arbitrary positive constant, ω equals $(3-4\nu)$ under plane-strain deformation and $(3-\nu)/(1+\nu)$ under plane-stress deformation.

8.3 A RECTANGULAR PLATE UNDER PURE BENDING

Let us use the stress function method to solve a rectangular plate subjected to pure bending, as illustrated in **Fig. 8.1**. When the plate thickness, t, is much smaller than its height, h, and length, L, the plate is under plane-stress deformation. The boundary conditions of the beam are

at $y = \pm h/2$, $0 \leq x \leq L$,

$$\sigma_{yy} = \sigma_{yx} = 0; \tag{a}$$

at $x = 0$ and $x = L$, $-h/2 \leq y \leq h/2$,

$$\sigma_{xy} = 0, \tag{b}$$

$$\int_{-h/2}^{h/2} \sigma_{xx} \, dy = 0, \quad \text{(c)}$$

$$\int_{-h/2}^{h/2} \sigma_{xx} y \, dy = M. \quad \text{(d)}$$

Conditions (c) and (d) are the result of the application of Saint-Venant's principle because we do not know the exact distribution of σ_{xx} at the two ends of the plate, although we know that σ_{xx} at the ends yields the resultant bending moment M but without the resultant force along the x-axis. For convenience, in the above we have let the plate thickness be a unit, i.e., t = 1.

Now we need to choose a proper stress function that can satisfy the boundary conditions. From **Table 8.3**, we can find that stress function No. 8 leads to a linear σ_{xx} at the two ends of the plate, which is positive when y is positive and becomes negative when y is negative. All the other stress components are zero. Thus if we use this stress function, i.e., if we let $\phi = ay^3$, boundary conditions (a), (b) and (c) will be satisfied automatically. To satisfy condition (d), substituting the corresponding σ_{xx} in **Table 8.3** into Eq. (d) we obtain

$$a = \frac{2M}{h^3}.$$

Hence, the stresses in the plate are

$$\sigma_{xx} = \frac{12M}{h^3} y, \quad \sigma_{yy} = 0, \quad \sigma_{xy} = 0.$$

8.4 PLATE BENDING UNDER A CONCENTRATED SHEAR FORCE

Consider a rectangular plate clamped at its left end and subjected to a concentrated shear force, P, at its right end, see **Fig. 8.2a**. Similarly we assume that the thickness of the plate is a unit and that its length L is much larger than its height h. The boundary conditions of the plate are therefore

at $y = \pm h/2$, $0 \le x \le L$,

$$\sigma_{yy} = \sigma_{yx} = 0; \quad \text{(a)}$$

at x = 0, -h/2 ≤ y ≤ h/2,
$$u = v = 0;\qquad(b)$$

at x = L, -h/2 ≤ y ≤ h/2,
$$\sigma_{xx} = 0;\qquad(c)$$

$$\int_{-h/2}^{h/2} \sigma_{xy}\,dy = P.\qquad(d)$$

Again, condition (d) above is due to the application of Saint-Venant's principle to match the resultant shear force P because we do not know the distribution of shear stress σ_{xy} applied externally.

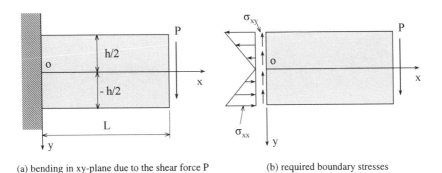

(a) bending in xy-plane due to the shear force P   (b) required boundary stresses

Figure 8.2 A cantilever plate under plane-stress bending.

It is easy to understand that the bending under P will cause, at the clamped end (x = 0), a normal stress σ_{xx}, whose distribution must bring about a resultant bending moment PL but a vanishing axial force, and a shear stress σ_{xy}, whose resultant shear force must be P. Hence, we need to select a stress function that can lead to consistent boundary stresses. When reviewing all the stress functions listed in **Table 8.3**, we find that none of the individual ones can satisfy the boundary conditions shown in **Fig. 8.2b**. Fortunately, we can use the superposition principle to combine some of them to obtain a satisfactory stress function. Stress function No. 4 in **Table 8.3** produces uniform shear stress on all the plate surfaces, while stress function No. 10 yields uniform shear stress on the top and bottom surfaces of the plate but brings about non-uniform shear stress at the two ends. Thus the superposition of stress functions No. 4 and No. 10 may cancel the shear stresses on the top and bottom surfaces but still leave shear stresses at the two ends. Function No. 10 also leads to a normal stress σ_{xx} at the right end (x = L) but the boundary conditions of our problem only require a normal stress at the left end. This means that we must find another stress function that can cancel the normal stress at the right end and in the meantime gives rise to a normal stress at the left end (x = 0). In fact, stress function No. 8 is just what we need. As a result, we can use the summation of No.4, No. 8 and No. 10 as the stress function for our current beam bending problem, *i.e.*,

$$\phi = axy + bxy^3 + cy^3,$$

where a, b and c are constants to be determined by boundary conditions. The stresses can then be calculated by substituting the above ϕ into Eq. (8-1). In this way, we obtain

$$\begin{cases} \sigma_{xx} = 6bxy + 6cy, \\ \sigma_{yy} = 0, \\ \sigma_{xy} = -\left(a + 3by^2\right). \end{cases}$$

Substituting the above into boundary conditions (a), (c) and (d), we get

$$a = -\frac{Ph^2}{8I}, \quad b = \frac{P}{6I}, \quad c = -\frac{PL}{6I},$$

where $I = h^3/12$. Thus the stresses are

$$\begin{cases} \sigma_{xx} = -\frac{P}{I}(L-x)y, \\ \sigma_{yy} = 0, \\ \sigma_{xy} = \frac{P}{2I}\left(\frac{h^2}{4} - y^2\right). \end{cases}$$

Now let us try to find the displacements. Using the generalised Hooke's law, Eq. (6-11), and geometrical relations between strains and displacements, Eq. (6-9), for plane-stress deformation, we can easily obtain

$$\begin{cases} \varepsilon_{xx} = \dfrac{\partial u}{\partial x} = -\dfrac{P(L-x)}{EI} y, \\ \varepsilon_{yy} = \dfrac{\partial v}{\partial y} = v\dfrac{P(L-x)}{EI} y, \\ \varepsilon_{xy} = \dfrac{1}{2}\left(\dfrac{\partial u}{\partial y} + \dfrac{\partial v}{\partial x}\right) = \dfrac{(1+v)P}{2EI}\left(\dfrac{h^2}{4} - y^2\right). \end{cases} \quad \text{(e, f, g)}$$

By integrating Eqs. (e) and (f), we get

$$\begin{cases} u = -\dfrac{P}{EI}\left(Lyx - \dfrac{1}{2}x^2 y\right) + \dfrac{P}{EI} f(y), \\ v = \dfrac{vP}{2EI}\left(Ly^2 - xy^2\right) + \dfrac{P}{EI} g(x). \end{cases} \quad \text{(h, i)}$$

The substitution of Eqs. (h) and (i) into Eq. (g) gives rise to

$$\left\{ g'(x) - \left(Lx - \dfrac{1}{2}x^2\right) \right\} + \left\{ f'(y) - \dfrac{1}{2}vy^2 + (1+v)y^2 \right\} = \dfrac{1}{4}(1+v)h^2. \quad \text{(j)}$$

In Eq. (j), the first term at the left hand side is a function of x, the second term is that of y, but the right-hand side of the equation is a constant. To make the equation valid, both the first and second terms in the left-hand side must be constants. Thus,

$$\begin{cases} g'(x) - \left(Lx - \dfrac{1}{2}x^2\right) = m, \\ f'(y) - \dfrac{1}{2}vy^2 + (1+v)y^2 = n, \\ m + n = \dfrac{1}{4}(1+v)h^2, \end{cases} \quad \text{(k, l, m)}$$

where m and n are constants to be determined. The integration of Eqs. (k) and (l) yields

$$\begin{cases} g(x) = \dfrac{1}{2}Lx^2 - \dfrac{1}{6}x^3 + mx + c_1, \\ f(y) = -\dfrac{2+v}{6}y^3 + ny + c_2, \end{cases} \qquad (o, p)$$

where c_1 and c_2 are integration constants. Hence, Eqs. (h) and (i) can be rewritten as

$$\begin{cases} u = \dfrac{P}{EI}\left(-Lxy + \dfrac{1}{2}x^2 y - \dfrac{2+v}{6}y^3 + ny + c_2\right), \\ v = \dfrac{P}{EI}\left(\dfrac{1}{2}vLy^2 - \dfrac{1}{2}vxy^2 + \dfrac{1}{2}Lx^2 - \dfrac{1}{6}x^3 + mx + c_1\right), \end{cases} \qquad (q, r)$$

where c_1, c_2, m, and n should be determined by the displacement boundary conditions of the beam. Clearly, the displacements given by Eqs. (q) and (r) cannot satisfy the clamping condition at the left end of the beam described by Eq. (b). Thus the above solution cannot be the exact solution of the problem. However, according to Saint-Venant's principle, we can let the solution satisfy the displacement boundary conditions at one point, *e.g.*, let $u = v = 0$ at $x = y = 0$, which eliminates rigid-body translation of the beam in the x- and y-directions, and $\dfrac{\partial v}{\partial x} = 0$ at $x = y = 0$, which eliminates the rigid-body rotation of the beam around the origin. Together with Eq. (m) we finally get

$$\begin{cases} c_1 = c_2 = m = 0, \\ n = \dfrac{1}{4}(1+v)h^2. \end{cases}$$

Hence, the final solution of the displacements becomes

$$\begin{cases} u = \dfrac{P}{EI}\left(-Lxy + \dfrac{1}{2}x^2 y - \dfrac{2+v}{6}y^3 + \dfrac{1+v}{4}h^2 y\right), \\ v = \dfrac{P}{EI}\left(\dfrac{1}{2}vLy^2 - \dfrac{1}{2}vxy^2 + \dfrac{1}{2}Lx^2 - \dfrac{1}{6}x^3\right). \end{cases}$$

This is of course an approximate solution based on Saint-Venant's principle.

8.5 STRESS CONCENTRATION AROUND A CIRCULAR HOLE

Many engineering structures have holes and we often see that some of them have failed around a small hole. The failure is related to the stress fields in the neighbourhood of the holes. In the following, let us see what problems can be caused by a small hole in a plate even under a very simple loading condition.

8.5.1 Modelling and Solution

Consider a thin but large plate subjected to a uniaxial uniform stress T in x-direction, as shown in **Fig. 8.3a**. We know that the stresses in the plate must be

$$\sigma_{xx} = T, \; \sigma_{yy} = \sigma_{xy} = 0. \tag{a}$$

However, if we drill a small hole of radius a at the centre of the plate, as shown in Fig. 8.3b, the stresses in the plate around the hole must become very different from those of Eq. (a). Since the hole is circular, it is more convenient to use a polar coordinate system. Unfortunately, the outer boundary is in x- and y-directions so a Cartesian coordinate system is more convenient. To resolve this problem, a simple way is to transfer the outer boundary to a circular one too so that the polar coordinate system can be used for both the inner and outer boundaries.

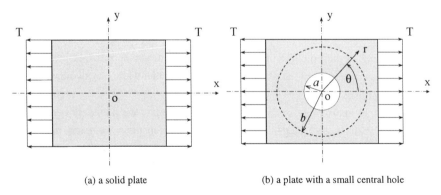

(a) a solid plate   (b) a plate with a small central hole

Figure 8.3 Thin and large plates with a central hole under uniaxial uniform tension.

Since the dimension of the plate is much larger than the hole radius, according to Saint-Venant's principle, stresses in the area far away from the hole must not be affected by the appearance of the hole. That is, in the zone away from the hole, stresses in the plate are still described by Eq. (a). Hence, if we imagine isolating an annular plate with an outer radius b, where b is much larger than a, as illustrated in **Fig. 8.3b**, the stresses on the outer radius, σ_{rr} and $\sigma_{r\theta}$, can be obtained by stress transformation formulae, Eqs. (3-14) and (3-15), by using Eq. (a). This gives rise to

$$\begin{cases} \sigma_{rr} = \dfrac{1}{2}T(1+\cos 2\theta), \\ \sigma_{r\theta} = -\dfrac{1}{2}T\sin 2\theta. \end{cases} \tag{b}$$

Therefore, we have converted the original problem of a rectangular plate with a small central hole to that of an annular plate with inner radius a and outer radius b, whose inner hole is stress-free but whose outer surface is subjected to the boundary stresses described by Eq. (b). In other words, we are to solve the annular plate with the following boundary conditions:

at $r = a$,
$$\sigma_{rr} = \sigma_{r\theta} = 0; \tag{e}$$

at $r = b$,
$$\sigma_{rr} = \frac{1}{2}T + \frac{1}{2}T\cos 2\theta, \quad \sigma_{r\theta} = -\frac{1}{2}T\sin 2\theta. \tag{d}$$

According to the principle of superposition, Eq. (d) can be resolved into two parts. Part one has σ_{rr} = constant and $\sigma_{r\theta} = 0$ when r = constant while part two has $\sigma_{rr} \propto \cos\theta$ and $\sigma_{r\theta} \propto \sin\theta$ at a given r.

To find an appropriate stress function that can satisfy the boundary conditions, we can also do it in two steps, *i.e.*, find the stress functions for parts one and two individually. From **Table 8.4**, we can see that stress functions Nos. 2 to 4 all give constant σ_{rr} with $\sigma_{r\theta} = 0$ at a given r. However, function No. 3 produces a circumferential displacement v that is a linear increasing function of θ. Because deformation inside the plate must be continuous and displacements at a point must be unique, $v|_{\theta=0}$ must be equal to $v|_{\theta=2\pi}$. Thus function No. 3 cannot be used.

With respect to part two, the appropriate stress function must produce $\cos 2\theta$ for σ_{rr} but $\sin 2\theta$ for $\sigma_{r\theta}$ at a given r. Since $\zeta = 2$, we can easily see from **Table 8.4** that only stress functions No. 15 to No. 17 are the possible ones. Function No. 14 is not suitable because when $\zeta = 2$ it leads to $\sigma_{rr} = 0$ that does not meet the boundary condition of part two in terms of σ_{rr}. Hence, the final stress function for our problem is the summation of functions No. 2, No. 4, No. 15, No. 16 and No. 17. That is,

$$\phi = c_1 r^2 + c_2 \ln r + \left(c_3 + c_4 r^2 + c_5 r^{-2}\right)\cos 2\theta,$$

where c_i (i = 1, ..., 5) are constants to be determined by the boundary conditions, Eqs. (c) to (d), when stresses are calculated by Eq. (8-4). The process is straightforward and we are not going to show the details. Finally we get the following stress components:

$$\begin{cases} \sigma_{rr} = \frac{T}{2}\left(1 - \frac{a^2}{r^2}\right) + \frac{T}{2}\left(1 + \frac{3a^4}{r^4} - \frac{4a^2}{r^2}\right)\cos 2\theta, \\ \sigma_{\theta\theta} = \frac{T}{2}\left(1 + \frac{a^2}{r^2}\right) - \frac{T}{2}\left(1 + \frac{3a^4}{r^4}\right)\cos 2\theta, \\ \sigma_{r\theta} = -\frac{T}{2}\left(1 - \frac{3a^4}{r^4} + \frac{2a^2}{r^2}\right)\sin 2\theta. \end{cases} \quad (8\text{-}6)$$

In deriving the above solution, we have let $b \to \infty$ because b is much larger than a.

8.5.2 Analysis of the Solution

Solution (8-6) represents a very important phenomenon, *stress concentration*, which must be considered in engineering practice. We can see from the solution that at r = a and $\theta = \pi/2$ and $3\pi/2$, $\sigma_{\theta\theta}$ reaches its maximum value, *i.e.*,

$$\sigma_{\theta\theta}\big|_{max} = 3T. \quad (e)$$

In general, if the maximum stress in a component is k times greater than the *nominal stress* σ_{nom}, that is, if

$$\sigma\big|_{max} = k\sigma_{nom}$$

we call k the *stress concentration factor* in the component, where the nominal stress is the stress that would exist if the hole (or fillet, notch, crack, etc.) were not there. For the above plate with a central hole under uniaxial tension, $\sigma_{nom} = T$ and thus the stress concentration factor $k = 3$. This means that the drilling of a hole in the plate raises the maximum stress locally three times over the nominal stress value T. That is why many structures with small holes fail in the vicinity of the holes. On the other hand, when $\theta = \pi/2$ and $3\pi/2$,

$$\sigma_{\theta\theta} = T\left\{1 + a^2/(2r^2) + 3a^4/(2r^4)\right\}.$$

The first term in the parentheses is the stress in the plate without the hole. The second and third terms are due to the effect of the hole. These terms decrease quickly as r increases and their contributions are only about 1/200 of the first when $r = 10a$. We can therefore consider that at a distance of $10a$ from the hole, the stresses are no longer influenced by the hole.

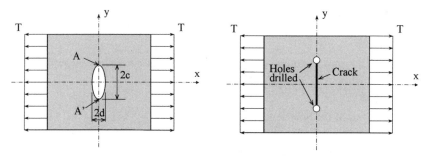

Figure 8.4 A plate with an elliptical hole under a uniaxial tension.

Figure 8.5 Two holes drilled at the tips of a crack to reduce the value of stress concentration factor k and thus to arrest the crack propagation.

If we make an elliptical hole in a plate under the uniform tension T, with one of the principal axes of the ellipse parallel to the tension as shown in **Fig. 8.4**, we can show[8.3] that the stresses at point A and A' reach the maximum of

$$\sigma_{xx}\big|_{max} = T\left(1+2\frac{c}{d}\right), \tag{f}$$

where c and d are the half-lengths of major and minor axes of the elliptical hole. When c = d, it becomes a circular hole and Eq. (f) reduces to Eq. (e). However, when d approaches zero, *i.e.*, when the elliptical hole becomes a crack with length 2c, the stress concentration factor $k = 1+2\frac{c}{d}$ will approach infinity and thus cause the crack to propagate. In practice, drilling holes at the ends of a crack can sometimes be used to increase the crack-tip radius and therefore arrest the crack propagation, as illustrated in **Fig. 8.5**. Some handbooks (*e.g.*, that by Roark, 1965) have worked out the stress concentration factors for quite a number of components and structures with fillets, holes, notches and other defects under various loading conditions.

8.5.3 Alternative Methods of Solution

There are some other ways to solve the above problem, *i.e.*, a large plate with a central small hole under uniaxial tension, as described by the boundary conditions (c) and (d). These are discussed below.

First, the problem can be resolved into two simpler problems with boundary conditions as follows, according to the principle of superposition.

Problem (1): An annular plate subjected to a uniform tension T/2 on its outer edge r = b. The boundary conditions of the problem can be described as

at r = a,
$$\sigma_{rr}^{(1)} = \sigma_{r\theta}^{(1)} = 0;$$
at r = b,
$$\sigma_{rr}^{(1)} = \frac{T}{2}, \ \sigma_{r\theta}^{(1)} = 0.$$

Problem (2): An annular plate subjected to the following boundary conditions:

at r = a,

[8.3] In the case with an elliptical hole, we may use the complex function method to obtain the solution of Eq. (f). Details can be found in the book by Muskhelishvili (1953).

$$\sigma_{rr}^{(2)} = \sigma_{r\theta}^{(2)} = 0; \tag{g}$$

at $r = b$,

$$\sigma_{rr}^{(2)} = \frac{T}{2}\cos 2\theta, \quad \sigma_{r\theta}^{(2)} = -\frac{T}{2}\sin 2\theta. \tag{h}$$

The solution of Problem (1) above can be obtained directly from **Table 6.1**, where we discussed the solution of a hollow circular disk under outer pressure, p_0. By changing p_0 to $-T/2$, R_i to a, R_o to b and by letting a/b be zero because b is much greater than a, we get the stress solution as

$$\begin{cases} \sigma_{rr}^{(1)} = \dfrac{T}{2}\left(1 - \dfrac{a^2}{r^2}\right), \\ \sigma_{\theta\theta}^{(1)} = \dfrac{T}{2}\left(1 + \dfrac{a^2}{r^2}\right), \\ \sigma_{r\theta}^{(1)} = 0. \end{cases} \tag{i}$$

To get a solution to Problem (2), we need to find a proper stress function. We can certainly follow the same method used in the previous section, *i.e.*, to select it from **Table 8.4**. But now let us try to solve it directly from the compatibility equation (8-2).

The relationship between stress function ϕ and stresses in the polar coordinate system, Eq. (8-4), indicates that to satisfy boundary condition (h), ϕ must have the form of

$$\phi = f(r)\cos 2\theta, \tag{j}$$

where f(r) is an undetermined function of variable r. Since ϕ must satisfy the compatibility equation (8-2), it requires that

$$\left(\frac{d^2 f}{dr^2} + \frac{1}{r}\frac{df}{dr} - \frac{4f}{r^2}\right)\cos 2\theta = 0.$$

The above equation must hold for any θ, thus

$$\frac{d^2f}{dr^2} + \frac{1}{r}\frac{df}{dr} - \frac{4f}{r^2} = 0. \tag{k}$$

To obtain a solution to Eq. (k), let $f = \sum c_n r^n$, where c_n ($n = \pm 1, \pm 2, \pm 3, \cdots$) are constants. The substitution of f into Eq. (k) leads to

$$\sum n(n^2 - 4)(n - 4) c_n r^{n-4} = 0.$$

The above equation must be valid for any r, thus we have $n = 0$, $n = \pm 2$, or $n = 4$. Hence, the stress function (j) becomes

$$\phi = \left(c_4 r^4 + c_2 r^2 + c_0 + c_{-2} r^{-2}\right) \cos 2\theta.$$

By using this stress function, getting stresses from Eq. (8-4), determining constants c_i (i = 0, 4, ±2) by satisfying the boundary conditions, Eqs. (g) and (h), and then superposing them with the solution of problem (1), Eq. (i), we will get the same stresses as in Eq. (8-6).

In a similar way, we can also obtain the stress function for problem (1). Because the problem is axisymmetric, φ must be a function of r only. By assuming

$$\phi = g(r),$$

we can obtain

$$\frac{1}{r}\frac{d}{dr}\left[r\frac{d}{dr}\left\{\frac{1}{r}\frac{d}{dr}\left(r\frac{dg}{dr}\right)\right\}\right] = 0$$

when substituting φ into the compatibility equation, Eq. (8-2). A direct integration gives rise to

$$g = Ar^2 \ln r + Br^2 + C \ln r + D,$$

where A, B, C and D are undetermined constants. A solution of stresses for problem (1) as shown in Eq. (i) can then be obtained. The reader can try to complete the process.

8.6 PURE BENDING OF A CURVED BEAM

Curved beams are common structural elements. Here we consider the simplest case that its inner surface is a circular arc of radius a and its outer surface is that of radius b, as shown in **Fig. 8.6**. The two end surfaces of the beam with an included angle θ^* are subjected to a bending moment M. The boundary conditions of the beam can therefore be described as

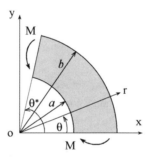

Figure 8.6 A curved beam under pure bending.

at $r = a$ and b, $0 \leq \theta \leq \theta^*$,

$$\sigma_{rr} = 0, \quad \sigma_{r\theta} = 0; \tag{a, b}$$

at $\theta = 0$ and θ^*, $a \leq r \leq b$,

$$\sigma_{\theta r} = 0,$$
$$\int_a^b \sigma_{\theta\theta} dr = 0, \tag{c, d, e}$$
$$\int_a^b \sigma_{\theta\theta} r dr = M.$$

Clearly, conditions (d) and (e) are due to the application of Saint-Venant's principle.

Now let us try to find a stress function that can satisfy the above boundary conditions. Condition (e) indicates that we must have non-vanishing $\sigma_{\theta\theta}$ at the two ends to balance the external bending moment M. Since M is a constant and the integration in Eq. (e) is about coordinate r, $\sigma_{\theta\theta}$ cannot be a function of coordinate θ. On the other hand, conditions (b) and (c) suggest that $\sigma_{r\theta}$ could be zero throughout the beam. Thus from **Table 8.4**, we find that stress functions No. 2 to No. 4 can be used, i.e., we can take

$$\phi = c_1 r^2 + c_2 r^2 \ln r + c_3 \ln r.$$

Equation (8-4) then gives rise to

$$\begin{cases} \sigma_{rr} = 2c_1 + c_2(2\ln r + 1) + c_3 \dfrac{1}{r^2}, \\ \sigma_{\theta\theta} = 2c_1 + c_2(2\ln r + 3) - c_3 \dfrac{1}{r^2}, \\ \sigma_{r\theta} = 0. \end{cases} \quad (f)$$

These stress components must satisfy boundary conditions (a) to (e), which leads to

$$\begin{cases} c_1 = -\dfrac{M}{A}\left(b^2 - a^2 + 2b^2 \ln b - 2a^2 \ln a\right), \\ c_2 = \dfrac{2M}{A}\left(b^2 - a^2\right), \\ c_3 = \dfrac{4M}{A} a^2 b^2 \ln\left(\dfrac{b}{a}\right), \end{cases}$$

where

$$A = \left(b^2 - a^2\right)^2 - 4a^2 b^2 \left[\ln\left(\dfrac{b}{a}\right)\right]^2.$$

The stresses in the curved beam are then determined by substituting the above constants back into Eq. (f).

Comparing solution (f) with those obtained by the engineering bending theory in the Mechanics of Materials, we can see that the engineering bending theory does not give good prediction of $\sigma_{\theta\theta}$ when a/b is small. The numerical comparison made by Timoshenko and Goodier (1970) showed that the engineering bending theory can no longer accurately predict stresses in curved beams when a/b becomes smaller that 0.5. When this happens, the above solution, Eq. (f), must be used.

8.7 EFFECT OF BODY FORCES

The stress function defined by Eq. (8-1) is for problems without body forces. When a problem is with *constant body forces*, i.e., f_x and f_y in the equations of equilibrium are independent of coordinates x and y, we should use

$$\begin{cases} \sigma_{xx} = \dfrac{\partial^2 \phi}{\partial y^2} - x\rho f_x, \\[4pt] \sigma_{yy} = \dfrac{\partial^2 \phi}{\partial x^2} - y\rho f_y, \\[4pt] \sigma_{xy} = -\dfrac{\partial^2 \phi}{\partial x \partial y}, \end{cases} \quad (8\text{-}7)$$

to replace Eq. (8-1). It is an exercise for the reader to show that the stress function defined by Eq. (8-7) satisfies the equations of equilibrium with body forces and is also a biharmonic function. Thus all the solution methods discussed in the above examples are still valid. If the body forces are potential forces, a similar stress function definition can be used. Further details can be found in more advanced books.

8.8 ANOTHER SCALAR METHOD

Another important scalar method in solving mechanics problems is the *energy method*, including methods using the principle of virtual work, the theorem of minimum potential energy and the first and second Castigliano's theorems that we studied in Strength of Materials. This is because energy is also a scalar quantity. The fundamental advantage of the energy method is its ability to yield approximate solutions for problems that are too complicated to solve by the methods discussed previously. Also importantly, it forms the theoretical basis of many powerful numerical methods, such as the finite element method.

The energy method and its applications can only be understood rigorously through a more systematic reading of particular textbooks and monographs. The reader is highly recommended to read works by Dym and Shames (1973), Washizu (1975) and Hu (1984). In Appendix C of this book, we review briefly the theorem of minimum potential energy that will be used in the formulation of the finite element method in Chapter 10.

References

Barber, JR (1992), *Elasticity*, Kluwer Academic Publishers, Dordrecht.
Dym, CL and Shames, JM (1973), *Solid Mechanics: A Variational Approach*, McGraw-Hill, New York.
Fung, YC (1965), *Foundations of Solid Mechanics*, Prentice-Hall, Inc., Englewood Cliffs, NJ.
Hu, H (1984), *Variational Principles of Theory of Elasticity with Applications*, Science Press, Beijing.
Kuske, A and Robertson, G (1974), *Photoelastic Stress Analysis*, John Wiley & Sons, London.
Muskhelishvili, NI (1953), *Some Basic Problems of the Mathematical Theory of Elasticity* (English translation by JRM Radok), Noordhoff, Groningen.
Roark, KC (1965), *Formulas for Stress and Strain*, McGraw-Hill, New York.
Shigley, JE (1972), *Mechanical Engineering Design*, McGraw-Hill, New York.
Timoshenko, SP and Goodier, JN (1970), *Theory of Elasticity*, McGraw-Hill, Singapore.
Washizu, K (1975), *Variational Methods in Elasticity and Plasticity*, 2nd edition, Pergamon Press, Oxford.

Important Concepts

Stress function
Stress concentration factor

Stress concentration
Arrest of crack propagation

Questions

8.1   When using the stress function method, why don't we need to check the basic equations in the solution?

8.2   What is the general way of selecting a proper stress function for a given problem?

8.3   Can you use some examples to show the importance of the principle of superposition to the selection of stress functions?

8.4   How can we possibly arrest the propagation of a crack according to solution (f) in section **8.5**?

Problems

8.1   What plane-stress problem can be solved by using

$$\phi = \frac{3F}{4c}\left(xy - \frac{xy^3}{3c^2}\right) + \frac{q}{2}y^2$$

as a stress function?

8.2   A thin cantilever beam is under a uniform pressure, q, on its top surface, as shown in **Fig. P8.1**. (a) Show that $\phi = ay^5 + bx^2y^3 + cy^3 + dx^2 + ex^2y$ can be a stress function. (b) Use the function to find the stress components in the cantilever beam. (c) Use the principle of superposition and the corresponding boundary conditions to explain why these terms are necessary in the stress function for this beam bending problem.

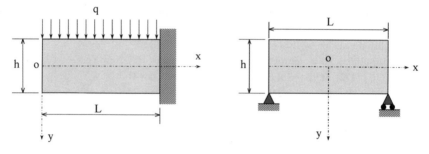

Figure P8.1   Figure P8.2

8.3   A simply supported plane-stress beam is subjected to its own weight, as illustrated in **Fig. P8.2**. The density of the beam material is ρ. (a) Show that $\phi = ax^2y^3 + by^5 + cy^3 + dx^2y$ can be used as a stress function, where a, b, c and d are constants. (b) Use the stress function to find the stress components of the beam.

8.4   When the rotating disk of **Fig. P7.3** is wrapped fully by the brake, the friction and normal pressure around the disk edge are approximated to be uniform and when the angular velocity is considered to be a constant, the mechanics model of the disk can be described by that of **Fig. P8.3a**. According to the principle of superposition, it can be further resolved into three simpler models, a disk subjected to external pressure (**Fig. P8.3b**), a disk under uniform rotation (**Fig. P8.3c**) and a disk subjected to a uniform surface shear (**Fig. P8.3d**). The solutions of models of **Fig. P8.3b** and **Fig. P8.3c** can be obtained easily by using the results listed in **Table 6.1**. (a) Show that the stress and displacements of the model in **Fig. P8.3d** can be obtained when using $\phi = A\theta$, where A is a constant. (b) Find the stress and displacement components in the disk of **Fig. P8.3a**. The radius of the spindle is R_s and that of the disk is R_d. (*Hint*: The initial stresses due to disk-spindle assembly can be superposed if required and therefore can be ignored in the present solution. The disk can be considered to be rigidly mounted on the spindle.)

8.5 The flange part of the workpiece during axisymmetric deep-drawing, as illustrated in **Fig. P7.4**, can be modelled as an annular plate subjected to a uniform tension at its inner edge, as shown in **Fig. P8.4**, when friction and compression from the flange holder are ignored. (a) Use stress function $\phi = A\ln r + Br^2 \ln r + Cr^2 + D$, where A, B, C and D are constants, to find the stresses and displacements. (b) Make use of the results in **Table 6.1** to get the stresses directly.

8.6 Use the following stress function to find the stresses in the dam discussed in Example 6.3 (**Fig. E6.3**): $\phi = r^3(A\cos\theta + B\sin\theta + C\cos3\theta + D\sin3\theta)$, where A, B, C and D are undetermined constants. (a) The weight of the dam itself is ignored. (b) The density of the dam material is ρ and the weight must be considered in the stress calculation.

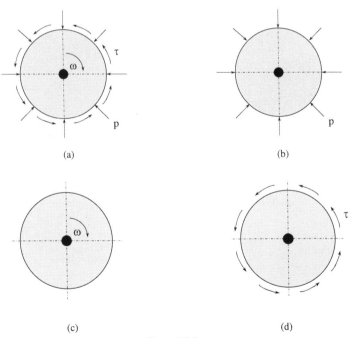

Figure P8.3

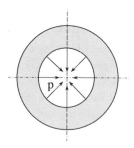
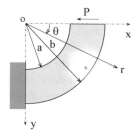

Figure P8.4 **Figure P8.5**

8.7 A curved beam is subjected to a concentrated shear force P as shown in **Fig. P8.5**, find the stresses. (*Hint*: (a) Take ϕ = (Ar³ +Br⁻¹ + Cr + Drlnr)sinθ where A, B, C and D are undetermined constants, or (b) Take ϕ = f(r)sinθ and solve for f(r) by satisfying Eq. (8-2).)

8.8 A large thin plate is subjected to pure shear on its edges, as shown in **Fig. P8.6**. (a) If there is a small hole of radius R at the centre of the plate, determine the stresses in the neighbourhood of the hole. (b) Find the stress concentration factor according to the solution. (*Hint*: Follow a similar method to that in section **8.5**.)

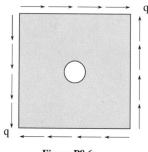
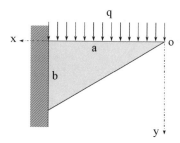

Figure P8.6 **Figure P8.7**

8.9 A plane-stress beam with a triangular cross-section is subjected to a uniform pressure q on its top surface, as illustrated in **Fig. P8.7**. Find the maximum principal stress in the beam. The following stress function can be used:

$$\phi = \frac{q}{2(1-\pi/4)}\left[-x^2 + xy + (x^2+y^2)\left(\frac{\pi}{4} - \tan^{-1}\frac{y}{x}\right)\right].$$

Chapter 9

PLASTICITY AND FAILURE

This chapter introduces some fundamentals of plasticity, discusses the basic methods for carrying out plastic deformation analysis and demonstrates the difference in the solution of problems with purely elastic deformation to those with elastic-plastic deformation. It also establishes the theory of plasticity and theories for predicting component failure via either plastic yielding or brittle fracture. Use of plastic deformation or prevention from failure is equally important in engineering practice.

In many cases, plastic deformation needs to be avoided but in many others it is desirable. The mirror surface machining by a diamond cutting tool shown above is a manufacturing process in industry, where plastic deformation of the workpiece plays a central role (Courtesy of MEL, Japan).

9.1 INTRODUCTION

We have understood in Chapter 5 that the regime of linear elastic deformation in a solid component is normally small and upon the removal of external loads such deformation will recover completely and stresses will vanish. Physically, this is because the atomic bonds in the solid are only stretched slightly within its elastic deformation regime, the structure of the atomic lattice of the material is not damaged and atoms can return to their original equilibrium positions after unloading. Most engineering components must work safely in the regime of elastic deformation without plastic deformation or cracking. For example, in a power transmission system we certainly do not allow the teeth of a gear to deform permanently or a gear box to fracture as this will break the whole transmission system and result in a serious accident. That is why the solutions of stresses, strains and displacements developed in the last three chapters are of great importance to the understanding of the reliability of structures and machine components. When proper failure theories are established, the solutions will offer critical guidelines towards the economic selection of materials, optimal design and effective maintenance of these structures and components.

On the other hand, however, we often need to make use of the permanent deformation of a solid. In manufacturing, we must deform or machine a piece of material to obtain a required shape. Typical examples are forging, rolling and stamping of metal components, which introduce plastic deformation via the application of external forces. The question raised in Chapter 1 on plate rolling needs to be answered by plastic deformation analysis.

We have known from the discussion in Chapter 5 that plastic deformation occurs in a solid under certain external loads when atomic bonds in part of the material are broken, some atoms move to their new equilibrium positions, new bonds form, the original atomic lattice deforms and hence permanent deformation takes place. In metals, such deformation is mainly due to the activation of slip systems such as dislocations. In some other materials, plasticity may happen via other mechanisms such as phase transformation. Clearly, the mechanisms of plastic deformation are much more complex than those of elasticity. Plasticity on the dimensional scale of our interest in this elementary text can be regarded as the macroscopic resultant of all the permanent movements of atoms.

From the elastic analysis in previous chapters, we know that stresses and strains have linear relationships represented by the generalised Hooke's law, Eq. (5-6). When we consider the deformation of a solid beyond its elastic limit, certainly, we must find other relationships that reflect the material's deformation against the stresses applied. Hence, before we can really carry out any stress and deformation

analysis when plasticity or failure occurs, we must answer the following basic questions:
(1) Are the basic equations established previously still valid when a solid or a part of it experiences plastic deformation?
(2) How can we know when plasticity or failure will occur? This means that we must have criteria that enable us to predict the occurrence of plastic deformation or failure.
(3) What are the stress-strain relationships in the zone where plasticity appears?
(4) Can we still use the methodology developed, such as the displacement approach, for studying deformation problems with plasticity?

In stress and strain analyses in Chapters 3 and 4, equations of motion, Eq. (3-46), strain-displacement relations, Eq. (4-7), and compatibility conditions, Eq. (4-15), were derived independently with respect to material properties. Thus these equations are valid for both elastic and inelastic regimes. In the next few sections we shall find the answers to questions (2), (3) and (4). To have a deeper discussion, however, let us introduce some new concepts first, that is, the concepts of octahedral shear stress and distortion energy.

9.2   OCTAHEDRAL SHEAR STRESS

In discussing stress analysis in Chapter 3, we found the stress transformation rule that gives rise to the stresses in any direction n, *i.e.*, σ_{nn}, $\sigma_{nt'}$ and $\sigma_{nt''}$, in terms of principal stresses. Now, we consider a special plane whose external normal n makes the same angle with the three principal stress directions, as shown in **Fig. 9.1**. Because of its special orientation, we call the plane the *octahedral plane*. From **Fig. 9.1**, it is easy to find that eight octahedral planes exist through a point.

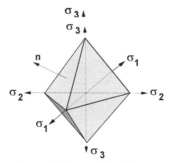

Figure 9.1 The octahedral planes.

The *octahedral shear stress*, τ_0, is the resultant shear stress on an octahedral plane, *i.e.*, $\tau_0 = \sqrt{(\sigma_{nt'})^2 + (\sigma_{nt''})^2}$. Using the formula in Chapter 3 for stress transformation, we have

$$\tau_0 = \frac{1}{3}\sqrt{(\sigma_1 - \sigma_2)^2 + (\sigma_2 - \sigma_3)^2 + (\sigma_3 - \sigma_1)^2}. \qquad (9\text{-}1)$$

9.3 DISTORTION ENERGY

As we can understand, when loads are applied to a solid, they will deform the solid. Provided no energy is lost in the form of heat, the external work done by the loads will be converted into internal work called *strain energy*. This energy, which is always positive, is stored in the deformed solid and is the summation of the work done by all stress components during straining. Hence, the *strain energy per unit volume* in an elastic solid (often called *strain energy density*), U, can be expressed as

$$U = \frac{1}{2}\left\{\sigma_{xx}\varepsilon_{xx} + \sigma_{yy}\varepsilon_{yy} + \sigma_{zz}\varepsilon_{zz} + 2\left(\sigma_{xy}\varepsilon_{xy} + \sigma_{yz}\varepsilon_{yz} + \sigma_{zx}\varepsilon_{zx}\right)\right\}.$$

Under linear elastic deformation, strains and stresses are related by the generalised Hooke's law. Thus if we express all the strains in terms of stresses, or all the stresses in terms of strains, U can also be written alternatively as

$$U = \frac{1}{2E}\left\{\left(\sigma_{xx}^2 + \sigma_{yy}^2 + \sigma_{zz}^2\right) - 2\nu\left(\sigma_{xx}\sigma_{yy} + \sigma_{yy}\sigma_{zz} + \sigma_{zz}\sigma_{xx}\right) + \right.$$
$$\left. + 2(1+\nu)\left(\sigma_{xy}^2 + \sigma_{yz}^2 + \sigma_{zx}^2\right)\right\}$$

or

$$U = G\left\{\frac{\nu}{1-2\nu}\left(I_1^\varepsilon\right)^2 + \left(\varepsilon_{xx}^2 + \varepsilon_{yy}^2 + \varepsilon_{zz}^2\right) + 2\left(\varepsilon_{xy}^2 + \varepsilon_{yz}^2 + \varepsilon_{zx}^2\right)\right\},$$

where G is the shear modulus of the material defined by Eq. (5-15). When x-, y- and z-directions are the principal directions, U can be written in terms of principal stresses, *i.e.*,

$$U = \frac{1}{2E}\left\{\left(\sigma_1^2 + \sigma_2^2 + \sigma_3^2\right) - 2\nu\left(\sigma_1\sigma_2 + \sigma_2\sigma_3 + \sigma_3\sigma_1\right)\right\}$$

$$= \frac{1}{2}K\left(I_1^\varepsilon\right)^2 + \frac{1+\nu}{6E}\left\{(\sigma_1 - \sigma_2)^2 + (\sigma_2 - \sigma_3)^2 + (\sigma_3 - \sigma_1)^2\right\},$$

where K is the bulk modulus of the material defined by Eq. (5-16). The first strain invariant, I_1^ε, represents the volume change rate of an infinitesimal material element, as shown in section **5.3.4**. Hence the first term in the above equation stands for the part of strain energy density for the volume change of the element, U_v, caused by the hydrostatic stress indicated by Eq. (k) of section **5.3.4**. The second term in U above is clearly corresponding to shear stresses presented by Eq. (3-44). It therefore stands for the part of strain energy density that causes the shape change of the element and is called the *distortion energy density*, U_d. Hence, the total strain energy density in a deformed linear elastic solid is

$$U = U_v + U_d,$$

where

$$U_v = \frac{1}{2}K\left(I_1^\varepsilon\right)^2,$$

and

$$U_d = \frac{1+\nu}{6E}\left\{(\sigma_1 - \sigma_2)^2 + (\sigma_2 - \sigma_3)^2 + (\sigma_3 - \sigma_1)^2\right\}. \tag{9-2}$$

9.4 ONSET OF PLASTICITY AND DEFORMATION AFTER INITIAL YIELDING

9.4.1 Plasticity under Simple Tension

Under a uniaxial tension, the stress-strain behaviour of a ductile material, such as mild steel, is generally described by the curve shown in **Fig. 9.2**. Plasticity occurs when the stress reaches the yield stress Y of the material, which is the initial yielding. When the stress increases further, different material may behave differently, as shown in **Fig. 9.3**. For the convenience of analytical studies, we often simplified models to approximate the actual behaviour according to our requirement. The models shown in **Fig. 9.4** are popular in engineering

Especially when hardening during plastic deformation is minor, the model of elastic/perfectly plastic deformation (**Fig. 9.4a**) provides reasonable accuracy and is of great simplicity. This is particularly true when forming metal at high temperatures, where hardening does not appear. On the other hand, when plastic deformation is very large, the contribution of elastic strains becomes negligible. In this case, the rigid/plastic models of **Figs. 9.4c** and **9.4d** can be used to simplify the process of plasticity analysis.

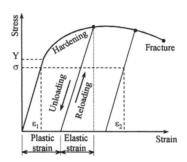

Figure 9.2 Schematic of the stress-strain curve of a material under simple tension.

When unloading takes place after plasticity has occurred, the deformation will be completely elastic as shown in **Figs. 9.2, 9.3c** and **9.4**. *This means that we can use the method of elastic analysis to study the stress and deformation during unloading.* The stress-strain behaviour at reloading after an unloading at any plastic deformation state with stress Y' will first be elastic. If the material has a hardening behaviour, it will not yield until the reloading stress reaches Y', see for example, **Fig. 9.3c**. Since Y' is greater than the material's initial yield stress Y, the material in reloading shows a hardened behaviour. This indicates that the yield stress of a work hardening material can be increased simply by loading it to its plastic deformation regime followed by unloading. The above phenomenon is called *work hardening*.

The total strain at an elastic-plastic deformation state is the summation of elastic strain and plastic strain, as illustrated in **Fig. 9.2**. It is also important to note that the stress-strain relationship in an elastic state is unique, *i.e.*, the strain is uniquely determined when the corresponding stress is known and vice versa, as the Hooke's law implies. However, in the elastic-plastic deformation regime, such a correspondence between stress and strain does not exist. For example, corresponding to stress σ, the strain can be either ε_1 or ε_2, depending on the loading history of the material.

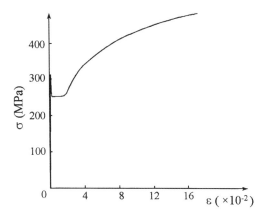

(a) mild steel

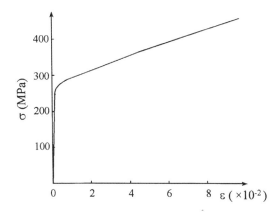

(b) A316 stainless steel

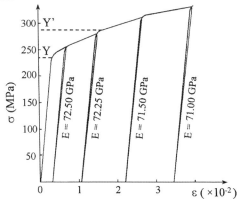

(c) loading-unloading-reloading curve of 2024 age-hardened aluminium alloy

Figure 9.3 Examples of stress-strain behaviour of metals under uniaxial tension. Tests were conducted at room temperature.

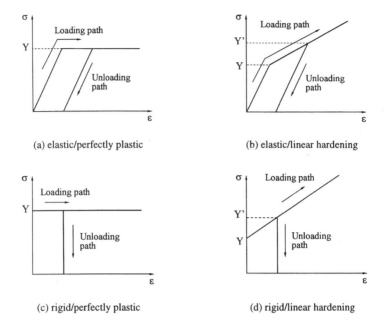

Figure 9.4 Some idealised models for describing the stress-strain behaviour of materials.

The simplified models of **Fig. 9.4** facilitate a plastic analysis at the cost of losing accuracy. It is sometimes necessary to represent the real stress-strain curve of a material, such as that of **Fig. 9.3**, by an equation obtained by empirically fitting the experimental data. Two commonly used equations are the power law equation

$$\sigma = A\varepsilon^n,$$

where A is called the *strength coefficient* and n is called the *strain-hardening exponent* to be determined by fitting the experimental data, and the *Ramberg-Osgood equation*,

$$\varepsilon = \frac{\sigma}{E} + k\left(\frac{\sigma}{E}\right)^n,$$

where k and n, similar to the A and n in the power law equation, are also material constants to be determined by empirical data fitting. Sometimes, it is also possible to fit the plastic part of a stress-strain curve as accurately as desired by a polynomial of arbitrary degree, *i.e.*,

$$\sigma = \begin{cases} E\varepsilon, & \text{in the elastic regime} \\ A_0 + A_1\varepsilon + \cdots + A_k\varepsilon^k, & \text{beyond the elastic regime} \end{cases}$$

where A_i ($i = 1,\ldots,k$) are constants determined by experimental data. When we use a numerical method with the aid of a powerful computer to analyse the stress and deformation of a problem, the above curve fitting will be handy.

When a solid is subjected to a complex stress state, *i.e.*, when all six stress components exist, the above simple stress-strain models can no longer describe the material's behaviour. In the next few sections, we shall introduce a very primary understanding of plasticity for solids under general three-dimensional stress states, including the prediction of the onset (or initiation) of plastic deformation, which is described by a yield criterion, and some basic plasticity analyses after initial yielding. More advanced theories and treatments, such as the flow rule, a general criterion of loading and unloading, anisotropic yielding, effects of strain rate, temperature and hydrostatic pressure, true stress and true strain under large straining and so on, can be found in textbooks and monographs on the theory of plasticity (*e.g.*, Hill, 1950; Mendelson, 1968; Chakrabarty, 1987).

Example 9.1 A cylindrical bar made of steel is subjected to a uniform tension P along its axis. The maximum axial strain was measured to be 0.0005. It is known that the Young's modulus of this type of steel is E = 200 GPa, its yield stress is Y = 600 MPa and the material's stress-strain curve shows almost no hardening. (a) Calculate the plastic strain in the bar at the above maximum strain. (b) If the bar is unloaded at this maximum strain, what is the residual strain remaining in the bar after complete unloading?

Solution: Because there is almost no hardening, the steel can be idealised as an elastic/perfectly plastic material as illustrated in **Fig. 9.4a**. The maximum elastic strain in the bar, *i.e.*, the strain at initial yielding or yield strain, is therefore

$$\varepsilon_Y = Y/E = 600/(200 \times 10^3) = 0.003.$$

Hence, at the maximum strain of $\varepsilon_{max} = 0.005$, the plastic strain in the bar is

$$\varepsilon^p = \varepsilon_{max} - \varepsilon_Y = 0.005 - 0.003 = 0.002.$$

When the bar is completely unloaded, all the elastic strain will disappear and the plastic strain remains. Thus the residual strain after complete unloading is equal to ε^p.

Example 9.2 A simple frame consisting of three uniform bars, made of the same material and of the same cross-sectional area A, as illustrated in **Fig.E9.1**, is subjected to a concentrated force P at its tip hinge. All the hinges can be assumed to be frictionless. The bar material is elastic/perfectly plastic. Its yield stress is Y and Young's modulus is E. Find (a) the *elastic limit load*, P^e, beyond which plastic yielding occurs in the frame material, (b) the *plastic limit load*, P^p, which represents the maximum load-carrying capacity of the frame, and (c) the *residual stresses* in the bars when the frame is completely unloaded from the loading level P that $P^e < P < P^p$.

Solution: The frame is clearly under symmetric deformation about the axis of bar (2). The initial length of bars (1) and (3) is L' = L/cosθ.

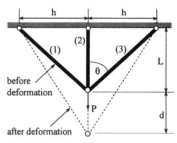

Figure E9.1

(i) *Equilibrium of stresses*
The internal resultant forces, $N_{(i)}$ (i = 1, 2, 3), can be obtained by the equilibrium of the frame in both horizontal and vertical directions, as shown in **Fig. E9.2**, which gives rise to

$$N_{(1)} = N_{(3)}, \quad (a)$$
$$N_{(1)}\cos\theta + N_{(2)} + N_{(3)}\cos\theta = P . \quad (b)$$

Because we are considering a small deformation, the change of θ after deformation is negligible. Since each bar is under uniform tension, the stress in a bar is simply

$$\sigma_{(i)} = N_{(i)}/A . \quad (i = 1, 2, 3) \quad (c)$$

Thus the substitution of Eq. (c) into Eqs. (a) and (b) results in the equilibrium equations of the frame, *i.e.*,

$$\sigma_{(1)} = \sigma_{(3)} . \quad (d)$$
$$2\sigma_{(1)}\cos\theta + \sigma_{(2)} = P/A . \quad (e)$$

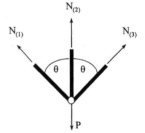

Figure E9.2

(ii) *Compatibility of deformation*
The tip hinge will have a vertical displacement, d, after deformation. Assume that the strains in the bars are $\varepsilon_{(i)}$ (i = 1, 2, 3). Because of the symmetry of deformation, $\varepsilon_{(1)} = \varepsilon_{(3)}$. As the bars are joined together by the tip hinge of the frame their deformation must be compatible. The elongation of bar (1) or (3) means that dcosθ + L' = L' (1 + $\varepsilon_{(1)}$) while that of bar (2) requires that d = L$\varepsilon_{(2)}$. Hence,

$$\varepsilon_{(1)} = \varepsilon_{(2)}\cos^2\theta . \quad (f)$$

(iii) *Elastic deformation and initial yielding*
In the regime of elastic deformation, the generalised Hooke's law leads to

$$\sigma_{(1)} = E\varepsilon_{(1)}, \quad \sigma_{(2)} = E\varepsilon_{(2)}.$$

Together with Eqs.(d) to (f), we can easily obtain

$$\sigma_{(2)} = \frac{P}{A} \frac{1}{\left(1 + 2\cos^3\theta\right)},$$

$$\sigma_{(1)} = \sigma_{(2)} \cos^2\theta.$$

(g, h)

The above solution shows that $\sigma_{(2)}$ is the largest so that if $\sigma_{(2)}$ reaches the yield stress of the bar material, Y, yielding occurs in bar (2) of the frame. The corresponding load is the elastic limit load of the frame. Let $\sigma_{(2)}$ be Y in Eq.(g), we get

$$P^e = YA(1 + 2\cos^3\theta).$$

(i)

(iv) *Elastic-plastic deformation and plastic limit load*

Although bar (2) is yielded when P reaches P_e, bars (1) and (3) are still in elastic deformation, because $\sigma_{(1)} = \sigma_{(3)}$ is less than Y. Thus when we increase P further, the frame will be in the regime of elastic-plastic deformation. In this regime, we always have $\sigma_{(2)} = Y$ because the bar material is elastic/perfectly plastic. Hence the equilibrium equation, Eq. (e), results in

$$\sigma_{(1)} = \frac{P/A - Y}{2\cos\theta}.$$

(j)

When $\sigma_{(1)}$ also reaches Y as P increases, bars (1) and (3) will also yield. The corresponding load P is therefore the plastic limit load of the frame, *i.e.*,

$$P^p = AY(1 + 2\cos\theta).$$

(k)

If we compare the magnitude of the plastic limit load with that of the elastic limit load, we find that

$$\frac{P^p}{P^e} = \frac{1 + 2\cos\theta}{1 + 2\cos^3\theta},$$

which is greater than unity. When $\theta = 45°$, for instance, $P^p = 1.41 P^e$. This means that if we allow the frame to work in its elastic-plastic deformation regime, the load-carrying capacity of the frame is much increased.

(v) *Unloading and residual stresses*

Now, we unload the frame at an elastic-plastic stage, *i.e.*, unload when $P^e < P < P^p$. As we have seen, unloading is elastic and the stresses and strains caused by unloading can be obtained by an elastic analysis. Thus the residual stresses and strains in the bars after a complete unloading from $P = P^*$ to $P = 0$ can be obtained by superposing the elastic stresses and strains caused by $P = -P^*$ to the elastic-plastic ones at $P = P^*$ obtained before. The elastic stresses due to unloading are replacing the P in Eqs.(g) and (h) by $-P^*$,

$$\sigma_{(2)} = \frac{-P*}{A} \frac{1}{\left(1 + 2\cos^3\theta\right)},$$

$$\sigma_{(1)} = \sigma_{(2)} \cos^2\theta.$$

(l, m)

Thus the residual stress in bar (1), $\sigma_{(1)}^*$, is the summation of Eqs.(j) and (m), *i.e.*,

$$\sigma_{(1)}^* = \frac{P*-AY}{2A\cos\theta} - \frac{P*}{A\left(1 + 2\cos^3\theta\right)}.$$

Similarly

$$\sigma_{(2)}^* = Y - \frac{P*}{A\left(1 + 2\cos^3\theta\right)}.$$

The reader can try to find the residual strains in the bars.

This example, though simple, shows the complete process of a general elastic-plastic analysis.

9.4.2 Plasticity under Complex Stress States

9.4.2.1 *Initial Yielding: Yield Criterion*

Now let us see how to predict the onset of plastic deformation under a complex stress state. As mentioned previously, the occurrence of plastic deformation in a solid on the macroscopic scale is the resultant effect of permanent movements of atoms. Thus to predict the initiation of plastic deformation, two factors must be considered. One is the stress in the solid that is the stimulus of the atomic motion and the other is the property of the solid material, determined by the type and strength of the atomic bond and lattice structure of the material, that governs the response of the atoms to an external stimulus. Therefore it is reasonable to assume that at a point in a solid when the principal stresses and some property parameters of the solid material representing all the possible permanent movements of atoms satisfy a critical condition, plastic yielding occurs at this point. Mathematically, such a condition can be written as

$$F\left(\sigma_1, \sigma_2, \sigma_3, k_1, k_2, \cdots, k_n\right) = 0, \tag{9-3}$$

where F is a function to be determined and $k_1, k_2, \cdots, k_n$ are n parameters that can describe all possible behaviour of permanent deformation of the material. Equation (9-3) is called a *yield criterion*, or *initial yield condition*. Here, 'initial' means that before the critical state described by Eq. (9-3) is reached, deformation is purely elastic. If presenting it graphically, F = 0 is a surface in the space of coordinates σ_1, σ_2 and σ_3. Therefore, the surface defined by Eq. (9-3) is also called an *initial yield surface*.[9.1] Any point inside the surface corresponds to an elastic deformation state and that on or beyond the surface represents a plastic deformation state.

Condition (9-3) is phenomenological, macroscopic and approximate. On the atomic or even microscopic scale, a single surface that divides elastic deformation and plastic deformation absolutely does not exist, because plastic deformation is the accumulation of a huge number of permanent atomic movements and macroscopically the process from an elastic deformation to a noticeable plastic deformation is gradual. However, on the macroscopic scale for normal engineering analysis, as we will see later in the chapter, condition (9-3) will be accurate enough to describe the initiation of any effective plastic deformation and is reliable and feasible.

The specific expression of function F in Eq. (9-3) depends on specific material properties. For metals, various criteria have been proposed in the past to predict the onset of yielding under complex stresses. However, most of them are only of historical interest, because they conflict with experimental findings. The two satisfactory and widely used criteria in engineering are those credited to Tresca (1864) and von Mises (1913).

9.4.2.2 *Tresca Criterion*

From a series of experiments on the extrusion of metals, Tresca observed that material flow seems to be along the direction of the maximum shear stress. He therefore proposed that yielding occurred at a point when the maximum shear stress at that point reached a critical value. According to the relationship between the maximum shear stress and principal stresses, Tresca's argument can be expressed as

$$\sigma_1 - \sigma_3 = 2k_{Tresca} \qquad (9\text{-}4)$$

and is called the *Tresca criterion*[9.2] of initial yielding. Compared with the general form of yielding conditions, Eq. (9-3), it is clear that the Tresca criterion uses $F = \sigma_1 - \sigma_2 - 2k_{Tresca}$, which ignores the intermediate principal stress, σ_2, and involves only

---

[9.1] A general yield surface must satisfy a number of conditions, such as convexity. More details can be found in the books by Hill (1950) and Chakrabarty (1987).
[9.2] Tresca was probably influenced by a more general criterion for the failure of soils proposed earlier by Coulomb (1773).

one material parameter, k_{Tresca}, to describe the properties of metals related to plastic deformation. Let us find out what k_{Tresca} means in the Tresca criterion.

If a material is under a uniaxial tension along the first principal stress direction, σ_3 vanishes. Equation (9-4) becomes

$$\sigma_1 = 2k_{Tresca}. \tag{9-5}$$

Since under a uniaxial tension, a metal will yield when $\sigma_1 = Y$, where Y is the yield stress of the metal, thus k_{Tresca} in Eq. (9-4) must be Y/2 and hence the Tresca criterion becomes

$$\sigma_1 - \sigma_3 = Y. \tag{9-6}$$

9.4.2.3 *von Mises Criterion*

Experiment has shown that metals do not yield under high hydrostatic stress. In the discussion above on strain energy, we understand that hydrostatic stress only causes a volume change of a material element with no change in shape. Huber (1904) therefore proposed that plastic yielding in a ductile material occurs when the distortion energy density in the material, Eq. (9-2), equals or exceeds that of the same material when it is yielding under uniaxial tension. From a theoretical consideration, von Mises (1913) suggested that when a combination of the principal stresses at a point, $(\sigma_1 - \sigma_2)^2 + (\sigma_2 - \sigma_3)^2 + (\sigma_3 - \sigma_1)^2$, reached a critical value, initial plastic yielding occurred at the point. Hencky (1924) explained, independently of Huber, that the yielding criterion proposed by von Mises actually corresponds to the distortion energy of Eq. (9-3).

Analytically, the above proposal on the initial yielding of a metal can be expressed as

$$(\sigma_1 - \sigma_2)^2 + (\sigma_2 - \sigma_3)^2 + (\sigma_3 - \sigma_1)^2 = 2k_{Mises}^2 \tag{9-7}$$

and is called the *von Mises criterion*.[9.3] Compared with the general yielding criterion of Eq. (9-3), we see that the von Mises criterion takes F =

[9.3] Although Huber's paper was published much earlier than that of von Mises, his work was not known for a long time since it was published in Polish. Moreover, Huber's suggestion was slightly different from the current understanding discussed above. He suggested that when the hydrostatic stress at a point in a solid is negative, the plastic yielding of the material at the point was determined by the distortion energy density; however, when the hydrostatic stress is positive, the total strain energy density determined the yielding.

$(\sigma_1 - \sigma_2)^2 + (\sigma_2 - \sigma_3)^2 + (\sigma_3 - \sigma_1)^2 - 2(k_{Mises})^2$ and involves the effect of the intermediate principal stress σ_2. But again, this criterion also includes only one parameter, k_{Mises}, to describe the properties of metals against yielding. Similar to the Tresca criterion, under uniaxial tension in the first principal stress direction, σ_2 and σ_3 vanish and Eq. (9-7) reduces to

$$\sigma_1^2 = k_{Mises}^2 . \tag{9-8}$$

Thus $k_{Mises} = Y$ if yielding takes place under uniaxial tension. Hence, the von Mises criterion becomes

$$(\sigma_1 - \sigma_2)^2 + (\sigma_2 - \sigma_3)^2 + (\sigma_3 - \sigma_1)^2 = 2Y^2 . \tag{9-9}$$

In the Cartesian coordinate system xyz, this criterion can be written in terms of the stress components as

$$(\sigma_{xx} - \sigma_{yy})^2 + (\sigma_{yy} - \sigma_{zz})^2 + (\sigma_{zz} - \sigma_{xx})^2 + 6(\sigma_{xy}^2 + \sigma_{yz}^2 + \sigma_{zx}^2) = 2Y^2 . \tag{9-10}$$

Example 9.3 A thin plate is subjected to a set of uniform stresses on its edges as shown in **Fig. E9.3**, where μ is a positive proportional factor. The plate material is the same as that in Example 9.1. Find the maximum stress σ beyond which plastic deformation appears.

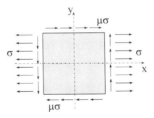

Figure E9.3

Solution: Based on the loading condition, we can easily understand that the stress state at any point in the plate is $\sigma_{xx} = \sigma$, $\sigma_{yy} = 0$, $\sigma_{xy} = \mu\sigma$, $\sigma_{zz} = \sigma_{zx} = \sigma_{zy} = 0$. σ_{zz} is clearly a principal stress. The other two principal stresses are in xy-plane and can be calculated by Eq. (3-21), i.e.,

$$\frac{1}{2}(\sigma_{xx} + \sigma_{yy}) \pm \sqrt{\left(\frac{\sigma_{xx} - \sigma_{yy}}{2}\right)^2 + \sigma_{xy}^2} = \frac{1}{2}\sigma\left(1 \pm \sqrt{1 + 4\mu}\right).$$

We therefore have

$$\sigma_1 = \frac{1}{2}\sigma\left(1 + \sqrt{1+4\mu}\right), \quad \sigma_2 = 0, \quad \sigma_3 = \frac{1}{2}\sigma\left(1 - \sqrt{1+4\mu}\right).$$

To determine the maximum elastic stress σ, we need to examine the stress state by a yield criterion, e.g., the Tresca or von Mises criterion. According to the Tresca criterion, Eq. (9-6), yielding occurs when

$$\sigma_1 - \sigma_3 = \frac{1}{2}\sigma\left(1 + \sqrt{1+4\mu}\right) - \frac{1}{2}\sigma\left(1 - \sqrt{1+4\mu}\right) = \sigma\sqrt{1+4\mu} = Y.$$

Hence, the maximum elastic stress depends on both σ and μ. For instance, when $\mu = 0.5$, the above equation gives rise to

$$\sigma = \frac{\sqrt{3}}{3}Y = 346.4 \text{ MPa},$$

because the yield stress of the material is $Y = 600$ MPa according to the data in Example 9.1. We can also use the von Mises criterion to determine the maximum elastic stress. In a similar way, the von Mises criterion, Eq. (9-9), leads to

$$(\sigma_1 - \sigma_2)^2 + (\sigma_2 - \sigma_3)^2 + (\sigma_3 - \sigma_1)^2 = 2\sigma^2(1+3\mu) = 2Y^2.$$

That is

$$\sigma = \frac{Y}{\sqrt{1+3\mu}} = \frac{\sqrt{10}}{5}Y = 379.5 \text{ MPa}.$$

The above result shows that in this case the Tresca criterion gives a more conservative estimation.

Example 9.4 A thin solid disk is rotating with a uniform angular velocity ω. The yield stress of the disk material is Y and the disk radius is R. Use the Tresca criterion to find the elastic limit of the angular velocity.

Solution: The stresses of a rotation disk have been obtained in Chapter 6 and listed in **Table 6.1**, i.e.,

$$\sigma_{rr} = \frac{3+\nu}{8}\rho\omega^2\left(R^2 - r^2\right), \quad \sigma_{\theta\theta} = \frac{3+\nu}{8}\rho\omega^2\left(R^2 - \frac{1+3\nu}{3+\nu}r^2\right),$$

$$\sigma_{r\theta} = \sigma_{zz} = \sigma_{zr} = \sigma_{z\theta} = 0.$$

Clearly, the principal stresses are $\sigma_1 = \sigma_{\theta\theta}$, $\sigma_2 = \sigma_{rr}$ and $\sigma_3 = \sigma_{zz} = 0$. Thus the Tresca criterion brings about

$$\frac{3+v}{8}\rho\omega^2\left(R^2 - \frac{1+3v}{3+v}r^2\right) = Y.$$

Yielding occurs first at the centre of the disk because the smaller the r the greater the left-hand side of the equation. Hence, we have

$$\omega = \frac{2}{R}\sqrt{\frac{2Y}{(3+v)\rho}}.$$

This relationship indicates that with a given disk material, if the disk radius R is greater, then the limit of angular velocity is smaller. When the disk radius is fixed, on the other hand, a material with higher yield stress can have larger limit velocity. If the material has a greater density ρ, however, the limit of angular velocity will be smaller. The function of Poisson's ratio is similar to that of ρ but its effect is minor. These are useful conclusions to a disk design in terms of geometry and material selections. The reader should try to use the von Mises criterion to repeat the above analysis and see if similar conclusions can be drawn.

9.4.3 Experimental Verification

As the Tresca criterion was based on the observation of extrusion tests and the von Mises criterion was from theoretical considerations, they need to be verified by experiments under different stress conditions, although both the criteria have clear physical meanings. The following tests were carried out specifically for this purpose.

9.4.3.1 *Lode's Test*

A series of well-recognised tests was carried out by Lode in 1926 on thin-walled tubes of steel, copper and nickel. In these tests, the specimens were strained progressively under a variety of stress states obtained by various combinations of longitudinal tension and internal hydrostatic pressure.

As mentioned before, the Tresca criterion does not include the effect of the intermediate principal stress, σ_2, while the von Mises criterion does. Thus a highly sensitive method of discerning between the Tresca and von Mises criteria is to examine the influence of σ_2 upon yielding. Lode introduced the parameter, which is called *Lode's parameter* after his work,

$$\mu = \frac{2\sigma_2 - \sigma_3 - \sigma_1}{\sigma_1 - \sigma_3} \tag{9-11}$$

to characterise the influence of the intermediate principal stress. Using his parameter, the von Mises criterion of Eq. (9-9) becomes

$$\frac{\sigma_1 - \sigma_3}{Y} = \frac{2}{\sqrt{3+\mu^2}} \tag{9-12}$$

and the Tresca criterion can be rewritten as

$$\frac{\sigma_1 - \sigma_3}{Y} = 1. \tag{9-13}$$

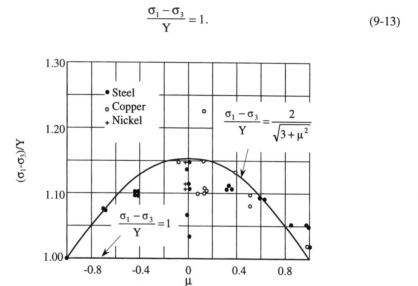

Figure 9.5 Lode's test results.

Figure 9.5 shows the experimental values of $(\sigma_1 - \sigma_3)/Y$ plotted versus the Lode's parameter μ. It indicates that they are in close agreement with the von Mises criterion. A number of additional results of tests on tubular steel specimens by Ros and Eichinger in 1929 confirm Lode's results and indicate a rather close consistency again with the von Mises criterion.

9.4.3.2 Test by Taylor and Quinney

In 1931, Taylor and Quinney carried out a series of tests by combining axial tension with torsion. As tests on solid cylinders are not reliable,[9.4] they used thin-walled

---

[9.4] This is because of the uncertainty of the stress distribution due to torsion after a small amount of plastic flow has taken place.

tubular specimens in which a uniform state of plane stress is assumed across the wall thickness. With respect to the polar coordinate system having the z-axis oriented parallel to the tube axis, the only two non-zero stress components in the tube are σ_{zz} caused by tension and $\sigma_{z\theta}$ introduced by torsion. Thus the three principal stresses can be expressed as

$$\sigma_1 = \frac{\sigma_{zz}}{2} + \sqrt{\left(\frac{\sigma_{zz}}{2}\right)^2 + \sigma_{z\theta}^2},$$
$$\sigma_2 = 0, \tag{9-14}$$
$$\sigma_3 = \frac{\sigma_{zz}}{2} - \sqrt{\left(\frac{\sigma_{zz}}{2}\right)^2 + \sigma_{z\theta}^2},$$

and in turn, the Tresca and von Mises criteria become

$$\left(\frac{\sigma_{zz}}{Y}\right)^2 + 4\left(\frac{\sigma_{z\theta}}{Y}\right)^2 = 1, \quad \text{according to Tresca criterion, Eq. (9-6)}$$

$$\left(\frac{\sigma_{zz}}{Y}\right)^2 + 3\left(\frac{\sigma_{z\theta}}{Y}\right)^2 = 1, \quad \text{according to von Mises criterion, Eq. (9-9)}$$

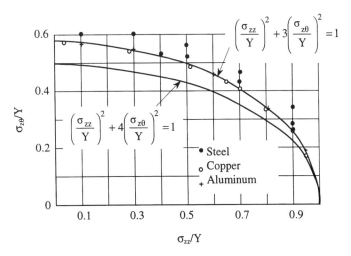

Figure 9.6 Taylor and Quinney's test results.

The equations above define two ellipses as shown in **Fig. 9.6**. A comparison with the corresponding experiment shows that the von Mises criterion, again, gives better predictions.

Although the von Mises criterion gives better prediction of initial yielding in general, the differences between the von Mises and Tresca criteria are rarely more than 15%. Because of its simplicity, the Tresca criterion has thus been widely used in engineering practice, particularly for analytical solutions and failure analysis as to be discussed later in this chapter.

9.4.4 Some Remarks

The physical meaning of the Tresca criterion is clear, as indicated by Tresca himself. That is, when the maximum shear stress at a point in a metal under complex stresses reaches a critical value, which is half of the yield stress under uniaxial tension, plastic yielding happens.

If we compare the left-hand side of the von Mises criterion, Eq. (9-9), with Eq. (9-1) of the octahedral shear stress, we see that the left-hand side of the criterion is proportional to the square of the octahedral shearing stress, τ_0^2. Thus the von Mises criterion can be elucidated in two ways: Under complex stresses, plastic yielding at a point takes place when the distortion energy defined by Eq. (9-2) reaches a critical value, $(1+\nu)Y^2/(3E)$, or when the octahedral shear stress reaches a critical value, $\sqrt{2}\,Y/3$ (Nadai, 1950).

In addition to the tests described above, the plasticity criteria have also got certain metallurgical supports on the microscopic scale. For instance, it has been established that each single crystallite deforms by slippage along certain crystallographic planes in certain crystallographic directions when the shear stress component along those directions reaches a critical value. In 1928, Sachs proposed to evaluate the yield strength of a polycrystalline aggregate, subjected to simple stress states, by averaging the shear stress components along the critical slip directions for the various orientations of the principal directions with respect to the crystallographic axes. This averaging principle, applied to aggregates of face-centred cubic crystals and to two stress states, pure tension and pure torsion, yielded a ratio of 1:1.15 for the maximum shear stress at the beginning of plasticity. This is very closely the same ratio as that furnished by the distortion energy condition ($2/\sqrt{3} = 1.155$). Moreover, Dehlinger (1943) showed that this conclusion can be generalised and that the distortion energy criterion of plasticity can be obtained, for any state of stress, by the process of averaging the yield strength of the cubic crystals that constitute the polycrystalline aggregate. The above results provide a physical evidence of the criteria of plastic yielding (Hoffman and Sachs, 1953). However, these studies were based on the argument that plastic deformation is due to slippage along certain crystallographic planes. Very recently, it was found that the criteria based on octahedral or maximum shear stresses can also be used to predict the inelastic deformation caused by phase transformation in mono-crystalline silicon (Zhang and Tanaka, 1998, 1999). Thus the applicability of the yield criteria may be wider than previously thought.

The plastic yielding criteria discussed above can only describe the initiation of plasticity. Further development of plastic deformation must be characterised by plastic flow rules. The reader can consult with many other books on plasticity listed in the References at the end of this chapter.

Example 9.5 A thin circular disk of inner radius a and outer radius b is under an inner pressure p, as shown in **Fig. E9.4**. The disk material that has an initial yield stress Y can be considered to be elastic-perfectly plastic. Find the elastic limit pressure, plastic limit pressure and residual stresses in the disk when it is completely unloaded after initial yielding.

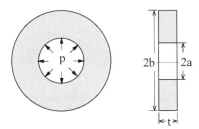

Figure E9.4 A thin disk under an inner pressure.

Solution: The deformation is in a plane-stress state when the thickness of the disk, t, is much smaller than the other dimensions, a and b. Using the polar coordinate system, the stress boundary conditions of the disk can be written as

$$\text{At } r = a, \ \sigma_{rr} = -p, \ \sigma_{r\theta} = 0. \tag{a}$$
$$\text{At } r = b, \ \sigma_{rr} = 0, \ \sigma_{r\theta} = 0. \tag{b}$$

(i) Stresses during elastic deformation
When the disk is in its elastic state, we know from the solution obtained in Chapter 6 that the only non-zero stresses are

$$\sigma_{rr} = p^* \left(1 - \frac{b^2}{r^2}\right),$$
$$\sigma_{\theta\theta} = p^* \left(1 + \frac{b^2}{r^2}\right), \tag{c, d}$$

where

$$p^* = \frac{pa^2}{b^2 - a^2}. \tag{e}$$

In this case, the principal stresses in the disk are

$$\sigma_1 = \sigma_{\theta\theta}, \; \sigma_2 = \sigma_{zz} = 0, \; \sigma_3 = \sigma_{rr}. \tag{f}$$

(ii) Onset of Plastic Deformation

When the inner pressure p increases to a critical value, plastic deformation will occur.[9.5] As we have known the principal stresses, Eq. (f), we can use either the Tresca criterion or the von Mises criterion to predict the critical value of p, denoting by p^e. From Eqs. (f) and (9-6), the Tresca criterion leads to

$$\sigma_{\theta\theta} - \sigma_{rr} = Y. \tag{g}$$

The substitution of Eqs. (c) and (d) into Eq. (g) above immediately brings about

$$2p^* \frac{b^2}{r^2} = Y. \tag{h}$$

It shows that the smaller the r the larger the value of the left-hand side of Eq. (h). Because the smallest value of r in the disk is r = a, plastic deformation will first appear on the inner surface of the disk. Hence, the critical value of inner pressure, p^e_{Tresca}, at which onset of plastic deformation happens, is

$$p^e_{Tresca} = \frac{Y}{2}\left(1 - \frac{a^2}{b^2}\right). \tag{i}$$

The above result indicates that for a given disk material, *i.e.*, for a given Y, we can increase the elastic limit by increasing the outer radius of the disk, b. However, there is a limit of p^e_{Tresca} in doing so. The maximum p^e_{Tresca} achievable in this way is $p^e_{Tresca} = Y/2$ when b approaches to infinite. According to Eq. (i), the other way of increasing p^e_{Tresca} is to use a material with a higher yield stress.

Similarly, the von Mises criterion can also be used to predict the yielding. Again, by substituting Eq. (f) into Eq. (9-9), we obtain

$$\sigma_{\theta\theta}^2 - \sigma_{rr}\sigma_{\theta\theta} + \sigma_{rr}^2 = Y^2. \tag{j}$$

Using Eqs. (c) and (d) leads to

$$p^* \sqrt{1 + 3b^4/r^4} = Y. \tag{k}$$

The left-hand side of Eq. (k) becomes larger when r becomes smaller. Thus the von Mises criterion also indicates that yielding first occurs on the inner surface (r = a) and this gives rise to the critical inner pressure

$$p^e_{Mises} = \frac{Y}{\sqrt{3 + \dfrac{a^4}{b^4}}} \left(1 - \frac{a^2}{b^2}\right). \tag{l}$$

The p^e_{Mises} in Eq. (l) is slightly larger than the p^e_{Tresca} given by Eq. (i). When b approaches to infinite, $p^e_{Mises} = Y/\sqrt{3}$. The elastic limit pressure, p^e_{Tresca} or p^e_{Mises}, indicates the limit of elastic deformation of the disk.

The form of the Tresca criterion is much simpler. Thus for convenience, we shall use the Tresca criterion in the following analysis and simply use p^e instead of p^e_{Tresca}.

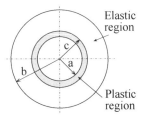

Figure E9.5 Elastic and plastic deformation regions in the disk.

(iii) *Elastic-Plastic Deformation*
If the inner pressure increases further, *i.e.*, if $p > p^e$, plastic deformation expands from r = a to r = c (a ≤ c ≤ b), that is, plastic deformation will take place in the region of a ≤ r ≤ c while that of c ≤ r ≤ b is still elastic, see **Fig. E9.5**. Now we need to obtain the stresses in both the plastic and elastic regions individually. It must be pointed out that stress solutions listed in Eqs. (c) and (d) no longer hold even in the elastic region, since the inner boundary of the current elastic region is at r = c and σ_{rr} at r = c is now unknown.

(a) *Stresses in the plastic region (a ≤ r ≤ c)*
Equilibrium equations hold for both elastic and plastic deformation. Thus in the plastic region, we still have

$$\frac{d\sigma_{rr}}{dr} + \frac{\sigma_{rr} - \sigma_{\theta\theta}}{r} = 0. \tag{m}$$

On the other hand, yielding must have occurred at the points inside the plastic region. Thus the stress state at any point in a ≤ r ≤ c must follow the yield criterion, Eq. (9-6). Equation (m) can then be rewritten as

---

[9.5] For convenience, we shall use the elastic/perfectly plastic model in the following analysis.

$$\frac{d\sigma_{rr}}{dr} = \frac{Y}{r}. \tag{n}$$

Integrating Eq. (n) directly and using the boundary condition, Eq. (a), we get

$$\sigma_{rr} = -p + Y \ln\left(\frac{r}{a}\right). \tag{o}$$

Using the above solution, Eq. (9-6) brings about

$$\sigma_{\theta\theta} = -p + Y\left\{1 + \ln\left(\frac{r}{a}\right)\right\}. \tag{p}$$

Equations (o) and (p) are the stress solution to the region with plastic deformation, because the compatibility equation in terms of stresses is also satisfied.

(b) *Stresses in the elastic region* $(c \le r \le b)$
When solving the circular tube subjected to inner pressure in the previous chapters, we found the general solutions of stresses before applying specific boundary conditions. They are

$$\sigma_{rr} = A - \frac{B}{r^2},$$
$$\sigma_{\theta\theta} = A + \frac{B}{r^2}, \tag{q}$$

where A and B are undetermined constants. Solution (q) is still valid in our current elastic deformation region $c \le r \le b$ provided that A and B are determined by the current boundary conditions.

We understand that the boundary condition at the outer surface of the disk is still described by Eq. (b). However, at the inner boundary of the elastic region, *i.e.*, at $r = c$, stresses $\sigma_{\theta\theta}$ and σ_{rr} must be equal to those in the plastic region, since across the elastic-plastic border $r = c$, stresses must be continuous. In addition, the stresses of Eq. (q) at $r = c$ must also satisfy Tresca criterion. Bearing all these in mind, we have

$$\sigma_{rr}\big|_{r=b} = A - \frac{B}{b^2} = 0,$$
$$\sigma_{rr}\big|_{r=c} = A - \frac{B}{c^2} = -p + Y \ln\left(\frac{c}{a}\right), \tag{r, s, t}$$
$$[\sigma_{\theta\theta} - \sigma_{rr}]_{r=c} = \frac{2B}{c^2} = Y.$$

Constant B can be obtained directly from Eq. (t) and then A from Eq. (r) in terms of c. Thus according to Eq. (q), the stresses in the elastic deformation region become

$$\sigma_{rr} = \frac{c^2 Y}{2b^2}\left(1 - \frac{b^2}{r^2}\right),$$

$$\sigma_{\theta\theta} = \frac{c^2 Y}{2b^2}\left(1 + \frac{b^2}{r^2}\right).$$ (u, v)

The radius of the plastic region, c, is controlled by the magnitude of the inner pressure applied, p, and can be determined by Eq. (s) above, *i.e.*,

$$\ln\left(\frac{c}{a}\right) + \frac{1}{2}\left(1 - \frac{c^2}{b^2}\right) = \frac{p}{Y}.$$ (w)

When the whole disk yields, *i.e.*, when the whole disk enters into plastic deformation, c = b. In this case, the disk cannot bear any further increment of inner pressure. Thus the inner pressure given by Eq. (w) when c = b is the maximum pressure, *i.e.*, the plastic limit pressure of the disk, p^P. That is

$$p^P = Y \ln\left(\frac{b}{a}\right).$$ (x)

Equation (x) implies that for a given material, the plastic limit pressure p^P can be increased with increasing the outer radius b.

(c) *Residual stresses after complete unloading*
As discussed above, when the disk is loaded up to a pressure p^r that is larger than the elastic limit pressure p^e but smaller than the plastic limit pressure p^s (*i.e.*, $p^e \le p^r \le p^P$), the disk will have an elastic region $c \le r \le b$, where stresses are given by Eqs. (u) and (v), and a plastic deformation region $a \le r \le c$, where stresses are specified by Eqs. (o) and (p). The radius of the elastic-plastic boundary, c, is determined by Eq. (u).

Let us see now what will happen if we unload from this deformation stage. Similar to the frame deformation discussed in Example 9.2, the unloading process from p^r to zero is equivalent to applying a negative p^r. Since an unloading process is completely elastic, the elastic stresses due to unloading can be obtained directly from the elastic solution, Eqs. (c) and (d), by replacing the p there with $-p^r$. Thus the elastic stresses due to unloading from p^r to zero are

$$\sigma_{rr} = \frac{-p^r a^2}{b^2 - a^2}\left(1 - \frac{b^2}{r^2}\right),$$

$$\sigma_{\theta\theta} = \frac{-p^r a^2}{b^2 - a^2}\left(1 + \frac{b^2}{r^2}\right).$$ (y)

The disk has had stresses before unloading. Thus the residual stresses in the disk after unloading are the summation of Eq. (y) with Eqs. (u) and (v) and Eqs. (o) and (p), respectively, in regions $a \le r \le c$ and $c \le r \le b$, according to the principle of superposition. This gives rise to the following residual stresses:

$$\sigma_{rr} = \begin{cases} -p^r + Y \ln\dfrac{r}{a} - \dfrac{p^r a^2}{b^2 - a^2}\left(1 - \dfrac{b^2}{r^2}\right), & a \le r \le c \\[2ex] \left(\dfrac{c^2 Y}{2b^2} - \dfrac{p^r a^2}{b^2 - a^2}\right)\left(1 - \dfrac{b^2}{r^2}\right), & c \le r \le b \end{cases}$$

$$\sigma_{\theta\theta} = \begin{cases} -p^r + Y\left(1 + \ln\dfrac{r}{a}\right) - \dfrac{p^r a^2}{b^2 - a^2}\left(1 + \dfrac{b^2}{r^2}\right), & a \le r \le c \\[2ex] \left(\dfrac{c^2 Y}{2b^2} - \dfrac{p^r a^2}{b^2 - a^2}\right)\left(1 + \dfrac{b^2}{r^2}\right), & c \le r \le b \end{cases}$$

where c is determined by

$$\ln\left(\dfrac{c}{a}\right) + \dfrac{1}{2}\left(1 - \dfrac{c^2}{b^2}\right) = \dfrac{p^r}{Y}.$$

The solutions of the above residual stresses lead to a significant conclusion that is very useful to our design. According to the yield criterion, Eq. (9-6), the disk with such residual stresses does not yield again if the inner pressure under a working condition is less than p^r ($< 2p^e$). Hence, by introducing plastic deformation in the manufacturing of the disk, with a given material and inner and outer radii, we can increase the elastic working regime from the original $p < p^e$ to $p < 2p^e$. This conclusion also applies to a wide range of component designs in engineering practice.

Example 9.6 A uniform beam with a rectangle cross-section, as shown in **Fig. 7.5** (page 176), is under pure bending with bending moment M. (a) Find the elastic and plastic limit bending moments, M^e and M^p. (b) Determine the residual stresses in the beam after a complete unloading from a deformation state with $M^e < M < M^p$.

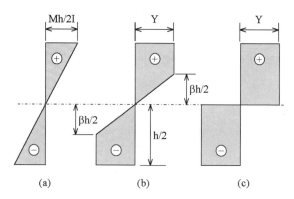

Figure E9.6 Stress distribution on the beam cross-section. (a) elastic state, (b) elastic-plastic state, (c) full plastic state.

Solution: The elastic solution of the beam has been obtained in section **7.2** as

$$\sigma_{zz} = My/I, \ \sigma_{xx} = \sigma_{yy} = \sigma_{zx} = \sigma_{zy} = \sigma_{xy} = 0, \tag{a}$$

where $I = bh^3/12$ is the moment of area of the beam cross-section. Obviously, a principal stress is σ_{zz} and the other two are σ_{xx} and σ_{yy} that are zero. The sign of σ_{zz} depends on that of coordinate y, which is positive when y is positive and becomes negative when y is negative, as shown in **Fig. E9.6a**. Thus σ_{zz} can be the largest or smallest principal stress, *i.e.*, σ_1 or σ_3. In this case, both the Tresca and von Mises criteria lead to

$$|\sigma_{zz}| = Y. \tag{b}$$

The left-hand side of the above equation reaches its largest value at $y = \pm h/2$ (**Fig. E9.6a**). Thus yielding in the beam will occur first at the top and bottom surfaces (**Fig. E9.6b**). By substituting Eq.(a) into Eq. (b) and taking $y = \pm h/2$, we obtain the elastic limit moment of bending as

$$M^e = Ybh^2/6. \tag{c}$$

When M increases further, the plastic zone will develop towards the central plane (neutral plane) of the beam, as illustrated in **Fig. E9.6b**, in which the core bounded by $-\beta h/2 < y < \beta h/2$ is still elastic where $0 \le \beta \le 1$ ($\beta = 1$ stands for the initial yielding at the top and bottom surfaces of the beam and $\beta = 0$ represents the full yielding of the beam as shown in **Fig. E9.6c**). Thus before full yielding, the bending moment can be calculated by

$$M(\beta) = 2b \left\{ \int_0^{\beta h/2} \frac{Yy}{\beta h/2} y \, dy + \int_{\beta h/2}^{h/2} Yy \, dy \right\}$$

$$= \frac{1}{12} Ybh^2 \left(3 - \beta^2\right). \tag{d}$$

Full yielding occurs when β approaches 0. Thus the plastic limit moment of bending is obtained when let $\beta = 0$ in Eq. (d), *i.e.*,

$$M^p = Ybh^2/4 = 3M^e/2 \tag{e}$$

which is 1.5 times larger than the elastic limit moment. Thus if a structure allows the beam to have a certain plastic deformation, the load-carrying capacity of the beam can be increased without changing its material and dimensions.

Unloading is elastic. The residual stress in the beam, when unloading from $M = M^*$ ($M^e < M^* < M^p$), is the superposition of the stress before unloading with

$$\sigma_{zz} = -M^*y/I, \ \sigma_{xx} = \sigma_{yy} = \sigma_{zx} = \sigma_{zy} = \sigma_{xy} = 0. \tag{f}$$

Note that at an elastic-plastic stage, the stress at different heights in the beam is different (**Fig. E9.6b**). Hence the residual stress should also be calculated correspondingly as

$$\sigma_{zz}^* = \begin{cases} Y - M*y/I, & \beta*h/2 \le y \le h/2 \\ \dfrac{Yy}{\beta*h/2} - M*y/I, & 0 \le y \le \beta*h/2 \end{cases}$$

where $\beta*$ is determined by Eq.(d), i.e.,

$$\beta* = \sqrt{3 - \dfrac{12M*}{Ybh^2}}.$$

All the other residual stress components are certainly zero. It would be beneficial if the reader can plot the distribution of the residual stress over the beam cross-section to understand its variation.

Example 9.7 Now let us revisit the plate rolling problem raised in section **1.2.2**, where we were requested to find a method to reduce the rolling force for the production engineer. The mechanics model of the problem shown in **Fig. 1.8** (page 8) can be further refined to that illustrated in **Fig. E9.7**, where h_0 is the inlet plate thickness, h_f is the outlet thickness, p and τ are the roll-plate interaction stresses normal and tangential to the deformed plate surface, respectively. The roll of radius R is rotating with a constant surface velocity V_r. We need (a) to find the rolling force and (b) from the formula of rolling force obtained to explore a method that can reduce the force without changing the rolling mill configuration and rolling parameters.

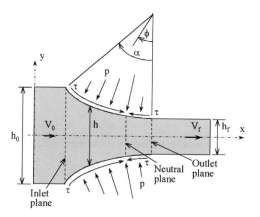

Figure E9.7 The mechanics model of a plate under plane-strain rolling.

Solution:
(i) *Modelling*

Let us recall briefly some modelling issues addressed in Chapter 1. To simplify our analysis, we assume that a cross-section in the plate normal to the x-axis remains a plane normal to the axis during deformation. To keep the volume rate of metal flow constant, the velocity of the plate must increase as it moves through the rolling zone (similar to fluid flow through a converging channel). As a result, when the roll surface is moving at a constant speed, the velocity of the plate in the inlet zone, V_0, must be smaller than V_r, and that in the outlet zone V_f, must be greater than V_r. Inside the roll-plate contact zone, there must exist a plane at which the plate speed is the same as the roll surface speed. This plane is called the *neutral plane*. Thus in both sides of the neutral plane, sliding occurs between the roll and plate surfaces and a shear stress τ applies. The shear stress changes its direction at the neutral plane because slip between roll and plate surfaces changes its direction there. In addition, it is reasonable to assume that $\tau = \mu p$, where μ is the coefficient of friction between the roll and plate surfaces when slip occurs. Rolls of a mill are much harder than the plates, thus when the plate is not too thin, the deformation of rolls can be ignored (Fleck *et al.*, 1992; Zhang, 1995). Furthermore, as discussed in section **6.2.3.3**, the plate deformation can be considered as a plane-strain problem.

(ii) *Equilibrium*
Consider the equilibrium of an isolated element of the plate in the deformation zone, as shown in **Fig. E9.8**. Since τ is usually much smaller than p in a rolling operation, the shear stress on the cross-section plane is negligible. (This simplification is also consistent with the plane cross-section assumption made above.) The difference between the two elements in the inlet and outlet zones lies in the direction of shear stress. Thus we only need to study one of them, say that of **Fig. E9.8a**.

The equilibrium in the vertical direction is satisfied automatically. The horizontal equilibrium of the elements requires that

$$- (\sigma_{xx} + d\sigma_{xx})(h + dh) - 2pR\, d\phi \sin\phi + \sigma_{xx} h + 2\mu pR\, d\phi \cos\phi = 0.$$

If the second order small quantities are neglected, the above equation becomes

$$- d(\sigma_{xx} h)/d\phi = 2pR(\sin\phi - \mu \cos\phi).$$

In rolling, angle α in **Fig. E9.7** is usually a few degrees. Hence, $\sin\phi \approx \phi$ and $\cos\phi \approx 1$. Thus the above equilibrium equation can be rewritten as

$$- d(\sigma_{xx} h)/d\phi = 2pR(\phi - \mu). \tag{a}$$

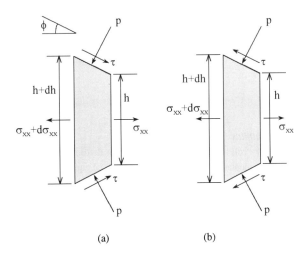

(a)   (b)

Figure E9.8 Stresses on the isolated elements of plate in the deformation zone. (a) element in the inlet zone, (b) element in the outlet zone.

(iii) Plastic deformation

Now let us consider the deformation of the plate during rolling. To reduce the thickness of the plate, stresses must be great enough to cause plastic deformation. Let us assume that the plate material is elastic/perfectly plastic and has a yield stress Y. σ_{xx} is a principal stress as we have ignored the shear stress on the cross-sectional plane. Once again, because τ is much smaller than p and ϕ is small, we can assume that $\sigma_{yy} \approx -p$ and it is also a principal stress. Hence, $\sigma_1 = \sigma_{xx}$, $\sigma_3 = \sigma_{yy}$ and $\sigma_2 = \nu(\sigma_{xx} + \sigma_{yy})$ because it is a plane-strain problem, where ν is the Poisson's ratio of the plate material. Thus the application of the Tresca criterion gives rise to

$$\sigma_{xx} + p = Y. \tag{b}$$

With this, Eq. (a) can be rewritten as

$$\frac{d\{(p-Y)h\}}{d\phi} = 2pR(\phi - \mu),$$

i.e.,

$$Yh\frac{d}{d\phi}\left(\frac{p}{Y}\right) + \left(\frac{p}{Y} - 1\right)\frac{d(Yh)}{d\phi} = 2pR(\phi - \mu).$$

Since Y is a constant for an elastic/perfectly plastic material and $dh/d\phi$ is small, the second term in the above equation can be ignored when compared with the first one. Thus

$$\frac{\frac{d}{d\phi}(p/Y)}{p/Y} = \frac{2R}{h}(\phi - \mu). \tag{c1}$$

Similarly, by considering the element in the outlet zone shown in **Fig. E9.8b**, we have

$$\frac{\frac{d}{d\phi}(p/Y)}{p/Y} = \frac{2R}{h}(\phi + \mu). \tag{c2}$$

(iv) Deformation compatibility

We have assumed that the roll is rigid, i.e., the roll does not deform. Thus the plate thickness change in the rolling gap can be calculated by the following geometrical equation:

$$h = h_f + 2R(1 - \cos\phi).$$

Expanding $\cos\phi$ by a power series of ϕ and ignoring the high-order small quantities because ϕ is small, the above relationship can be approximately rewritten as

$$h = h_f + R\phi^2. \qquad (d)$$

(v) *Rolling pressure*
A direct integration of Eq. (c) by using Eq. (d) brings about the following solution:

$$p = \begin{cases} Y\dfrac{h}{h_0} e^{\mu(H_0-H)}, & \text{in the inlet zone} \\ Y\dfrac{h}{h_f} e^{\mu H}, & \text{in the outlet zone} \end{cases} \qquad (e)$$

where

$$H = 2\sqrt{R/h_f}\ \tan^{-1}\!\left(\phi\sqrt{R/h_f}\right)$$

and $H_0 = H|_{\phi=\alpha}$. The neutral plane is the boundary of the inlet and outlet zones. Thus at the neutral section, the rolling pressure p calculated by the formula for inlet and outlet zones must be the same. This determines the position of the neutral plane, *i.e.*,

$$\phi_n = \sqrt{h_f/R}\ \tan\!\left\{\left(H_n\sqrt{h_f/R}\right)\!/2\right\},$$

where

$$H_n = \frac{1}{2}\left[H_0 - \frac{1}{\mu}\ln\!\left(\frac{h_0}{h_f}\right)\right].$$

The total rolling force is therefore the integration of p over the whole roll-plate interaction zone. If p of Eq. (e) can be reduced, the rolling force can also be reduced.

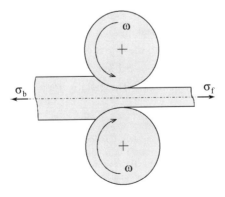

Figure E9.9 Plate rolling with back and front tensions.

(vi) *A way of rolling force reduction*

The solution, Eq. (e), shows that the rolling pressure p is determined by a number of variables, such as the amount of thickness reduction $h_0 - h_f$, yield stress of the plate material Y, roll radius R and coefficient of friction μ. The problem raised in Chapter 1 was to reduce the rolling force without changing all the above parameters. It seems to be an impossible task. However, if we revisit Eq. (b), which is the yield criterion, we find that it is possible to reduce Y equivalently by changing σ_{xx}. Practically, we can easily apply a back tension, σ_b, or a front tension, σ_f, or both of them, to change σ_{xx}, as shown in **Fig. E9.9**. By doing so, we can get a modified solution similar to Eq. (e), that is,

$$p = \begin{cases} (Y - \sigma_b)\dfrac{h}{h_0}e^{\mu(H_0 - H)}, & \text{in the inlet zone} \\ (Y - \sigma_f)\dfrac{h}{h_f}e^{\mu H}. & \text{in the outlet zone} \end{cases}$$

Obviously, the rolling pressure and thus the total rolling force is reduced because of the application of σ_b and σ_f.

(v) *A remark*

In the solution process demonstrated above, we can understand that after the initial modelling one may introduce more assumptions to simplify the model to avoid mathematical difficulties, provided that they are reasonable. The assumptions of rigid roll, uniform friction, plane cross-section and so on are examples. An in-depth understanding of the analytical solution is important to practical application and sometimes can bring about unexpected methods such as the rolling force reduction method discussed above, to improve the application.

9.5 FAILURE THEORIES

As mentioned in section **9.1**, most engineering components must work safely in the regime of elastic deformation. Thus in designing a component or structure using a specific material, it is important to place a *failure limit* on the stress state that the structure can sustain. Generally speaking, if a structural material is *ductile* (*e.g.*, mild steel), failure is most likely due to the initiation of yielding, as we have understood throughout the discussion in sections **5.1** and **9.4**; if the material is *brittle* (*e.g.*, cast iron), failure is often by *fracture*.[9.6] The failure limit is clear if a component is

---

[9.6] Usually, brittle materials are approximately considered to be those whose maximum strain at fracture (sometimes called *fracture strain*) in a uniaxial tensile test is less than about 5%. The fracture of a brittle material is due to the flaws or imperfections in the material that raise the stress in their localised regions to a high enough value that the material's cohesive strength is exceeded. Cracks will therefore grow and spread to bring about fracture. The failure theories to be introduced here define failure in terms of the applied stresses and tensile and compressive strengths of a material. The field of *fracture mechanics* provides a more advanced approach to failure analysis, which is concerned with the stress and deformation in the local region of a crack tip in a material, and is useful to account for the effect of rate of loading. More details about fracture mechanics can

subjected to a uniaxial state of stress, such as in the case of simple tension where the yield stress of a ductile material or the stress at fracture of a brittle material is the failure limit. However, the *mode of failure* of a structural material, yielding or fracture, also depends on the stress field, temperature variation, rate of loading, chemical environment, manufacturing process of the structure and so on. For example, ceramics and monocrystalline silicon are regarded as brittle materials because under simple loading conditions, such as uniaxial tension and simple bending, they will fail by fracture; however, under complex loading conditions, such as indentation and scratching, plastic deformation via dislocation or phase transformation will occur (Zarudi, Zhang and co-workers, 1996, 1998, 1999, 2000).

The general guideline on the development of a yield surface discussed in section **9.4.2.1** is also suitable for establishing a failure criterion. As we will see shortly, the Tresca and von Mises criteria of initial yielding are straightforward the failure criteria for ductile materials because plastic yielding is one of the failure modes.

In the following, we shall introduce five theories of failure. It is necessary to emphasise that these theories are applicable to the cases that are under static loading, at room temperature and without the consideration of other environmental factors, such as chemical factors.

9.5.1 Ductile Materials

9.5.1.1 *The Maximum Shear Stress Theory*

This is the Tresca criterion discussed previously and described by Eq. (9-6). The theory assumes that when the maximum shear stress in a material reaches the yield stress in uniaxial tension, failure takes place due to the initiation of plastic deformation. If the safety factor of a design is α, the allowable stress must satisfy

$$|\sigma_1 - \sigma_3| \leq \frac{Y}{\alpha}. \tag{9-15}$$

In other words, to avoid the failure of a component, the theory requires that the difference of the first and third principal stresses *at any point* in the component must not reach Y/α. In discussing the failure mode of a circular shaft of a ductile material in section **7.1**, we have used the present theory.

---

be found in many textbooks and monographs, *e.g.*, those by Ewalds and Wanhill (1984) and Broek

9.5.1.2 *The Maximum Distortion Energy Theory*

This is the von Mises criterion shown in Eq. (9-9) or (9-10), which states that failure occurs when the distortion energy at any point in a material reaches a critical value, i.e.,

$$(\sigma_1 - \sigma_2)^2 + (\sigma_2 - \sigma_3)^2 + (\sigma_3 - \sigma_1)^2 \leq \frac{2Y^2}{\alpha} \qquad (9\text{-}16a)$$

or

$$(\sigma_{xx} - \sigma_{yy})^2 + (\sigma_{yy} - \sigma_{zz})^2 + (\sigma_{zz} - \sigma_{xx})^2 + 6(\sigma_{xy}^2 + \sigma_{yz}^2 + \sigma_{zx}^2) \leq \frac{2Y^2}{\alpha}. \qquad (9\text{-}16b)$$

9.5.2 Brittle Materials

9.5.2.1 *The Maximum Normal Stress Theory*

This theory assumes that a brittle material will fail when the magnitude of the maximum normal stress (the first principal stress) in the material, $|\sigma_1|$, reaches the normal stress of the material at failure in simple tension, σ_f. This means that failure happens when

$$|\sigma_1| = \sigma_f. \qquad (9\text{-}17a)$$

To avoid failure, the allowable stress at any point in a component must satisfy the inequality

$$|\sigma_1| \leq \frac{\sigma_f}{\alpha}. \qquad (9\text{-}17b)$$

Experimentally, it has been found that the above theory gives good predictions for brittle materials having similar stress-strain behaviour under tension and compression. Clearly, the theory does not involve the effect of the other two principal stresses.

(1985).

In discussing the strength of a rotating disk in section **6.3.2.2** and the failure mode of a circular shaft of a brittle material under torsion in section **7.1**, we have used the present maximum normal stress theory.

9.5.2.2 *The Maximum Normal Strain Theory*

This theory considers that a brittle material will fail when the maximum normal strain (the first principal strain) in the material, ε_1, is tensile and reaches the normal strain of the material at failure in simple tension, ε_f, *i.e.*, failure happens when

$$\varepsilon_1 = \varepsilon_f. \tag{9-18a}$$

Similarly, to avoid failure the allowable strain at any point in a component must follow

$$\varepsilon_1 \leq \frac{\varepsilon_f}{\alpha}. \tag{9-18b}$$

Since the theory assumes that a brittle material does not have plastic deformation before failure, both ε_1 and ε_f above can be replaced by stresses using the generalised Hooke's law. If the failure stress in simple tension is σ_f, the corresponding failure strain is $\varepsilon_f = \sigma_f/E$ and using Eq. (5-13) we can find that the first principal strain ε_1 is

$$\varepsilon_1 = \frac{1}{E}[\sigma_1 - \nu(\sigma_2 + \sigma_3)].$$

Thus the criterion (9-18b) can be rewritten as

$$\sigma_1 - \nu(\sigma_2 + \sigma_3) \leq \frac{\sigma_f}{\alpha}. \tag{9-19}$$

This theory takes the second and third principal stresses into account and can explain well the failure of brittle materials under compression.

9.5.3 The Mohr Theory

The failure theories based on the concept of maximum shear stress and maximum distortion energy were developed for ductile materials, although they have sometimes been applied to brittle materials and their applicability seems to be even wider as pointed out in section **9.4.4**. On the other hand, the maximum normal stress and strain theories are for brittle materials. Nevertheless, a good prediction by the maximum normal stress theory requires that the stress-strain behaviour of a brittle material under tension and that under compression should be similar. Unfortunately, many brittle materials do not possess such a mechanical property.

The Mohr theory deals with both plastic yielding and brittle fracture for materials whose strengths in uniaxial tension and compression may be different. Theoretically, it is an extension of the maximum shear stress theory, by defining failure in terms of a limiting combination of normal and shear stresses in a material in a graphical manner. As we have been familiar with in Chapter 3 (see also Appendix A), corresponding to any pair of principal stresses, (σ_1, σ_2), (σ_1, σ_3) and (σ_2, σ_3), we can construct a Mohr's circle. Hence, if the three Mohr's circles with respect to the three pairs of principal stresses are drawn together, we get **Fig. 9.7**. It can be shown (see, e.g., Westergaard, 1952) that the normal and shear stresses on any plane can be represented by a point lying within the shaded area. The Mohr theory assumes that on any plane failure is governed by the maximum shear stress. This implies that the largest circle defined by the pair (σ_1, σ_3) determines the critical failure state.

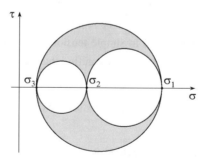

Figure 9.7 Mohr's stress circles representing three-dimensional stress states.

Based on the above theory, we can now easily determine the failure stress of a component, either plastic yielding or brittle fracture depending on the material of the component. We need to do three tests on the material to determine the failure limit boundary, a uniaxial tension for obtaining the tensile failure stress σ_{ft}, a uniaxial compression for getting the compressive failure stress σ_{fc} and a pure torsion test for determining the failure shear stress σ_{fs}. In tension, the principal stresses are $\sigma_1 = \sigma_{ft}$ and $\sigma_2 = \sigma_3 = 0$; in compression, they are $\sigma_1 = \sigma_2 = 0$ and $\sigma_3 = \sigma_{fc}$; while in torsion they become $\sigma_1 = \sigma_{fs}$, $\sigma_2 = 0$ and $\sigma_3 = -\sigma_{fs}$. The Mohr's circle for each of these stress states can be easily plotted, as shown in **Fig. 9.8**. An envelope tangent to these circles, as indicated in the figure, represents the failure curve of the material. If the Mohr's circle of the stress state at any point in a component extends beyond or has a point of tangency with the envelope, failure is assumed to take place. Clearly, when the failure stresses of a material under tension and compression are the same, *i.e.*, $\sigma_{ft} = \sigma_{fc}$, the Mohr theory becomes the same as the maximum shear stress theory.

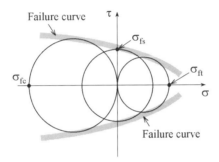

Figure 9.8 The failure curve determined by the Mohr theory.

All the above failure theories are commonly used in design or failure analysis in engineering practice. However, we should understand that their usefulness is limited and no single theory may apply to a material at all times, because a material may behave differently, ductile or brittle, depending on various conditions mentioned previously.

The use of the failure theories is straightforward and similar to the applications of the yielding criteria discussed in Examples **9.3** and **9.4**. The fundamental steps are to find the principal stresses at the most critical point in a component or structure and then select a proper failure criterion to work out the critical loading condition. In the design of a structure, a safety factor needs to be taken into account.

References

Broek, D (1985), *Elementary Engineering Fracture Mechanics*, 4th edition, Martinus Nijhoff, London.

Chakrabarty, J (1987), *Theory of Plasticity*, McGraw-Hill, New York.

Coulomb, CA (1773), *Mem. Math. Phys.*, **7**, 343.

DeGarmo, EP, Black, JT and Kohser, RA (1997), *Materials and Processes in Manufacturing*, 8th edition, Prentice-Hall, Inc., Englewood Cliffs, NJ.

Dehlinger, U (1943), Die Fliessbedingung bei mehrachsigem Spannungszustand vielkristalliner Metalle, *Z. Metallkunde*, **35** (9), 182-184.

Ewalds, HL and Wanhill, RJH (1984), *Fracture Mechanics*, Edward Arnold, Delft.

Fleck, NA, Johnson, KL, Mear, M and Zhang, L (1992), Cold rolling of thin foil, *Journal of Engineering Manufacture*, IMechE, **B206**, 119-131.

Hencky, H (1924), *Z. Angew. Math. Mech.*, **4**, 323.

Hill, R (1950), *The Mathematical Theory of Plasticity*, Oxford University Press, Oxford.

Hoffman, O and Sachs, G (1953), *Introduction to the Theory of Plasticity for Engineers*, McGraw-Hill, New York.

Lode, W (1926), *Z. Physik*, **36**, 913.

Mendelson, A (1968), *Plasticity: Theory and Application*, Macmillan, New York.

Nadai, A (1950), *The Theory of Flow and Fracture of Solids*, McGraw-Hill, New York, p.402.

Sachs, G (1928), Zur Ableitung einer Fleissbedingung, *Z. Ver. Deut. Ing.*, **72**, 734.

Taylor, GI and Quinney, H (1931), *Phil. Trans. Roy. Soc.*, **A230**, 323.

Tresca, H (1864), *Comptes Rendus Acad. Sci., Paris*, **59**, 754; and Tresca, H (1868), *Mem. Sav. Acad. Sci., Paris*, **18**, 733.

Von Mises, R (1913), *Gottinger Nachrichten Math. Phys. Klasse*, 582.

Westergaard, HM (1952), *Theory of Elasticity and Plasticity*, Harvard University Press, Cambridge, Mass., pp.61-64.

Zarudi, I and Zhang, L (1999), Structural Changes in Mono-Crystalline Silicon Subjected to Indentation – Experimental Findings, *Tribology International*, **32**, 701-712.

Zarudi, I and Zhang, L (2000), On the Limit of Surface Integrity of Alumina by Ductile-Mode Grinding, *Trans ASME, Journal of Engineering Materials and Technology*, **122**, 129-134.

Zarudi, I, Zhang, L and Cockayne, D (1998), Subsurface structure of alumina associated with single-point scratching, *Journal of Materials Science*, **33**, 1639-1654.

Zarudi, I, Zhang, L and Mai, Y-W (1996), Subsurface damage in alumina induced by single point scratching, *Journal of Materials Science*, **31**, 905-914.

Zhang, L (1995), On the mechanism of cold rolling thin foil, *International Journal of Machine Tools and Manufacture*, **35**, 363-372.

Zhang, L and Tanaka, H (1998), Atomic deformation induced by two-body and three-body contact sliding, *Tribology International* **31**, 425-433.

Zhang, L and Tanaka, H (1999), On the mechanics and physics in the nano-indentation of silicon monocrystals, *JSME International Journal*, **A42** (4), 546-559.

Zhang, L and Zarudi, I (1999), An understanding of the chemical effect on the nano-wear deformation in mono-crystalline silicon components, *Wear*, **225-229**, 669-677.

Zhang, L and Zarudi, I (2000), Plasticity in monocrystalline silicon under complex loading conditions, *Key Engineering Materials*, **177-180**, 121-128.

Important Concepts

Octahedral plane
Strain energy
Distortion energy density
Initial yielding surface
Unloading
von Mises criterion
Elastic limit
Residual stress
Failure mechanisms

Octahedral shear stress
Strain energy density
Yielding
Loading
Reloading
Tresca criterion
Plastic limit
Failure theory
Failure curve

Questions

9.1   What is initial yielding? What is work hardening? Physically, what causes plastic deformation in a ductile material?

9.2   If we load a hardening material to its plastic deformation regime and unload it, then during reloading, we will find that the material's yield stress becomes greater. Why is this?

9.3   Do we have a plastic limit load for a structure with a work hardening material?

9.4   What are the physical indications of the two most commonly used criteria?

9.5   Among the basic equations developed in elasticity, which of them becomes invalid when plasticity occurs?

9.6   When a material is under elastic deformation, does loading and unloading follow the same path? What will happen when unloading a structure in its elastic-plastic regime?

9.7   What is a yield criterion? What are the two major groups of variables that must be considered in developing a yield criterion? Can you think of a new yield criterion for a material newly developed?

9.8   What are the major steps of carrying out an elastic-plastic analysis? What is the difference between an elastic-plastic analysis and an elastic analysis studied in previous chapters?

9.9   Generally, how do we determine the elastic limit load and plastic limit load of a structure? If the material of a component is elastic/perfectly plastic, what will happen when the external load reaches the plastic limit load?

9.10 Why are the failure modes of brittle and ductile materials different? What are the basic procedures for a failure analysis using the criteria discussed?

Problems

Unless otherwise specified, the material properties required in solving the following problems should be taken from **Table 5.2**.

9.1 Find the octahedral shear stress and distortion energy density at the points in a mild steel component that has the stress states given in Problem 3.1.

9.2 A standard specimen of 2024-O aluminium alloy for a simple tensile test has a diameter of 12.65 mm. The tension experiment was done with a gauge length of 50.8 mm. The results were recorded and are listed in the following table. Plot the stress-strain curve and determine (a) the yield stress of the material (0.2% offset) and (b) modulus of elasticity.

Load (kN)	Elongation within the gauge length (mm)	Load (kN)	Elongation within the gauge length (mm)	Load (kN)	Elongation within the gauge length (mm)
7.47	0.042418	17.53	0.508	28.02	6.096
9.03	0.051308	18.68	0.762	29.37	8.128
11.70	0.1016	19.22	1.016	30.42	10.160
14.23	0.2032	21.04	1.524	31.54	12.192
15.83	0.3048	22.64	2.032	32.56	13.919
17.08	0.4064	24.95	3.048		

9.3 A long prismatic steel bar is loaded by a uniform tension along its longitudinal axis to a stress of $\sigma = 450$ MPa. The bar material can be idealised by an elastic/linear hardening model with a hardening modulus $E_t = 20$ GPa, as shown in **Fig. P9.1**. The Young's modulus of the material E is 207 GPa and the yield stress is 350 MPa. Find the residual strain and residual stress after complete unloading. If the material is rigid/linear hardening, i.e., E is infinite, what is the residual strain after complete unloading?

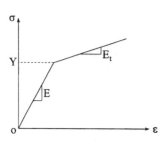

Figure P9.1

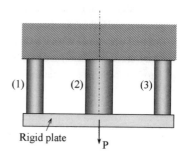

Figure P9.2

9.4 A structure shown in **Fig. P9.2** has three uniform bars whose materials can be approximated as elastic/perfectly plastic. The bars are of the same length and are aligned vertically in the loading direction. Bars (1) and (3) are identical with a cross-sectional area A, Young's modulus E and yield stress Y_{13}. Bar (2) has a cross-sectional area of B, Young's modulus E and yield stress Y_2. (a) Find the elastic limit load, P^e, the plastic limit load of the structure, P^p, and the residual stresses in all the bars after the complete unloading from P^p. (b) If the structure is reloaded in the same way, determine the new elastic limit load, P^e_{new}, and compare it with P^e.

9.5 If the simple frame of Example 9.2 is subjected to a horizontal force, Q, at the tip hinge, find the elastic limit load, Q^e, plastic limit load Q^p and the residual stresses in the bars after a complete unloading from a force Q that is larger than Q^e but smaller than Q^p. If the frame is subjected to a vertical load P and horizontal load Q at the same time, what are the corresponding solutions?

9.6 An axisymmetric assembly of a solid bar and a tube is loaded in the way shown in **Fig. P9.3**. The cross-sectional areas of the bar and tube are A_{bar} and A_{tube}, respectively, and the Young's moduli of them are E_{bar} and E_{tube}, respectively. It is known that the yield stress of the bar is smaller than that of the tube, $i.e.$, $Y_{bar} < Y_{tube}$. Both the bar and tube materials are elastic/perfectly plastic. (a) Find the elastic limit load P^e, plastic limit load P^p and the residual stresses in the bar and tube after complete unloading from a load P that $P^e < P < P^p$. (b) If the assembly is reloaded in the same way, what is the second elastic limit load?

9.7 Two solid cylinders of the same length L are joined together as illustrated in **Fig. P9.4**. All the joints are rigid in the axial direction but the radial expansion of the cylinders is free. At the initial environmental temperature of $T = 0°C$, the cylinders are without any stress. When the temperature changes to $T = 150\,°C$, it is found that cylinder (1) is yielded plastically but cylinder (2) is still elastic. Find the residual stresses in both the bars when the environment temperature goes back to $0\,°C$. The material of the bars is elastic/perfectly plastic with Young's modulus $E = 210$ GPa, coefficient of thermal expansion $\alpha = 0.000011\ /\,°C$ and yield stress $Y = 250$ MPa. The cross-sectional area of bar (2) is twice that of bar (1), $i.e.$, $A_{(2)} = 2A_{(1)}$.

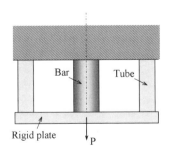

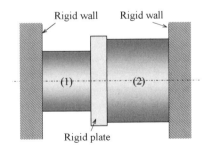

Figure P9.3　　　　　　　　　　Figure P9.4

9.8 Use both the Tresca and von Mises criteria to define the elastic limit pressure p^e of the thin-walled pressure vessel in Example 3.2 (page 41). If the inner radius of the vessel is 400 mm, thickness is 4 mm and the yield stress of the vessel material is 250 MPa, compare the difference of p^e obtained from the two criteria.

9.9 If the pin and collar for an interference-fit illustrated in **Fig. 1.6** is made of low-carbon steel, determine the maximum total interference δ before causing plastic deformation. The nominal radius of the pin is 20 mm and the nominal outer radius of the collar is 50 mm. (*Hint*: See the discussion at the end of section **6.3.2.2**.)

9.10 In the thermal fitting discussed in section **7.3.2**, if the required maximum octahedral shear stress in the disk after assembly and after complete cooling is 60% of the yield stress of the disk material, find the minimum radius of the disk before thermal assembly. The outer radius of the disk is 80 mm and that of the shaft is 30 mm. Both the shaft and disk are made of high-carbon steel.

9.11 A gun barrel needs to be assembled by shrinking an outer barrel over an inner barrel so that the maximum principal stress is 70 percent of the yield stress of the material. The material of both members is steel with yield stress Y = 537.93 MPa, Young's modulus E = 206.90 GPa and Poisson's ratio ν = 0.292. The nominal radii of the barrels are 4.76 mm, 9.53 mm and 14.29 mm. Use the von Mises criterion to find the maximum interference for assembly.

9.12 If the gun of Problem 9.11 is fired with an internal pressure of 275.86 MPa, will this pressure cause plastic yielding in any of the barrel members?

9.13 A simply supported beam is subjected to a uniform pressure on its top surface, as shown in **Fig. P9.5**. The support is at the central plane. Find the elastic limit pressure q^e and determine the boundary between the elastic deformation zone and the plastic deformation zone when q is larger than q^e but less than the plastic limit pressure q^p.

9.14 When the beam of Problem 9.13 is a cantilever clamped at the end of x = 0, answer the same questions.

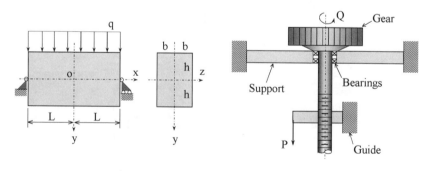

Figure P9.5 **Figure P9.6**

9.15 **Fig. P9.6** shows a gear-driven power screw that is used to pull up an eccentric load of P = 50 kN. The screw is supported in bearings that take out the radial- and thrust-load components. The load is suspended from a nut, as shown, guided to prevent rotation with the screw. The eccentricity of the load induces a bending moment into the screw of magnitude M = 845 N·m. The friction in the threads and between the nut and guide, together with the lead angle of the threads, requires that a torque of Q = 450 N·m be applied to the gear to raise the load. Experiments with a similar structure indicate that failure of such a system can be predicted if the mean diameter of the threads is used for calculating cross-sectional areas, moments of inertia, stresses, etc. In the present problem, the mean diameter is 48 mm. The screw material is a cold-drawn steel having the following mechanical properties: Y = 456 MPa, E = 200 GPa and ν = 0.3. Examine whether plastic deformation will occur in the power screw.

9.16 A very long solid circular shaft has a radius R and is subjected to an axial force T and a torque Q at its two ends, as shown in **Fig. P9.7**. For the following cases, determine the critical radius with a safety factor α = 1.5 when T = 100 kN and Q = 20 kN·m:

(a) The shaft is made of cast iron, whose failure stress in uniaxial tension is 138 MPa. Compare the predictions of the maximum normal stress theory, the maximum normal strain theory and the Mohr theory.

(b) The shaft is made of low-carbon steel (see **Table 5.2** for mechanical properties). Compare the predictions of the maximum shear stress theory, the maximum distortion energy theory and the Mohr theory.

Figure P9.7

Chapter 10

AN INTRODUCTION TO THE FINITE ELEMENT METHOD

Throughout the study from Chapter 2 to Chapter 9, we have established necessary concepts for deformation and stress analysis and learned the fundamental theories in solid mechanics. In this chapter, we shall introduce the basics of a numerical method for stress and deformation analysis, the finite element method that has been one of the most powerful numerical tools for design, manufacturing and reliability assessment of various structures and components.

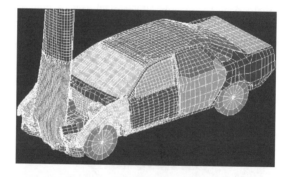

The above pictures show an example of the application of the finite element method, simulated deformation of a car body when subjected to an impact (Courtesy of Paul Briozzo, The University of Sydney, Australia). Top: with the finite element mesh. Bottom: without the mesh.

10.1 INTRODUCTION

In the last nine chapters, we have studied some basic skills for stress and deformation analyses in both the elastic and plastic regimes, learned how to model a practical engineering problem for mechanics analysis and understood some important principles and phenomena in solid mechanics, such as the principle of superposition, St Venant's principle, the concept of stress concentration and the mechanism of strengthening by introducing plastic deformation. In the meantime, we have also experienced significant mathematical difficulties in solution.

In this chapter, we shall see how the finite element method (often abbreviated as FEM), one of the most powerful numerical methods in engineering practice, is derived based on the fundamental theory of solid mechanics, how it can be used to overcome the mathematical difficulties encountered in our previous solution processes and what will be the major procedures, concerns and considerations in the application of the method. We shall then take a relatively complex engineering problem as an example to experience the whole process from the mechanics modelling and solution to the analysis of the results obtained. Accordingly, we shall further appreciate the importance of our fundamental understanding of stress, deformation, description of boundary conditions and modelling methodology for obtaining reliable numerical results.

10.2 FUNDAMENTALS

10.2.1 Introductory Examples

To understand the principle of the finite element method, let us first consider how to approximate a circle shown in **Fig. 10.1a** by using straight line elements. If we use three line elements, we can form a triangle that is clearly far away from the circle (**Fig. 10.1b**). However, if we use six line elements to form a hexagon, it approximates the circle much better (**Fig. 10.1c**). It is obvious that if we use an infinite number of line elements, we will get exactly the smooth circle of **Fig. 10.1a**. In practice, however, it is only possible to use a finite number of elements to get an approximate circle. Note that the approximation by a finite number of line elements in this simple problem is only geometrical.

Now consider a relatively more complex case, the deformation of the pipe system shown in **Fig. 10.2**, which is supported by rigid walls at its ends, but is subjected to concentrated loads P_1, P_2 and P_3 at the locations shown. The stresses in the system are often required for the selection of materials and for the determination of the pipe dimensions. If we use the solution method studied in previous chapters, we shall encounter tremendous difficulties because the pipe system is three-dimensional, the cross-sections of the pipes are not uniform and the flanges and pipes can be made from different materials.

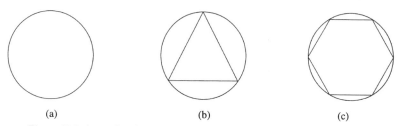

Figure 10.1 Approximation to a circle using a finite number of line elements.

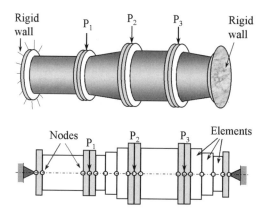

Figure 10.2 Finite element modelling of a piping system.

We can also use a finite number of pipe elements to approximate the system, as shown in the bottom part of **Fig. 10.2**, where each element can be independent in terms of dimensions and material properties but has a common joint end with its adjacent element. Such a common joint is called a *node*, which transmits displacements and stresses (or equivalent forces) from an element to its neighbour. Inside an element, stresses, strains and displacements are continuous.

In the above approximation using a finite number of elements, we have introduced two types of approximation errors. The first is the *geometrical discretisation error*, which is of the same nature as that encountered in the first example of approaching a continuous circle by a finite number of line elements. The second is the *piecewise approximation error of mechanics quantities*, *i.e.*, stresses, strains and displacements. In the original pipe system, all these mechanics quantities are continuous and thus the degrees of freedom of the system are infinite. After the approximation, by allowing the stresses and displacements to transmit through a finite number of nodes, the degrees of freedom of the approximate system become finite. In this way, we can then use a computer to solve the problem reasonably. When the number of elements approaches to infinite both the geometrical error and the piecewise error of mechanics quantities will vanish. Thus in general, if we model an engineering problem using a sufficient number of elements with correct boundary conditions and material properties, we will obtain a good approximate solution.

The above two examples demonstrate the original idea of *the finite element method* and the origin of its name.

10.2.2 Types of Elements

The element used for the above pipe system has only two nodes, one at each end of the element. When dealing with more complex problems under complex stress states, we need to use other types of elements. For example, for the dam of a reservoir under plane-strain problems, we can divide the body by two-dimensional triangle elements (**Fig. 10.3**). However, to study the stresses in a more complex structure, such as the car-body structure shown in **Fig. 10.4**, a combination of different types of elements must be used to capture the complex three-dimensional deformation of individual components of the structure. The selection of the types of elements is not unique and often depends on the convenience of mesh division, requirement of accuracy, computer capacity and computational cost. **Figure 10.5** shows some commonly used types of elements for two- and three-dimensional problems. A commercial FEM code usually possesses a large number of different elements for users to select.

Naturally, the reader may have had the following question in mind: 'For a given problem, how do we know which type of elements to use and what number of elements?' The first part of the question can only be answered by understanding the nature of the problem, which is indeed dependent on what we have learned in the last nine chapters. For instance, for a long beam under transverse bending, we must use a beam element to capture its nature of bending. If a bar element is selected instead, which cannot provide bending stresses, we are unable to obtain a correct solution. On the other hand, the selection of the type of element in many cases is not unique. For the same problem, different people may use different elements. One can use a triangle element for the deformation analysis of the dam, as shown in **Fig. 10.3**, but others may use rectangular elements. The considerations are complex. Further details can be found in more advanced texts, for example, those by Bathe (1982) and Zienkiewicz and Taylor (1991).

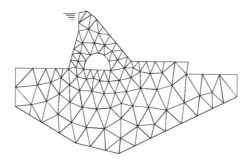

Figure 10.3 An examples of finite element meshing (division): the dam of a reservoir.

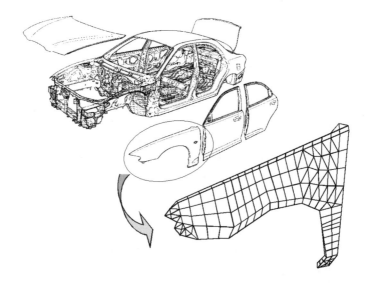

Figure 10.4 Finite element division of a car-body panel using different types of elements.

The second part of the question is relatively easier to answer in principle. To make sure that the number of elements is sufficient, one needs to carry out an error analysis based on a series of trial calculations by refining, *e.g.*, doubling, the number of elements each time and comparing the subsequent solutions. Usually, a finite element mesh division is usable only when the change of the solution (or convergence error) becomes acceptable for a further increment of element numbers. (In many engineering applications, for instance, a convergence error of 10% is acceptable.) However, if a special deformation zone exists, for instance, a stress concentration zone, the above simple refining method will become inefficient. A general consideration is that in the zone that may experience a sharp stress variation (*e.g.*, in the area with a concentrated force, a notch, a crack, or a hole) smaller-sized elements should be arranged. Some typical considerations will be discussed later in this chapter.

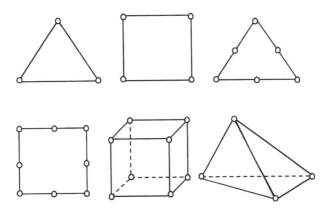

Figure 10.5 Some commonly used types of elements. Each hollow circle indicates a node.

10.2.3 Control Volume

The domain of a component or structure that is selected for finite element analysis is called *the control volume* of the finite element modelling. In the example given in **Fig. 10.3**, the control volume is the same as the whole cross-section of the dam. It is not always the case, however. In many circumstances, the geometry and boundary and loading conditions have a certain symmetry that enables us to use a smaller control volume to reduce computational cost and increase accuracy. The tension of the thin plate with a central hole discussed in section **8.5** is a good example. In this case, we know that deformation in the plate must be symmetrical to both its horizontal and vertical axes. Thus the deformation in the upper half and lower half as well as that in the left half and right half of the plate must be symmetrical. We can therefore take only a quarter of the plate as the control volume.

The selection of a control volume is an important aspect of a successful and efficient finite element analysis. It needs a sound understanding of the material studied in the previous chapters. We have discussed its mechanics in Chapter 6 on modelling and solution and will introduce more examples for the finite element modelling later in this chapter.

10.3 FORMULATION IN THE FINITE ELEMENT METHOD

As we have seen in the examples above, the primary idea of the finite element method is to use a finite number of elements to replace the originally continuous body approximately. This process is called *discretization*. After discretization, the connection between two adjacent elements is through their common nodes. Mathematically, discretization means that we change the originally continuous body, which has an infinite number of degrees of freedom, into a mechanics model, which is an assembly of some elements and has a finite number of degrees of freedom, although within each element, all the mechanics quantities, such as stresses, strains and displacements, are still continuous.

We can imagine that to carry out a piecewise discretization we need to have a special treatment of the basic equations and boundary conditions. On the other hand, we realise that if a control volume contains some thousands of elements, we are unable to handle them manually one by one. A feasible way is to find a standard procedure, which applies to all the elements and is easy to program so that computers can be used to do the job mechanically.

A general mathematical formulation of the finite element method can be found in many textbooks and monographs. In the following, we shall use a very simple example to present the characteristics and principle of the method to gain a good understanding. This example, though simple, possesses all the details of normal finite element formulation: *'The sparrow may be small but it has all the vital organs.'* Computer programming also follows the procedures given below.

10.3.1 Problem Description

Consider a bar with a variable cross-section in x-direction, as shown in **Fig. 10.6**, subjected to uniaxial concentrated external forces, P_1 and P_2, distributed surface stress $\tau = \tau(x)$ and its own weight, ρf_x, where ρ is the density of the bar material and $f_x = g$ is the acceleration of gravity.

According to the loading conditions, we know that when the length of the bar is much larger than its cross-sectional dimensions, the stresses, strains and displacements are the functions of coordinate x only and the only non-zero stress, strain and displacement will be the normal stress component, σ_{xx}, the normal strain component, ε_{xx}, and the axial displacement component, u. For this problem, therefore, we have the following unknowns,

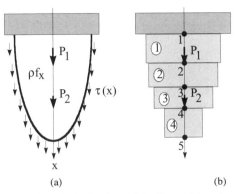

Figure 10.6 A long bar under uniaxial loading. (a) loading and boundary conditions, (b) finite element division. To demonstrate the variation of its cross-sectional dimension, the bar is not sketched proportionally.

$$\sigma_{xx} = \sigma(x), \quad \varepsilon_{xx} = \varepsilon(x), \quad u = u(x). \tag{10-1}$$

The external forces and stresses applying on the bar are

$$\begin{aligned}
&\text{surface stress:} \quad \tau = \tau(x), \\
&\text{body force:} \quad f = \rho f_x, \\
&\text{concentrated forces:} P_1 \text{ and } P_2.
\end{aligned} \tag{10-2}$$

The displacement boundary condition is that at the top end, *i.e.*, at $x = 0$,

$$u = 0. \tag{10-3}$$

For the present one-dimensional problem, the generalised Hooke's law becomes

$$\sigma = E\varepsilon \tag{10-4}$$

and the strain-displacement relationship reduces to

$$\varepsilon = \frac{du}{dx}. \tag{10-5}$$

According to the theorem of minimum potential energy (see also Appendix C of this book), the equilibrium equation is equivalent to the condition that the potential energy of the bar, V, takes as its minimum, that is,

$$\frac{dV}{du} = 0. \tag{10-6}$$

Our task now is to find a solution that satisfies the above stress and displacement boundary conditions, Eqs. (10-2) and (10-3), and all the basic equations, Eqs. (10-4) to (10-6). We are going to follow the displacement approach discussed in Chapter 6 in our finite element formulation.

10.3.2 Finite Element Solution

(1) *Element Division*

The first step in using the finite element method is to model the bar as a stepped one, consisting of a discrete number of elements, each having a uniform cross-section, as shown in **Fig. 10.6b**. For convenience, we use only four elements to demonstrate the principle and procedure. The cross-sectional area of each element is the average within the element region. Thus in total, we have four elements and five nodes, with each element connecting to two nodes.

In addition to the above, surface stress and body force are also treated approximately as constant within each element (it is not necessary though). However, element length, cross-sectional area, surface stress and body forces may all differ in magnitude from element to element. Better approximations will be obtained by increasing the number of elements, as we saw in the introductory examples in section **10.2.1**. It is worth mentioning that in dividing the elements, it is convenient to define a node at each location where a concentrated force is applied. In the model of **Fig. 10.6b**, we divide elements in the way that nodes 2 and 4 are at the application points of the concentrated forces P_1 and P_2.

(2) *Numbering*

In the present one-dimensional bar deformation, any point in the bar has only one displacement component, u, as stated before. Thus each node of an element is only permitted to displace in the positive or negative x-direction. In other words, each node has only one *degree of freedom*. The five-node finite element model therefore has five degrees of freedom.

We denote the displacements along each degree of freedom by U_1, U_2, U_3, U_4, U_5, respectively, called *nodal displacement*, and arrange them in a vector form as

$$\mathbf{U} = \{U_1, U_2, U_3, U_4, U_5\}^T. \tag{10-7}$$

This vector is called *the global displacement vector*. Correspondingly, we can also define *the global load vector*, **F**, as follows, which represents the equivalent loads on the nodes due to external forces including body forces, surface stresses and concentrated forces,

$$\mathbf{F} = \{F_1, F_2, F_3, F_4, F_5\}^T, \tag{10-8}$$

where F_i (i = 1, 2, 3, 4, 5) is called the *nodal load* or *nodal force* on node i.

The above notations are defined on the whole bar with x as the reference coordinate system and that is why we use the term 'global'. On the other hand, we need to discuss stress, strain and displacement within an individual element locally. When these are clear in an element, we should then connect them with the global quantities. It is therefore necessary to have a table of *element connectivity* to show the local-global correspondence of nodes. Keep in mind that in the present case each element has two nodes locally, thus we have the numbering correspondence of **Table 10.1**.

Table 10.1 Connectivity table of local and global node numbers

Element	Node		
ⓔ	1	2	Local
①	1	2	Global
②	2	3	
③	3	4	
④	4	5	

The concepts of degree of freedom, nodal displacements, nodal loads, and element connectivity are central to the finite element method and need to be understood clearly.

(3) *Coordinates and Shape Functions*

Now consider a typical element ⓔ. In the local numbering scheme, the first node is always 1 and the second is always 2, because each element has only two nodes in the present simple case. Let us use x_1 to represent 'the global x-coordinate of node 1' and x_2 to stand for 'the global x-coordinate of node 2'.

On the other hand, we define a local coordinate system, denoted by ξ, as

$$\xi = 2\frac{x - x_1}{x_2 - x_1} - 1. \tag{10-9}$$

In this way, we shall always have $\xi = -1$ at node 1 ($x = x_1$) and $\xi = +1$ at node 2 ($x = x_2$). The length of the element is covered when ξ changes from -1 to $+1$.

Assume that the displacement variation inside an element can be expressed by a linear function of ξ. (Note that we are introducing a third type of approximation error here by assuming a 'known' expression of displacement.) If the nodal displacements under the local coordinate system of element ⓔ are u_1 and u_2, respectively, for the local node 1 and node 2, the linear displacement within the element ⓔ can be written as

$$u = N_1 u_1 + N_2 u_2, \tag{10-10}$$

where

$$N_1(\xi) = \frac{1}{2}(1-\xi), \quad N_2 = \frac{1}{2}(1+\xi) \tag{10-11}$$

are called *element shape functions*. In a matrix form, Eq. (10-10) can be written as

$$u = \mathbf{N}\mathbf{u}_e, \tag{10-12}$$

where

$$\mathbf{N} = \{N_1, N_2\} \tag{10-13}$$

and

$$\mathbf{u}_e = \{u_1, u_2\}^T. \tag{10-14}$$

Vector $\mathbf{u}_e$ is called *the element displacement vector*.

It may be noted that the previous transformation from the global coordinate x to the local coordinate ξ, Eq. (10-9), can also be written in terms of the shape functions as

$$x = N_1 x_1 + N_2 x_2. \tag{10-15}$$

Compared with the displacement approximation, Eq. (10-10), we see that both the displacement, u, and the coordinate, x, are transformed within the element by the same shape functions, N_1 and N_2. Because of this, the above derivation is called *isoparametric formulation* in the finite element literature.

In the above formulation, we use linear shape functions. However, other choices of shape functions are possible provided that they satisfy the conditions of *compatibility* and *completeness*. The details of these conditions can be found in most textbooks on the finite element method.[10.1] It has been shown theoretically that when the number of elements approaches infinity, the approximate solution made by the shape function will approach the exact solution if the shape function satisfies the conditions of completeness and compatibility.

(4) Strain and Stress in Terms of Nodal Displacements

Using Eqs. (10-5), (10-9) to (10-11), we can easily express the strain in an element in terms of the element nodal displacements, q_1 and q_2. As shown above, displacement u in an element is defined by Eq. (10-10) and therefore is also a function of the local coordinate ξ. Thus Eq. (10-5) can be re-written as

$$\varepsilon = \frac{du}{dx} = \frac{du}{d\xi}\frac{d\xi}{dx}. \tag{a}$$

On the other hand, ξ is related to the global coordinate x by Eq. (10-9), which gives rise to

$$\frac{d\xi}{dx} = \frac{2}{x_2 - x_1}. \tag{b}$$

According to Eq. (10-10), we have

$$\frac{du}{d\xi} = \frac{1}{2}(-u_1 + u_2). \tag{c}$$

Thus the substitution of Eqs. (b) and (c) into Eq. (a) yields

---

[10.1] Simply speaking, the compatibility of a shape function requires that displacements specified by the function must be continuous across the element boundary while the completeness demands that the first derivatives of the shape function are finite within an element. Further details can be found in the book by Bathe (1982).

$$\varepsilon = \frac{1}{x_2 - x_1}(-u_1 + u_2), \tag{d}$$

or in its matrix form, similar to Eq. (10-12),

$$\varepsilon = \mathbf{B}\,\mathbf{u}_e \tag{10-16}$$

where

$$\mathbf{B} = \frac{1}{x_2 - x_1}\{-1,\ 1\}. \tag{10-17}$$

The stress-strain relationship, Eq. (10-4), can also be written in its matrix form as

$$\sigma = E\varepsilon = E\,\mathbf{B}\,\mathbf{q}_e. \tag{10-18}$$

In summary, when assuming that the displacement within an element follows a linear distribution, Eq. (10-12), we obtained the element strain, Eq. (10-16), by using the strain-displacement relationship, and the element stress, Eq. (10-18), by using the stress-strain relationship. All of them are now the explicit functions of the element displacement vector, $\mathbf{u}_e$. Since we are following the displacement approach, the compatibility equation within the element must have been satisfied automatically. Hence, the one left to examine is the equilibrium equation.

As mentioned previously, the satisfaction of the equilibrium equation is equivalent to that of Eq. (10-6) from the principle of minimum potential energy. In the following, we will calculate the potential energy of the bar by using the displacement, strain and stress obtained above.

(5) Element Stiffness Matrix

The potential energy for the present one-dimensional problem can be written as[10.2]

$$V = \frac{1}{2}\int_L \sigma\varepsilon A\,dx - \int_L uf A\,dx - \int_L u\tau\,dx - \sum_{i=1}^{2} P_i u_i, \tag{10-19}$$

[10.2] See Appendix C for the details of obtaining Eq. (10-19). The fundamentals of the energy method are addressed in many texts. Further details can be found in the books by Dym and Shames (1973), Washizu (1975) and Hu (1984). For more formulations and programming techniques of the finite element method, the reader may consult the books by Hinton (1977), Bathe (1982), Chandrupatla and Belegundu (1991) and Zienkiewicz and Taylor (1991).

where L is the total length of the bar, $A = A(x)$ is the cross-sectional area of the bar and u_i ($i = 1, 2$) is the displacement at the application points of the concentrated load, P_i. Since the bar has been divided into 4 elements, the potential energy is the summation of those individual elements. Thus Eq. (10-19) becomes

$$V = \sum_{e=1}^{4}\left\{\frac{1}{2}\int_{L_e}\sigma\varepsilon A dx\right\} - \sum_{e=1}^{4}\left\{\int_{L_e} ufA dx\right\} - \sum_{e=1}^{4}\left\{\int_{L_e} u\tau dx\right\} - (P_1 U_2 + P_2 U_4), \quad (10\text{-}20)$$

where L_e is the length of element ⓔ ($e = 1, 2, 3, 4$) and U_2 and U_4 are the global nodal displacements at the application points of the concentrated loads P_1 and P_2, as shown in **Fig. 10.6** and Eq. (10-7).

The first term of Eq. (10-20) represents the total strain energy, W, in the bar and that in the brackets of the term is the strain energy of element ⓔ, W_e. Using Eqs. (10-16) and (10-18), we have

$$W_e = \frac{1}{2}\int_{L_e}\sigma\varepsilon A dx = \frac{1}{2}\int_{L_e} \mathbf{u}_e^T \mathbf{B}^T \mathbf{E} \mathbf{B} \mathbf{u}_e A dx = \frac{1}{2}\mathbf{u}_e^T \mathbf{K}_e \mathbf{u}_e, \quad (10\text{-}21)$$

where

$$\mathbf{K}_e = \frac{EA_e}{L_e}\begin{pmatrix} 1 & -1 \\ -1 & 1 \end{pmatrix} \quad (10\text{-}22)$$

is called *the element stiffness matrix*. In Eq. (10-22), A_e is the cross-sectional area of element ⓔ and is a constant within the element as we assumed at the beginning. Thus the total strain energy becomes

$$W = \sum_{e=1}^{4}\left\{\frac{1}{2}\int_{L_e}\sigma\varepsilon A dx\right\} = \sum_{e=1}^{4}\left(\frac{1}{2}\mathbf{u}_e^T \mathbf{K}_e \mathbf{u}_e\right), \quad (10\text{-}23)$$

which is now related to the element nodal displacements explicitly.

Similar to the above procedures, we can also correlate the other terms in the total potential energy, Eq. (10-20), with the element nodal displacements. Using Eq. (10-12), the second term, *i.e.*, the body force term, becomes

$$\sum_{e=1}^{4}\left\{\int_{L_e} ufAdx\right\} = \sum_{e=1}^{4} u_e^T f_e, \qquad (10\text{-}24)$$

where

$$f_e = \frac{A_e L_e \rho g}{2}\begin{pmatrix}1\\1\end{pmatrix} \qquad (10\text{-}25)$$

is called the *element body force vector*. The third term of Eq. (10-20) can also be similarly re-written as

$$\sum_{e=1}^{4}\left\{\int_{L_e} u\tau dx\right\} = \sum_{e=1}^{4} u_e^T t_e, \qquad (10\text{-}26)$$

where

$$t_e = \frac{T_e L_e}{2}\begin{pmatrix}1\\1\end{pmatrix} \qquad (10\text{-}27)$$

is called the *element surface force vector*. In Eq. (10-27), T_e is the average value of τ in element ⓔ.

(6) Assembly of the Global Stiffness Matrix and Load Vector

The total potential energy is now in the form of

$$V = \sum_{e=1}^{4}\left(\frac{1}{2} u_e^T K_e u_e\right) - \sum_{e=1}^{4} u_e^T f_e - \sum_{e=1}^{4} u_e^T t_e - (P_1 U_2 + P_2 U_4). \qquad (10\text{-}28)$$

Note that the local nodal displacement vector, u_e, and the global nodal displacement vector U, as given in Eq. (10-7), are related by the element connectivity shown in **Table 10.1**. For instance, for element ① we have $u_① = \{U_1, U_2\}^T$ but for element ② we have $u_② = \{U_2, U_3\}^T$. Hence, by taking element connectivity into account, the total potential energy can be expressed as the global displacement vector, that is,

$$V = \frac{1}{2}\mathbf{U}^T\mathbf{K}\mathbf{U} - \mathbf{U}^T\mathbf{F}, \qquad (10\text{-}29)$$

where

$$\mathbf{K} = E \begin{pmatrix} \dfrac{A_1}{L_1} & -\dfrac{A_1}{L_1} & 0 & 0 & 0 \\ -\dfrac{A_1}{L_1} & \left(\dfrac{A_1}{L_1} + \dfrac{A_2}{L_2}\right) & -\dfrac{A_2}{L_2} & 0 & 0 \\ 0 & -\dfrac{A_2}{L_2} & \left(\dfrac{A_2}{L_2} + \dfrac{A_3}{L_3}\right) & -\dfrac{A_3}{L_3} & 0 \\ 0 & 0 & -\dfrac{A_3}{L_3} & \left(\dfrac{A_3}{L_3} + \dfrac{A_4}{L_4}\right) & -\dfrac{A_4}{L_4} \\ 0 & 0 & 0 & -\dfrac{A_4}{L_4} & \dfrac{A_4}{L_4} \end{pmatrix}$$
(10-30)

is the *global stiffness matrix*,[10.3] obtained by adding up all the element stiffness vectors $\mathbf{K}_e$ (e = 1, 2, 3, 4) shown in Eq. (10-22). The vector

$$\mathbf{F} = \begin{Bmatrix} \dfrac{A_1 L_1 \rho g}{2} + \dfrac{L_1 T_1}{2} \\ \left(\dfrac{A_1 L_1 \rho g}{2} + \dfrac{L_1 T_1}{2}\right) + \left(\dfrac{A_2 L_2 \rho g}{2} + \dfrac{L_2 T_2}{2}\right) + P_1 \\ \left(\dfrac{A_2 L_2 \rho g}{2} + \dfrac{L_2 T_2}{2}\right) + \left(\dfrac{A_3 L_3 \rho g}{2} + \dfrac{L_3 T_3}{2}\right) \\ \left(\dfrac{A_3 L_3 \rho g}{2} + \dfrac{L_3 T_3}{2}\right) + \left(\dfrac{A_4 L_4 \rho g}{2} + \dfrac{L_4 T_4}{2}\right) + P_2 \\ \dfrac{A_4 L_4 \rho g}{2} + \dfrac{L_4 T_4}{2} \end{Bmatrix} \qquad (10\text{-}31)$$

in Eq. (10-29) is called the *global nodal force* vector defined by Eq. (10-8), which is the summation of all the element body force vectors, Eq. (10-25), element surface force vectors, Eq. (10-27), and the external concentrated forces, P_1 and P_2.

[10.3] It is worthwhile to note that many components of the global stiffness matrix are zero and the matrix is tridiagonal. This is actually an important property of the stiffness matrix generated by FEM. The zero components do not need to be stored and can save a lot of memory space in computation. Many introductory books about FEM programming are available (*e.g.*, Hinton, 1977).

(7) Satisfaction of the Equilibrium Equation and Boundary Conditions

We have used all the basic equations except the equilibrium equation specified by the conditions of minimum potential energy, Eq. (10-6). Now, because the first global nodal displacement $U_1 = 0$, which is specified by the displacement boundary condition of Eq. (10-3), Eq. (10-6) becomes

$$\frac{\partial V}{\partial U_i} = 0, \quad (i = 2, 3, 4, 5). \tag{10-32}$$

The stress boundary conditions have been included in the force vector **F**. The displacement boundary condition, $U_1 = 0$, should also be considered in the calculation of V in Eq. (10-29). The substitution of Eq. (10-29) into Eq. (10-32) leads to

$$\tilde{\mathbf{K}}\tilde{\mathbf{U}} = \tilde{\mathbf{F}}, \tag{10-33}$$

where $\tilde{\mathbf{U}} = \{U_2, U_3, U_4, U_5\}^T$, $\tilde{\mathbf{F}} = \{F_2, F_3, F_4, F_5\}^T$ and $\tilde{\mathbf{K}}$ is the sub-matrix of **K** in Eq. (10-30) with its first row and first column deleted, because $U_1 = 0$ is known.

(8) The Finite Element Solution

After solving Eq. (10-33) for $\tilde{\mathbf{U}} = \{U_2, U_3, U_4, U_5\}^T$ numerically, we can easily obtain stress by Eq. (10-18), strain by Eq. (10-16) and displacement by Eq. (10-12).

10.4 USUAL PROCEDURES OF THE FINITE ELEMENT SOLUTION

By summarising the above, we can see that a finite element solution, when using the displacement approach, follows the procedures below:

(1) Modelling. This is to generate a mechanics model based on the original engineering problem for finite element analysis. Simplifications and boundary conditions must be described clearly. It is one of the most important steps to model the problem correctly.

(2) Element Division. Based on the mechanics model obtained, the continuous body is divided into a number of discrete elements which connect with each other via common nodes. The use of the type(s) and number of elements must be considered carefully. In most commercial FEM codes nowadays, the element division can be done automatically when the type and number of elements are specified by the users. However, automation does not mean optimisation. In this step, two types of approximation errors are introduced, *i.e.*, the geometrical discretization error and the piecewise approximation error of mechanics quantities.

(3) **Numbering.** With the element division, nodes and elements must be numbered in order to allow further analysis. Commercial codes can do the numbering automatically. Steps (1) to (3) are often called *pre-processing*.

(4) **Element Analysis.** This is to form an element stiffness matrix, *etc*. Here, the displacement distribution in the element is assumed. Thus another approximation error is introduced as emphasised previously.

(5) **Global Analysis.** This is to form the global stiffness matrix and global load vector, including the treatment of any displacement boundary conditions.

(6) **Solution of the Finite Element Equation.** Numerical solutions will be performed by the codes using pre-set numerical solvers. Now, the assumed displacement field has satisfied all the basic equations. All the stress and displacement boundary conditions have also been incorporated into Eq. (10-33) during the global assembly. Thus when the nodal displacements, U_i, are solved, they give the correct solution. Here, we assume that the approximation errors introduced are acceptable.

Steps (4) to (6) are carried out by FEM codes automatically.

(7) **Analysis of Results.** With the solution achieved, we need to analyse the results for our particular purpose. This includes (a) to understand and explain the way of variation of stresses, strains and displacements, and (b) to draw conclusions or guidelines for improving our design, manufacturing techniques and so on. If any of the results are not reasonable, we must go back to investigate our model to identify possible errors in the pre-processing stage. This process is called *post-processing* in a finite element analysis.

In the above, we only demonstrated the displacement approach to the finite element formulation of linear elastic problems when the material obeys the generalised Hooke's law. When plasticity happens, the derivation will become more complicated, because stress-strain relationships, as discussed in Chapter 9, are non-linear. It is also possible to follow other approaches, such as the stress approach. The reader can consult more advanced books for details.

As an example in practical application, **Fig. 10.7** shows the general procedure of finite element analysis in the metal forming process. We can see clearly that for such a complex system involving non-linear interaction among dies, punches, material properties, and so on, it is impossible to get any analytical solution using the method discussed in previous chapters. Thus the application of the FEM is of great importance. However, it must be emphasised again that the finite element method is only a numerical tool to avoid the mathematical difficulties of solving the basic equations with given boundary conditions.[10.4] For an engineer seeking a solution to a practical problem, the most important steps are modelling and interpretation of results, which rely on the knowledge of mechanics theory established in the last nine chapters. In section **10.6**, we shall discuss the solution of a complicated engineering problem to experience the whole process of practical modelling when using the finite element method.

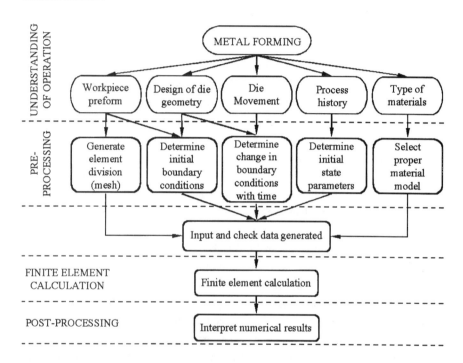

Figure 10.7 The general procedure of the finite element method in metal forming.

10.5 SOME CONSIDERATIONS IN FINITE ELEMENT MODELLING

All the considerations of mechanics modelling discussed in Chapter 6 are important for the solution of engineering problems using the finite element method, including the correct description of boundary conditions, symmetry of deformation and the understanding of special deformation characteristics of a problem, such as axisymmetric and plane deformation. In addition, we have understood from the engineering examples in previous chapters that concentrated forces, contact stresses

---

[10.4] When a problem to solve is time dependent, initial conditions at time equal to zero should also be specified.

and small holes and fillets in a components will normally cause stress concentration and must be taken into account in finite element modelling.

Figs. 10.8 and **10.9** illustrate some examples that involve the above considerations. The component in **Fig. 10.8** is a thin plate specimen for simple tensile testing. We easily realise that it is a plane-stress deformation problem and thus a plane element can be used. Furthermore, we see that when the specimen is elongated in the x-direction uniformly, its deformation is symmetrical to both x- and y-axes. Hence, only a quarter of the specimen needs to be analysed to reduce computational cost. (The control volume taken is therefore a quarter of the specimen.) Owing to deformation symmetry, the material points on the x-axis do not move in the y-direction and those on the y-axis do not move in the x-direction. Thus correspondingly displacement conditions must be applied onto the nodes on the x- and y-axes, as shown in **Fig. 10.8** and serve as the displacement boundary conditions of the model. The ends of the specimen are subjected to a uniform tensile stress σ. In the model, such stress must be applied onto each node as well. Moreover, we note that there are concave fillets of small radius. In the vicinity of the fillets there may exist a stress concentration. Thus the size of elements around the fillet must be much smaller. However, whether the element division used is fine enough or too fine depends on the accuracy required.

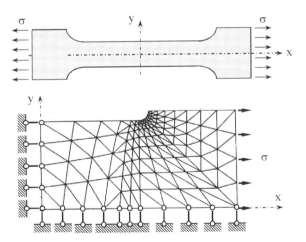

Figure 10.8 The use of deformation symmetry in finite element modelling. Note also the application of displacement and stress conditions onto the boundary nodes of the control volume that is only a quarter of the original specimen dimension.

In modern finite element codes commercially available, an adaptive approach to meshing is often used to automatically adjust the mesh size based on certain error criteria programmed. For example, in studying the deformation of a plate subjected to extrusion, as illustrated in **Fig. 10.9**, we can start with a coarse mesh shown in **Fig. 10.9a** in the finite element calculation.[10.5] Based on the error criteria embedded in the program and the computational accuracy specified, a re-meshing can be carried out by the software automatically. At the exit corner of the workpiece, high stress concentration requires much finer elements in the local zone. **Fig. 10.9b** is an automatically refined mesh generated during deformation analysis by a finite element software, which satisfies the accuracy requirement. However, it is worthwhile to mention that the adaptive re-meshing takes much longer CPU time.

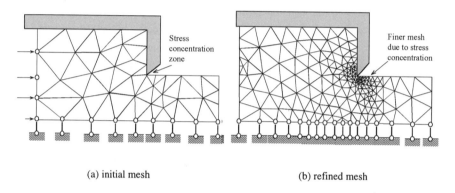

(a) initial mesh   (b) refined mesh

Figure 10.9 Automatic re-meshing of elements by a finite element software.

Example 10.1 A thin rectangular plate clamped firmly at its two ends, as shown in **Fig. E10.1** is subjected to a uniform pressure on its top edge. Provide its mechanics model for finite element analysis. Twelve rectangular elements (each element has four nodes) are to be used to divide the control volume uniformly.

Solution: Since the loading and boundary conditions are symmetrical to the y-axis, we can take only half of the plate as the control volume for the finite element analysis. Along the y-axis, material points in the plate must not move horizontally, *i.e.*, no displacement u will happen at any points on the y-axis. The finite element model of this problem is shown in **Fig. E10.2**.

[10.5] Note that the control volume is only a quarter of the original dimension of the workpiece. Equivalent displacement and stress conditions have been applied onto the boundary nodes.

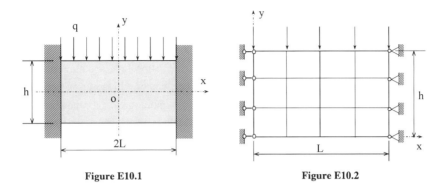

| Figure E10.1 | Figure E10.2 |

Example 10.2 A square plate is subjected to a pair of concentrated forces, P = 20 kN/m, along one of its diagonals, as shown in **Fig. E10.3**. Describe the mechanics model for finite element analysis. Use four triangular elements, each with three nodes, to divide the control volume.

Solution: The deformation symmetry of the plate is obvious, that is, the material points in the plate on the diagonals will displace only along the diagonals. The four triangular parts of the plate divided by the diagonals will have identical deformation status. Hence, only a quarter of the plate needs to be considered in the finite element analysis. The finite element model is given in **Fig. E10.4**.

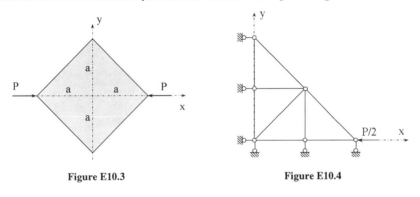

| Figure E10.3 | Figure E10.4 |

10.6 A CASE STUDY

Now let us take a relatively complex engineering problem as an example to see how to use the modelling skills developed in previous chapters and to use the finite element method to obtain a solution with sufficient accuracy.

10.6.1 Background of the Problem

Grinding has been a major manufacturing process for many decades and has been one of the most versatile methods of machining for providing precision components

(Andrew *et al.*, 1985; Malkin, 1989; Shaw, 1996). Traditionally, the achievements required of a grinding operation are low surface roughness and high dimensional accuracy. Recently, however, rapid developments in various areas of high technology, such as space, optics, atomic power, micro-machine and electronics, put further demands on the quality of a ground component. In addition to surface roughness and dimensional accuracy, a high-technological component needs guarantees such as optimal distributions of residual stresses and microstructures of the surface material in the ground subsurface layer. Thus investigations into the relationship between grinding conditions and the resultant deformation in a ground component, with the aid of mechanics analysis, are necessary.

There are many different types of grinding operations, such as surface grinding, cylindrical grinding, centreless grinding and so on. In the following, we shall only discuss a mechanics solution to the residual stresses in a metal workpiece generated by surface grinding. **Figure 10.10** shows a CNC surface grinder with a dynamometer installed for measuring three-dimensional grinding forces for mechanics analysis.

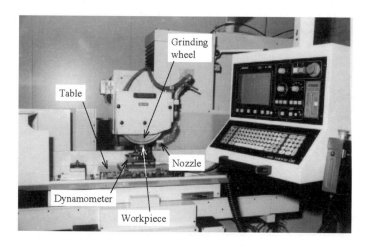

Figure 10.10 A CNC surface grinder.

10.6.2 Modelling

The schematic illustration of a surface grinding operation has been demonstrated in **Fig. 3.2d**, where we can see that the grinding wheel is rotating with a uniform angular velocity ω, driven by the spindle of the grinder, and the workpiece is moving with the machine table at a velocity V. When a depth of cut, d, is set, as shown in **Fig. 10.11**, the material on the workpiece surface will be removed due to the cutting of the grinding wheel.

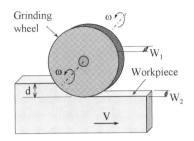

Figure 10.11 Schematic of a surface grinding operation.

Figure 10.12 Some grinding wheels.

At first glance, it seems to be straightforward to generate a mechanics model of the workpiece for stress and deformation analysis, because it looks like a contact problem of a circular disk (the wheel as illustrated in **Fig. 10.12**) with the workpiece. However, when we examine the details, we realise that it is an extremely complex system. For example, a grinding wheel consists of abrasives, such as alumina, cubic boron nitride (CBN) and diamond abrasives, held together by a suitable agent, called *bond*, such as resin and metal. The geometry of an abrasive is irregular and the distribution of abrasives in the grinding wheel is random, as shown in **Fig. 10.13**. Thus the grinding wheel as a whole is anisotropic and inhomogeneous, and its mechanical properties cannot be described by the theories studied so far. In addition, because the geometry of individual abrasives is different, the cutting behaviour of individual cutting edges, *e.g.*, due to the different rake angle α shown in **Fig. 10.14**, is different. Hence, if we try to model every detail of the wheel-workpiece interaction, it will be unfeasible.

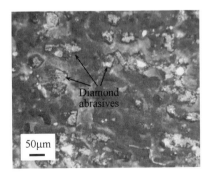

(a) the surface of a wheel with diamond abrasives   (b) the surface of a wheel with CBN abrasives

Figure 10.13 The microstructure of grinding wheels.

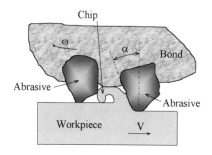
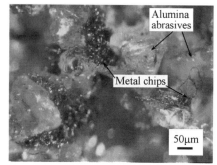

Figure 10.14 The micro-cutting of individual abrasives.

Figure 10.15 The chips due to micro-cutting that are filling the gaps among abrasives on a wheel surface.

Since our objective is to understand the stress and deformation of the workpiece during grinding, the case will become simpler if we only consider the workpiece itself in our mechanics model while treating the functions of the grinding wheel, such as the grinding forces on the workpiece surface, as externally applied boundary conditions. In this way, if we can describe such boundary conditions clearly, we can avoid the involvement of the anisotropy and inhomogeneity of the wheel material in the mechanics analysis.

In doing so, we are facing the problem of how to reasonably describe the boundary conditions of the workpiece. As we understood above, because the shapes of abrasives are irregular, their spacing over the wheel surface is random and the depths of cut of individual abrasives are different from each other, thus the stress distribution in the wheel-workpiece interaction zone could be irregular and discontinuous. However, by observing a wheel surface after grinding, see **Fig. 10.15** (also refer to **Fig. 10.14**), we can imagine that the bond material of the wheel and the chips in the gaps among the abrasives can also be in contact with the workpiece in grinding and because of this the interaction stress distribution can be more continuous. An experimental observation on a particular case (Okamura *et al.* 1984), as shown in **Fig. 10.16**, does support the consideration. Thus, approximately, the stress distribution in the wheel-workpiece interface can be assumed to be continuous. In addition, to further simplify the distribution of the interface stress to facilitate our mechanics analysis, it is reasonable to assume, as the experimental observation in **Fig. 10.16** indicates, that the stress follows a general triangular distribution over the wheel-workpiece contact area along the grinding direction but is uniform across the wheel thickness, as shown in **Fig. 10.17**, where W_1 is the wheel width. The base length of the triangle equals the contact length of the wheel with the workpiece, L_c, which is determined by a simple geometrical relation, $L_c = (Dd)^{1/2}$, when the deformation of the grinding wheel is ignored, where D is the diameter of the grinding wheel and d is the depth of cut as illustrated in **Fig. 10.11**. Furthermore, in a precision grinding, d is often much smaller than the dimensions of both the workpiece and wheel and thus in the mechanics model we can consider that the workpiece surface is flat and smooth. The height of the triangle, p_0, is determined by equating the resultant grinding force with the volume covered by the stress distribution. Because the workpiece moves at speed V, thus in the mechanics modelling, when we consider that the workpiece is static, the triangular surface stress moves with speed V along the grinding direction, as shown in **Fig. 10.17**. The apex position of the moving surface stress, L_0, is determined by experiment to suit grinding with different conditions. The normal and tangential components of the moving stress can also be determined experimentally because the resultant normal and tangential forces of grinding can be measured easily by embedding a dynamometer. Here, let us assume that the ratio of the resultant tangential force to the resultant normal force measured is µ.

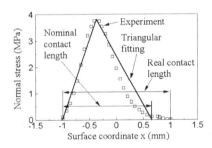

Figure 10.16 The stress distribution over the wheel-workpiece interface along the grinding direction. Note that the distribution can be reasonably approximated by a triangle to facilitate the modelling. In the experiment, the grinding wheel is with alumina abrasives (WA80L9V) and the workpiece is steel SK3 (Okamura *et al.*, 1984).

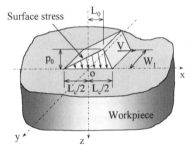

Figure 10.17 The mechanics model for a workpiece subjected to surface grinding. The function of the grinding wheel has been replaced by a triangular surface stress moving along the grinding direction with the table speed of the grinder, V. L_c is the length of the wheel-workpiece contact interface.

Fig. 10.17 is a mechanics model that can be used for analysing three-dimensional stress and deformation in a workpiece. However, it can be simplified further when the following features exist. If the thickness of a workpiece, *i.e.*, W_2 in **Fig. 10.11**, is very small, the case can be treated as a plane-stress problem, as we defined in Chapter 6. Contrary to this, if the thicknesses of the wheel and workpiece are much larger than the contact length L_c, then the deformation of the workpiece can be approximated by a plane-strain problem. Under these special conditions, we only need to solve a plane deformation model shown in **Fig. 10.18**, where 2L is the length of the workpiece in the grinding direction and h is its height above the fixture.

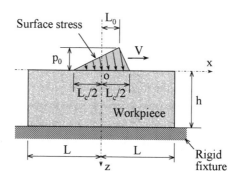

Figure 10.18 A simplified mechanics model for a workpiece under plane-stress or plane-strain grinding.

10.6.3 Solution

As an example, let us consider the solution to surface grinding under the simplest circumstance described by the mechanics model of plane-strain deformation in **Fig. 10.18**. Assume that at time $t = 0$, the base centre of the surface stress is at the origin of the coordinate system O. Thus at instant $t = t^*$, the boundary conditions can be written as

at $z = 0$, $-L \leq x \leq -L_c/2 + Vt^*$ and $L_c/2 + Vt^* \leq x \leq L$,
$$\sigma_{zz} = \sigma_{zx} = 0;$$
at $z = 0$, $-L_c/2 + Vt^* \leq x \leq L_0 + Vt^*$,
$$\sigma_{zz} = - p_0(2x + L_c - 2Vt^*)/(2L_0 + L_c - 2Vt^*),$$
$$\sigma_{zx} = \mu p_0(2x + L_c - 2Vt^*)/(2L_0 + L_c - 2Vt^*);$$
at $z = 0$, $L_0 + Vt^* \leq x \leq L_c/2 + Vt^*$,
$$\sigma_{zz} = - p_0(2x - L_c - 2Vt^*)/(2L_0 - L_c - 2Vt^*),$$
$$\sigma_{zx} = \mu p_0(2x - L_c - 2Vt^*)/(2L_0 - L_c - 2Vt^*);$$
at $x = \pm L$, $0 \leq z \leq h$,
$$\sigma_{xx} = \sigma_{xz} = 0;$$
at $z = h$, $-L \leq x \leq L$,
$$u = w = 0.$$

With such complex moving boundary conditions, an easier way[10.6] is to use the finite element method to obtain a numerical solution. Clearly, the loading condition is without any symmetry. Thus in the selection of control volume we cannot reduce its dimension by deformation symmetry. However, we understand that the deformation in a workpiece caused by grinding is very localised in the neighbourhood of the wheel-workpiece interaction zone. According to Saint-Venant's principle discussed in Chapter 6, the localised deformation will have a negligible effect on the stresses, strains and displacements in the area far away from the interaction zone. It is therefore possible to reduce the dimension of the control volume without affecting the solution accuracy in the vicinity of the interaction zone. Based on some trial and error tests, we can find that an economic yet sufficiently accurate control volume is with the dimension of $(L^*+L') \times h^* = 16L_c \times 6L_c$ as shown in **Fig. 10.19**. Of course, if the original dimension of the workpiece $2L \times h$ is smaller than $(L^*+L') \times h^*$, then the size of the control volume must be $2L \times h$.

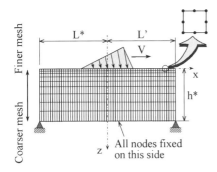

Figure 10.19 The finite element model.

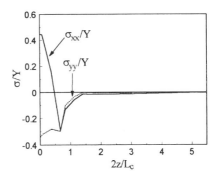

Figure 10.20 Major residual stresses in a workpiece.

The other problem related to the accuracy of the numerical solution is the division of the finite element mesh. An error analysis by increasing or decreasing the element size must therefore be carried out. Again, because the deformation in the specimen is localised in the surface area, we must use finer mesh there to accommodate the sharp variation of stresses and strains. The general configuration of the finite element mesh is shown in **Fig. 10.19** that demonstrates the overall transition of the elements from

---

[10.6] It is possible to get an analytical solution for an elastic deformation problem with the moving surface stress boundary, see Ling (1973). However, grinding involves significant plastic deformation. Thus the only feasible way is to use a numerical method such as the finite element method presented here.

fine to coarse. The result of an error analysis[10.7] indicates that 448 eight-node elements with 64 surface elements[10.8] will lead to a solution with an acceptable accuracy less than 5%. As we have understood in the finite element formulation above, boundary stresses must be converted to nodal forces that are applied to corresponding surface nodes. In the present case, the movement of the grinding stress on the surface must also be considered in element division to allow a nodal force to move from one node to another at the given velocity V.

10.6.4 Results and Analysis

With the above finite element model, we used a commercially available finite element software, ADINA, to study the residual stresses in an EN23 workpiece (plane-strain). The material is considered to be elastic/perfectly plastic and its properties are yield stress Y = 796 MPa, Young's modulus E = 214 GPa, Poisson's ratio v = 0.27. The grinding conditions are μ = 0.4, p_0 = 2.25Y, L_c = 2 mm and L_0 = 0.125L_c. The finite element analysis shows that the residual shear stress, σ_{xy}, is very small compared with the normal residual stresses, σ_{xx} and σ_{yy}. **Figure 10.20** presents the distributions of the two major residual stresses in the depth direction of the workpiece, *i.e.*, in the z-direction. (As indicated by the mechanics model of **Fig. 10.18**, the workpiece surface is at z = 0.) We can see that σ_{xx} in the neighbourhood of the workpiece surface is tensile and has a large magnitude. This is certainly undesirable in terms of the reliability of the ground component, because a large tensile residual stress may activate a surface flaw, such as a scratch or crack, and make it propagate. The large variation gradient of the residual stresses may also introduce surface distortion and thus degrade the precision of the surface profile. Thus the grinding conditions used are not very appropriate.

In the above, we only considered the mechanical stress on the workpiece surface by the grinding wheel. In a practical grinding process, material removal by the

---

[10.7] There are several ways of carrying out an error analysis. The most direct way is to compare the numerical result with the analytical solution of a similar problem. In the present case, as mentioned previously, we can get an analytical solution in the regime of elastic deformation. A comparison can then guide the division of elements. Alternatively, we can conduct a sequential calculation with different element sizes, for instance, to perform the calculations with 32, 64, 128 and 256 surface elements and compare the difference in stresses obtained. When the refinement of elements does not make considerable changes of stresses, the element division can be considered appropriate. In the present case, the relative error in stress components obtained by 64, 128 and 256 surface elements is within 5%. Thus the mesh with 64 surface elements can be regarded as an economic configuration with sufficient accuracy. Finally, if the finite element software used has the capacity of an adaptive meshing function, as mentioned previously, we can request the software to generate an acceptable element division by specifying an acceptable error.

grinding wheel will also generate significant heat and that is why coolant is often supplied in a grinding production. In many cases, the heat generated will cause thermal deformation or even the material's phase transformation. The deformations induced by mechanical loading, thermal heating and phase transformation, when all appear, will couple together. In this case, in addition to the moving surface stress modelled in **Fig. 10.18**, a moving heat source must be incorporated. Details can be found in the work by Zhang and co-workers carried out from 1992 to 1999. However, the principle and process of modelling is the same as discussed in the above.

10.6.5 Summary

The above discussion outlines a general process of mechanics modelling, solution and analysis of results in solving an engineering problem. We have seen that understanding the characteristics of the original problem is the most critical step. Otherwise, the model generated may not represent the problem. Experimental observations and measurements also play an important role in the understanding. For example, considerations concerning the grinding wheel and its interaction with the workpiece and the measurement of the distribution of interaction stress are key to the determination of the model of **Fig. 10.18**. After that, the simplification to a triangular moving surface stress, the application of Saint-Venant's principle and the consideration of plane-stress/strain all contribute to the efficient analysis of the model. The analysis of results is important because only a reasonable mechanics interpretation of the results can provide a useful guideline for improving the design of the grinding process.

Throughout the solution, we can see that a deep understanding of the principle, concepts and solution skills is central to optimise an engineering solution. Numerical methods are only tools to help overcome mathematical difficulties.

References

Andrew, C, Howes, TD and Pearce, TRA (1985), *Creep-Feed Grinding*, Industrial Press Inc., New York.

Bathe, K-J (1982), *Finite Element Procedures in Engineering Analysis*, Prentice-Hall, Inc., Englewood Cliffs, NJ.

[10.8] Here surface elements indicate the elements allocated on the workpiece surface. Since we are using uniform element division along the workpiece surface in this example (see **Fig. 10.19**), the use of 64 surface elements means that we have 64 elements in a row horizontally (in the x-direction).

Chandrupatla, TR and Belegundu, AD (1991), *Introduction to Finite Elements in Engineering*, Prentice-Hall, Inc., Englewood Cliffs, NJ.

Dym, CL and Shames, JM (1973), *Solid Mechanics: A Variational Approach*, McGraw-Hill, New York.

Hinton, E (1977), *Finite Element Programming*, Academic Press, London.

Hu, H (1984), *Variational Principles of Theory of Elasticity with Applications*, Science Press, Beijing.

Ling, FF (1973), *Surface Mechanics*, Wiley, New York.

Mahdi, M and Zhang, L (1996), A theoretical investigation on the mechanically induced residual stresses due to surface grinding, in: *Progress of Cutting and Grinding, Vol.3*, edited by N Narutaki, D Chen, Y Yamane and A Ochi (Proceedings of *The 3rd International Conference on Progress Cutting & Grinding*, Osaka, Japan, 19-22 November 1996), Japan Society for Precision Engineering, pp.484-487.

Malkin S (1989), *Grinding Technology*, Ellis Horwood, Chichester.

Okamura, K *et al.* (1984), Interface condition between grinding wheel and workpiece in plounge grinding, *Proceedings of the International Conference on Grinding*, Society of Manufacturing Engineers, MR84-522, pp.1-10.

Shaw, MC (1996), *Principles of Abrasive Processing*, Claredon Press, Oxford.

Washizu, K (1975), *Variational Methods in Elasticity and Plasticity*, 2nd edition, Pergamon Press, Oxford.

Zhang, L, Suto, T, Noguchi, H and Waida, T (1992), An overview of applied mechanics in grinding, *Manufacturing Review*, **5**, 261-273.

Zhang, L *et al.* (1993, 1995, 1997, 1998, 1999), Applied mechanics in grinding, Part II to Part VII, *International Journal of Machine Tools and Manufacture*, **33**, 245-255; **33**, 587-597; **35**, 1397-1409; **37**, 619-633; **38**, 1289-1304; **39**, 1285-1298.

Zienkiewicz, OC and Taylor, RL (1991), *The Finite Element Method*, 4th Edition, Vols. 1 and 2, McGraw-Hill, London.

Important Concepts

Approximation errors
Control volume
Global numbering
Global displacement
Shape function
Element stiffness matrix
Element load vector
Assembly
Post-processing

Type of element
Mesh
Local numbering
Local displacement
Isoparametric formulation
Global stiffness matrix
Global load vector
Pre-processing

Questions

10.1 What is the finite element method?

10.2 What approximation errors are usually introduced in the formulation of a finite element analysis?

10.3 Why do we need to introduce the shape function?

10.4 What are the basic considerations in finite element modelling?

10.5 What are the usual procedures of a finite element analysis?

10.6 Why does a deep understanding based on solid mechanics play a central role in the pre-processing and post-processing of the finite element analysis of a problem?

10.7 Why do experimental observations and measurements play an important role in the mechanics modelling of a complex engineering problem?

10.8 What are the anticipated problems in the surface grinding modelling discussed in this chapter if (a) the wheel-workpiece is treated as a contact system, *i.e.*, the wheel is also included in the analysis, (b) the interaction stress between the wheel and workpiece follows an instantly measured distribution, (c) the profile of the interaction stress is not replaced by the triangular distribution, or (d) the exact shape at the wheel-workpiece interface is taken into consideration, *i.e.*, the workpiece is not treated as having a smooth and flat surface at $z = 0$?

Problems

10.1 Provide mechanics models for a finite element analysis of the thin plates shown in **Fig. P10.1**. Use either four-node rectangular elements or three-node triangular elements or a combination of them depending on the requirements. State the reasons for the selection of mesh and control volume and the setting of boundary conditions.

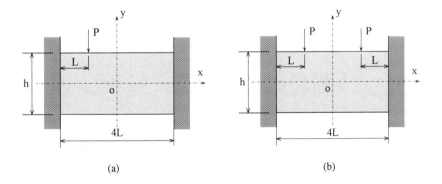

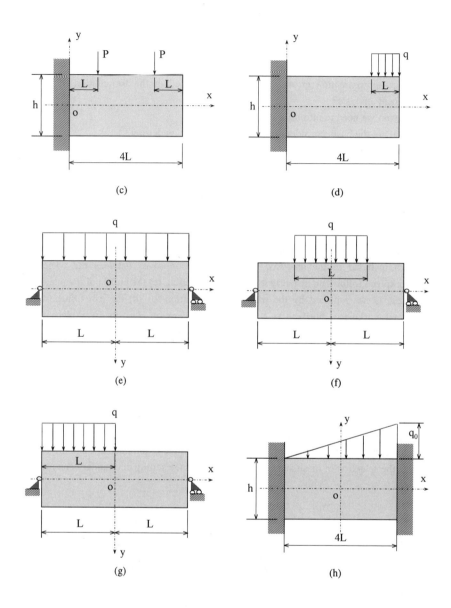

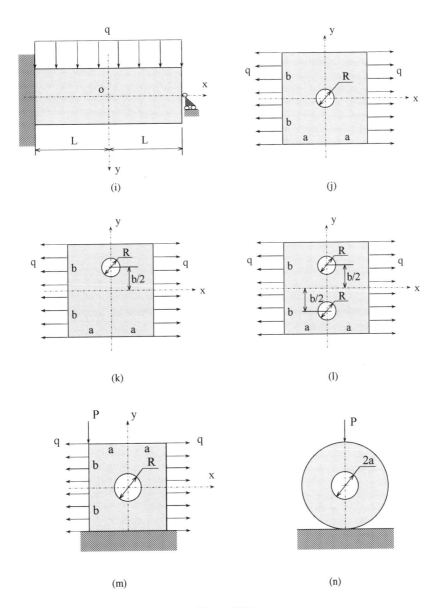

Figure P10.1

10.2 A thin rectangular plate is clamped at one end and subjected to a uniform tensile stress at the other end as shown in **Fig. P10.2**. Divide the plate by four bar elements. (The same element type of two nodes used in the example of section **10.3**.) Obtain the finite element solution of the bar in terms of stresses, strains and nodal displacements. Compare the finite element solution with the corresponding analytical solution. The bar material has the following properties: E = 210 GPa, ν = 0.3. The external stress q = 20 MPa and the dimensions of the bar are: L = 100 mm, h = 5 mm and t = 1 mm, where t is the bar thickness. (*Hint*: Follow the procedures and considerations in section **10.3** but use the dimensions and material properties given.)

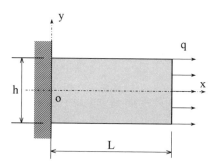

Figure P10.2

10.3 If the grinding problem discussed in this chapter must be treated as a three-dimensional problem, *i.e.*, neither plane-stress nor plane-strain, (a) describe the boundary conditions, (b) specify a corresponding finite element model, (c) work out a reasonable finite element mesh, and (d) explain the process of achieving an economic yet accurate enough control volume.

Appendix A

MOHR'S CIRCLE

A useful graphical method of analysing the state of stress was first proposed by a German engineer, Otto Mohr (1835-1918), in his papers in 1882 and 1900. Here we will introduce the method for representing two-dimensional stress states, which is most useful. A detailed discussion on the Mohr's circle for three-dimensional stress states can be found in many books such as those by Westergaard (1952), Sokolnikoff (1956) and Chakrabarty (1987).

Squaring Eqs. (3-14b) and (3-15b), we get

$$\left\{\sigma_{nn} - \frac{\sigma_{xx} + \sigma_{yy}}{2}\right\}^2 = \left(\frac{\sigma_{xx} - \sigma_{yy}}{2}\right)^2 \cos^2 2\theta + (\sigma_{xy})^2 \sin^2 2\theta + 2\sigma_{xy}\left(\frac{\sigma_{xx} - \sigma_{yy}}{2}\right)\cos 2\theta \sin 2\theta,$$

$$(\sigma_{nt})^2 = \left(\frac{\sigma_{xx} - \sigma_{yy}}{2}\right)^2 \sin^2 2\theta + (\sigma_{xy})^2 \cos^2 2\theta - 2\sigma_{xy}\left(\frac{\sigma_{xx} - \sigma_{yy}}{2}\right)\cos 2\theta \sin 2\theta.$$

The addition of the above two equations gives

$$\left\{\sigma_{nn} - \frac{\sigma_{xx} + \sigma_{yy}}{2}\right\}^2 + (\sigma_{nt})^2 = \left(\frac{\sigma_{xx} - \sigma_{yy}}{2}\right)^2 + (\sigma_{xy})^2.$$

For convenience, let $\sigma_{nn} = \sigma$ and $\sigma_{nt} = \tau$. Then the above equation can be rewritten as

$$\left\{\sigma - \frac{\sigma_{xx} + \sigma_{yy}}{2}\right\}^2 + \tau^2 = \left(\frac{\sigma_{xx} - \sigma_{yy}}{2}\right)^2 + (\sigma_{xy})^2. \tag{A-1}$$

Comparing Eq. (A-1) with the equation of the circle in the xy-coordinate plane shown in **Fig.A.1**, *i.e.*,

$$(x - C)^2 + y^2 = R^2,$$

which has a radius R with its centre at x = C, we find that Eq. (A-1) represents a circle in the $\sigma\tau$-coordinate plane with its radius

$$R = \sqrt{\left(\frac{\sigma_{xx} - \sigma_{yy}}{2}\right)^2 + \left(\sigma_{xy}\right)^2} \qquad (A-2)$$

and its centre at

$$C = \frac{\sigma_{xx} + \sigma_{yy}}{2}, \qquad (A-3)$$

as shown in **Fig.A.2**. This is called the *Mohr's circle* for stress representation.

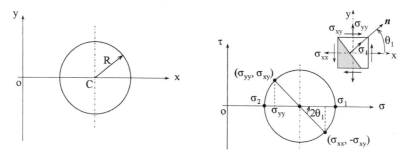

Figure A.1 A circle in the xy-coordinate plane.   **Figure A.2** Mohr's circle in the stress coordinate plane.

Notice that there are two points on the circle where $\tau = 0$, *i.e.*, shear stress $\sigma_{nt} = 0$. Thus we know that the coordinates of these points correspond to the two principal stresses of a two-dimensional stress state, *i.e.*, σ_1 and σ_2. Since we know the radius, Eq. (A-2), and the coordinate of the circle centre, Eq. (A-3), we can calculate graphically the values of σ_1 and σ_2, *i.e.*,

$$\left.\begin{array}{c}\sigma_1 \\ \sigma_2\end{array}\right\} = \frac{\sigma_{xx} + \sigma_{yy}}{2} \pm \sqrt{\left(\frac{\sigma_{xx} - \sigma_{yy}}{2}\right)^2 + \left(\sigma_{xy}\right)^2}. \qquad (A-4)$$

From **Fig.A.2**, it is clear that the directions of these principal stresses are at the angles θ and $90° + \theta$ to the original stress system in xy-plane (σ_{xx}, σ_{yy} and σ_{xy}), where θ is determined by

$$\tan 2\theta = \sigma_{xy} \bigg/ \left(\frac{\sigma_{xx} - \sigma_{yy}}{2}\right) = \frac{2\sigma_{xy}}{\sigma_{xx} - \sigma_{yy}}. \qquad (A-5)$$

Obviously, Eqs. (A-4) and (A-5) are reproductions of Eqs. (3-21) and (3-20), respectively. The Mohr's circle also shows graphically that the direction of the maximum (or minimum) shear stress bisects those of the two principal stresses, σ_1 and σ_2. The magnitude of the maximum (or minimum) shear stress in the xy-plane is the radius of the circle, *i.e.*,

$$\left.\begin{array}{c}\tau_{max}\\ \tau_{min}\end{array}\right\} = \pm\sqrt{\left(\frac{\sigma_{xx}-\sigma_{yy}}{2}\right)^2+(\sigma_{xy})^2},$$

where $\tau = \sigma_{nt}$ as we introduced above.

Example A.1 Use the Mohr's circle method to find the principal stresses and their directions in the following plane-stress state in the xy-plane.

$$\sigma_{xx} = 80 \text{ MPa}, \sigma_{yy} = -40 \text{ MPa}, \sigma_{xy} = 60 \text{ MPa}.$$

Solution: The Mohr's circle of the stress state can be drawn easily as shown in **Fig. EA.1a**. The two in-plane principal stresses can be measured approximately, or calculated by Eq. (A-4), as

$$\sigma_1 = 105 \text{ MPa}, \sigma_3 = -65 \text{ MPa}.$$

The tangent of the principal direction is defined by Eq. (A-5), that is,

$$\tan 2\theta = \frac{2\times 60}{80+40} = 1.$$

Thus $2\theta = 45°$ or $225°$. From the Mohr's circle, however, we see that the counterclockwise angle from x-axis to that of σ_1 is $45°$. Thus $2\theta_1 = 45°$ and $\theta_1 = 22.5°$ as shown in **Fig. EA.1b**. The direction of σ_2 is defined by $2\theta_2 = 225°$, ie, $\theta_2 = 112.5°$.

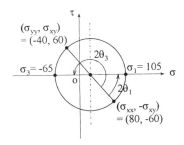

(a) Mohr's circle of the stress state

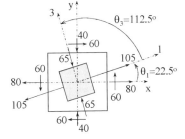

(b) principal directions in the xy-plane

Figure EA.1

References

Chakrabarty, J (1987), *Theory of Plasticity*, McGraw-Hill, Singapore, pp.30-32.
Sokolnikoff, IS (1956), *Mathematical Theory of Elasticity*, McGraw-Hill, New York, pp.52.
Westergaard, HM (1952), *Theory of Elasticity and Plasticity*, Harvard University Press, Cambridge, Mass., pp.61-64.

Appendix B

FORMULAE FOR SPECIAL CONTACT PROBLEMS

The following formulae are obtained by superposing the solution obtained in section **7.4.1**, *i.e.*, the solution of a half-space under a normal concentrated load (Boussinesq's solution). When friction on the contact interface becomes important, the solution of a half-space under a tangential concentrated force (Cerruti's solution, see section **7.4.2**) should also be used to accommodate the effect of friction that is tangential to the contact interface. In addition, it is easy to understand that the contact area in the following cases must be much smaller than the radius of curvature of the bodies in contact because Boussinesq's solution is based on a half-space with a flat surface.

In the formulae, δ is used to denote the relative displacement of the centres approach to each other under the external load applied and E* is called the *effective elastic modulus* of a corresponding contact system defined by

$$E^* = \left(\frac{1-v_1^2}{E_1} + \frac{1-v_2^2}{E_2} \right)^{-1}, \tag{B-1}$$

where E_i and v_i (i = 1, 2) represent Young's modulus and Poisson's ratio of elastic body i in contact.

B.1 Two Balls in Contact (Fig. B.1)

When two spheres are in contact, both under a load P applied axisymmetrically, the contact area is a circle. The contact compressive stress q (normal to the contact surface because friction is ignored), radius of contact area a and displacement δ can be obtained as

$$a^3 = \frac{3}{4} \frac{R_1 R_2}{(R_1 + R_2)} \frac{P}{E^*},$$

$$\delta^3 = \frac{9}{16} \frac{(R_1 + R_2)}{R_1 R_2} \left(\frac{P}{E^*} \right)^2, \tag{B-2}$$

$$q = q_0 \left(1 - \frac{r^2}{a^2} \right)^{1/2},$$

where

$$q_0^3 = \frac{6}{\pi^3}\left(\frac{R_1+R_2}{R_1 R_2}\right)^2 (E^*)^2 P. \tag{B-3}$$

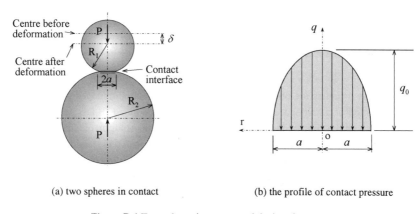

(a) two spheres in contact (b) the profile of contact pressure

Figure B.1 Two spheres in contact and the interface stress.

B.2 A Sphere in Contact with a Flat Half-Space (Fig. B.2)

In this case, the contact area is still a circle. If we let R_2 in formulae (B-2) and (B-3) be infinite, we obtain the corresponding formulae for the present case, that is,

$$a^3 = \frac{3}{4} R_1 \frac{P}{E^*},$$

$$\delta^3 = \frac{9}{16} \frac{1}{R_1}\left(\frac{P}{E^*}\right)^2, \tag{B-4}$$

$$q = q_0 \left(1 - \frac{r^2}{a^2}\right)^{1/2},$$

where

$$q_0^3 = \frac{6}{\pi^3}\left(\frac{1}{R_1}\right)^2 (E^*)^2 P. \tag{B-5}$$

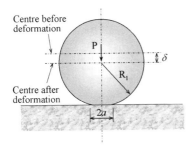
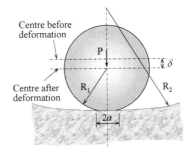

Figure B.2 A sphere in contact with a large flat surface.

Figure B.3 A sphere in contact with a large concave surface.

B.3 A Sphere in Contact with a Large Concave Spherical Surface (Fig. B.3)

Again, the contact area is a circle. Letting R_2 in formulae (B-2) and (B-3) be $-R_2$, the following formulae can be derived directly for the present contact system:

$$a^3 = \frac{3}{4} \frac{R_1 R_2}{(R_2 - R_1)} \frac{P}{E^*},$$

$$\delta^3 = \frac{9}{16} \frac{(R_2 - R_1)}{R_1 R_2} \left(\frac{P}{E^*} \right)^2, \quad \text{(B-6)}$$

$$q = q_0 \left(1 - \frac{r^2}{a^2} \right)^{1/2},$$

where

$$q_0^3 = \frac{6}{\pi^3} \left(\frac{R_2 - R_1}{R_1 R_2} \right)^2 \left(E^* \right)^2 P. \quad \text{(B-7)}$$

B.4 The Contact of Two Cylinders with Parallel Axes (Fig. B.4)

It is easy to imagine that the contact area is a rectangle of $2b \times L$, where L is the length of the cylinder and $2b$ is the contact length in the cross-sectional plane of the cylinder, as shown in **Fig. B.4**. Under a pair of uniformly distributed line loads of intensity p (unit: force per unit length, *e.g.*, N/m), which is loaded symmetrically as shown in the figure, the contact stress and contact area are determined by

$$b^2 = \frac{4}{\pi} \frac{R_1 R_2}{(R_1 + R_2)} \frac{p}{E^*},$$

$$\delta = \frac{2p}{\pi} \left\{ \frac{1-v_1^2}{E_1} \left(\ln \frac{2R_1}{b} + 0.407 \right) + \frac{1-v_2^2}{E_2} \left(\ln \frac{2R_2}{b} + 0.407 \right) \right\}, \quad \text{(B-8)}$$

$$q = q_0 \left(1 - \frac{x^2}{b^2} \right)^{1/2},$$

where

$$q_0^2 = \frac{1}{\pi} \left(\frac{R_1 + R_2}{R_1 R_2} \right) pE^*. \quad \text{(B-9)}$$

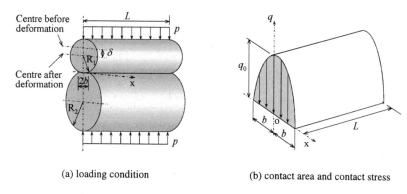

(a) loading condition   (b) contact area and contact stress

Figure B.4 Contact of two cylinders with parallel axes.

B.5   A Cylinder in Contact with a Flat Half-Space

Let R_2 in Eqs. (B-8) and (B-9) be infinite, we get

$$b^2 = \frac{4}{\pi} R_1 \frac{p}{E^*},$$

$$q = q_0 \left(1 - \frac{x^2}{b^2} \right)^{1/2}, \quad \text{(B-10)}$$

where

$$q_0^2 = \frac{pE^*}{\pi R_1}. \quad \text{(B-11)}$$

The general explicit expression of δ is not available, but we can have approximate formulae in special cases. For example, when $E_1 = E_2 = E$, $v_1 = v_2 = 0.3$ (the Poisson's values of many types of steel are around 0.3), we have:

$$b = 1.522\sqrt{\frac{pR_1}{E}},$$
$$\delta = 0.579\frac{p}{E}\left(\ln\frac{2R_1}{b} + 0.407\right), \quad \text{(B-12)}$$
$$q_0 = 0.418\sqrt{\frac{pE}{R_1}}.$$

B.6 The Contact of a Cylinder with a Large Concave Cylindrical Surface

Similar to the case of section **B.3**, assume that the axes of the cylinder and the concave cylindrical surface are parallel to each other and the uniform line load p is applied along the symmetrical plane of the system. Then the formulae can be obtained directly by replacing R_2 in Eqs. (B-8) and (B-9) with $-R_2$. This gives rise to

$$b^2 = \frac{4}{\pi}\frac{R_1 R_2}{(R_2 - R_1)}\frac{p}{E^*},$$
$$q = q_0\left(1 - \frac{x^2}{b^2}\right)^{1/2}, \quad \text{(B-13)}$$

where

$$q_0^2 = \frac{1}{\pi}\left(\frac{R_2 - R_1}{R_1 R_2}\right)pE^*. \quad \text{(B-14)}$$

B.7 The Contact of Two Cylinders with Perpendicular Axes (Fig. B.5)

Under this special contact condition, the contact area is an ellipse with major axis length $2a$ and minor axis length $2b$, as shown in **Fig. B.5**. The corresponding formulae are

$$a = \alpha\sqrt[3]{P\frac{2R_1 R_2}{R_1 + R_2}\left(E^*\right)^{-1}},$$
$$b = \beta a, \quad \text{(B-15)}$$
$$q_0 = \frac{3P}{2\pi ab},$$

$$\delta = \lambda \sqrt[3]{\frac{R_1+R_2}{2R_1R_2}P^2\left(E^*\right)^{-2}}.\qquad (B\text{-}16)$$

where the coefficients α, β and λ can be obtained by using **Fig. B.6** when R_1 and R_2 are known.

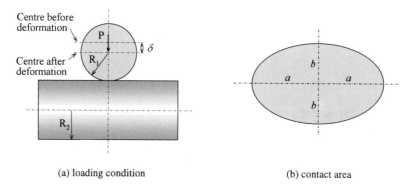

(a) loading condition (b) contact area

Figure B.5 The contact of two cylinders with perpendicular axes.

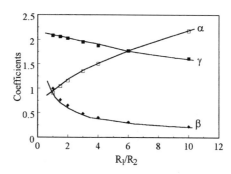

Figure B.6 Variation of coefficients in Eqs. (B-15) and (B-16) with the ratio of cylinder radii.

Appendix C

THE THEOREM OF MINIMUM POTENTIAL ENERGY

Our aim here is to use a very simple example to explain why Eq. (10-6) in the formulation of the finite element method is equivalent to the corresponding equilibrium equations. For this purpose, we are not going to review the general theory and applications of the theorem of minimum potential energy. The reader can find relevant details in the references recommended in section **8.8**.

The theorem of minimum potential energy can be simply stated as: 'Of all displacements satisfying the given boundary conditions of an elastic body, the true displacements make the potential energy of the body an absolute minimum.'

Consider the equilibrium of the spring-mass system shown in **Fig. C.1** subjected to an external force P. The displacement of the mass is δ. Thus when the spring constant is k, the equilibrium equation in x-direction of the system is

$$P - mg - k\delta = 0. \tag{C-1}$$

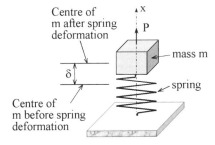

Figure C.1 The equilibrium of a mass-spring system.

Now let us apply the theorem of minimum potential energy to see if we can get Eq. (C-1). The strain energy in the spring is $(k\delta^2/2)$, the work done by gravity is $(-mg\delta)$ and the work done by the external force P is $(P\delta)$. Thus the total potential energy of the system, V, is

$$V = \frac{1}{2}k\delta^2 + mg\delta - P\delta. \qquad (C\text{-}2)$$

The theorem of minimum potential energy indicates that the true displacement δ makes the potential energy the minimum, i.e.,

$$\frac{dV}{d\delta} = 0.$$

Hence, the substitution of Eq. (C-2) into the above equation leads to

$$P - mg - k\delta = 0$$

that is exactly the equilibrium equation of the system, Eq. (C-1).

Variational theory has proved that the above conclusion is valid for an elastic solid under general deformation. However, it is understandable that in a general case the calculation of potential energy will become lengthy. For instance, if we consider an elastic-isotropic solid of volume Ω subjected to external surface stresses $\tau = (\tau_x, \tau_y, \tau_z)$ on a surface area S_σ (stress boundary conditions), concentrated forces $\mathbf{P} = (P_1, P_2, \ldots, P_n)$ acting on points 1, 2, … n and body forces $f = (f_x, f_y, f_z)$, then the total potential energy of the solid can be calculated by

$$V = \iiint_\Omega W d\Omega - \iint_{S_\sigma} \tau \bullet \mathbf{u} ds - \iiint_\Omega \rho f \bullet \mathbf{u} d\Omega - \mathbf{P} \bullet \overline{\mathbf{u}}, \qquad (C\text{-}3)$$

where W is the strain energy density of the solid, as discussed in section **9.3**, ρ is the density of the solid material and $\overline{\mathbf{u}} = (u_1, u_2, \cdots, u_n)^T$ are the displacements at the application points of P_i (i = 1, 2, …, n) in their loading directions. The physical meanings of individual terms in Eq. (C-3) are obvious. The first is the total strain energy in the solid due to deformation, the second is the work done by external surface stresses during deformation, the third is the work done by body forces and the last term is the work done by the concentrated forces.

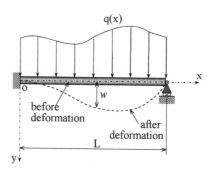

Figure C.2 A beam under bending and tension.

For simplicity, let us take the deformation of a beam (**Fig. C.2**) as an example to show the calculation of potential energy in a continuous body. The beam is clamped at x = 0, simply supported at x = L and is subjected to a transverse force q(x). According to Eq. (C-3), we have

(i) Strain energy:

$$\iiint_\Omega U d\Omega = \iiint_\Omega \frac{1}{2E}(\sigma_{xx})^2 dxdydz = \frac{1}{2}\int_0^L EI\left(\frac{d^2w}{dx^2}\right)^2 dx.$$

In the above derivation, the relationship of the normal stress σ_{xx} with the bending moment M on a beam cross-section

$$\sigma_{xx} = \frac{My}{I}$$

has been used, where $I = \iint y^2 dydz$.

(ii) Work done by external load q(x):

$$\iint_{S_\sigma} \tau \cdot u ds = \int_0^L qw dx.$$

The third and fourth terms in Eq. (C-3) vanish because there is no body force or concentrated force applied. Hence, the total potential energy of the beam is determined by

$$V = \frac{1}{2}\int_0^L EI\left(\frac{d^2w}{dx^2}\right)^2 dx - \int_0^L qw dx.$$

Appendix D

UNITS AND CONVERSION FACTORS

Length

$1\,\text{m} = 10^{10}\,\text{Å}$ $1\,\text{Å} = 10^{-10}\,\text{m}$

$1\,\text{m} = 10^{9}\,\text{nm}$ $1\,\text{nm} = 10^{-9}\,\text{m}$

$1\,\text{m} = 10^{6}\,\mu\text{m}$ $1\,\mu\text{m} = 10^{-6}\,\text{m}$

$1\,\text{m} = 10^{3}\,\text{mm}$ $1\,\text{mm} = 10^{-3}\,\text{m}$

$1\,\text{m} = 10^{2}\,\text{cm}$ $1\,\text{cm} = 10^{-2}\,\text{m}$

$1\,\text{mm} = 0.0394\,\text{in.}$ $1\,\text{in.} = 25.4\,\text{mm}$

$1\,\text{cm} = 0.394\,\text{in.}$ $1\,\text{in.} = 2.54\,\text{cm}$

$1\,\text{m} = 3.28\,\text{ft}$ $1\,\text{ft} = 0.3048\,\text{m}$

Angle (rad)

$1\,\text{rad} = \frac{180°}{\pi}$ $1° = \frac{\pi}{180}\,\text{rad}$

Mass

$1\,\text{Mg} = 10^{3}\,\text{kg}$ $1\,\text{kg} = 10^{-3}\,\text{Mg}$

$1\,\text{kg} = 10^{3}\,\text{g}$ $1\,\text{g} = 10^{-3}\,\text{kg}$

$1\,\text{kg} = 2.205\,\text{lb}_m$ $1\,\text{lb}_m = 0.4536\,\text{kg}$

$1\,\text{g} = 2.205 \times 10^{-3}\,\text{lb}_m$ $1\,\text{lb}_m = 453.6\,\text{g}$

Density

$1\,\text{kg/m}^{3} = 10^{-3}\,\text{g/cm}^{3}$ $1\,\text{g/cm}^{3} = 10^{3}\,\text{kg/m}^{3}$

$1\,\text{Mg/m}^{3} = 1\,\text{g/cm}^{3}$ $1\,\text{g/cm}^{3} = 1\,\text{Mg/m}^{3}$

$1\,\text{kg/m}^{3} = 0.0624\,\text{lb}_m/\text{ft}^{3}$ $1\,\text{lb}_m/\text{ft}^{3} = 16.02\,\text{kg/m}^{3}$

$1\,\text{g/cm}^{3} = 62.4\,\text{lb}_m/\text{ft}^{3}$ $1\,\text{lb}_m/\text{ft}^{3} = 0.01602\,\text{g/cm}^{3}$

$1\,\text{g/cm}^{3} = 0.0361\,\text{lb}_m/\text{in.}^{3}$ $1\,\text{lb}_m/\text{in.}^{3} = 27.7\,\text{g/cm}^{3}$

Force

$1\,N = 10^5\,\text{dynes}$ $1\,\text{dyne} = 10^{-5}\,N$
$1\,N = 1/9.80665\,\text{kgf}$ $1\,\text{kgf} = 9.80665\,N$
$1\,N = 0.2248\,\text{lb}_f$ $1\,\text{lb}_f = 4.448\,N$

Stress (Pa, defined as N/m^2)

$1\,\text{MPa} = 145\,\text{psi}$ $1\,\text{psi} = 6.90 \times 10^3\,\text{MPa}$
$1\,\text{MPa} = 0.102\,\text{kg/mm}^2$ $1\,\text{kg/mm}^2 = 9.806\,\text{MPa}$
$1\,\text{Pa} = 10\,\text{dynes/cm}^2$ $1\,\text{dyne/cm}^2 = 0.10\,\text{Pa}$
$1\,\text{kg/mm}^2 = 1422\,\text{psi}$ $1\,\text{psi} = 7.03 \times 10^{-4}\,\text{kg/mm}^2$

Energy (J, defined as N·m)

$1\,J = 10^7\,\text{ergs}$ $1\,\text{erg} = 10^{-7}\,J$
$1\,J = 6.24 \times 10^{18}\,\text{eV}$ $1\,\text{eV} = 1.602 \times 10^{-19}\,J$
$1\,J = 0.239\,\text{cal}$ $1\,\text{cal} = 4.184\,J$
$1\,J = 9.48 \times 10^{-4}\,\text{Btu}$ $1\,\text{Btu} = 1054\,J$
$1\,J = 0.738\,\text{ft} \cdot \text{lb}_f$ $1\,\text{ft} \cdot \text{lb}_f = 1.356\,J$
$1\,\text{eV} = 3.83 \times 10^{-20}\,\text{cal}$ $1\,\text{cal} = 2.61 \times 10^{19}\,\text{eV}$
$1\,\text{cal} = 3.97 \times 10^{-3}\,\text{Btu}$ $1\,\text{Btu} = 252.0\,\text{cal}$

Power (W, defined as J/s)

$1\,W = 0.239\,\text{cal/s}$ $1\,\text{cal/s} = 4.184\,W$
$1\,W = 3.414\,\text{Btu/h}$ $1\,\text{Btu/h} = 0.293\,W$
$1\,\text{cal/s} = 14.29\,\text{Btu/h}$ $1\,\text{Btu/h} = 0.070\,\text{cal/s}$

Temperature

$T(K) = 273 + T(^\circ C)$ $T(^\circ C) = T(K) - 273$
$T(K) = \frac{5}{9}\left[T(^\circ F) - 32\right] + 273$ $T(^\circ F) = \frac{9}{5}\left[T(K) - 273\right] + 32$
$T(^\circ C) = \frac{5}{9}\left[T(^\circ F) - 32\right]$ $T(^\circ F) = \frac{9}{5}T(^\circ C) + 32$

Moment or Torque (N·m)

Velocity (m/s)

Acceleration (m/s^2)

Angular velocity (rad/s)

Angular acceleration (rad/s^2)

Prefixes

Prefix	Symbol	Factor
yotta	Y	10^{24}
zetta	Z	10^{21}
exa	E	10^{18}
peta	P	10^{15}
tera	T	10^{12}
giga	G	10^{9}
mega	M	10^{6}
kilo	k	10^{3}
hecto	h	10^{2}
deka	da	10^{1}
deci	d	10^{-1}
centi	c	10^{-2}
milli	m	10^{-3}
micro	μ	10^{-6}
nano	n	10^{-9}
pico	p	10^{-12}
femto	f	10^{-15}
atto	a	10^{-18}
zepto	z	10^{-21}
yocto	y	10^{-24}

Index

A
Airy stress function, 211
anisotropy ratio, 103
approximation error, 282
axisymmetric deformation, 123

B
basic assumptions, 16
    absence of initial stresses, 18
    continuity, 16
    homogeneity, 17
    isotropy, 17
    no-initial stresses, 17
    small deformation, 18
basic equations, 115
biharmonic function, 211, 213
body force, 25, 54, 232
boundary conditions, 115
    displacement boundary conditions, 115
    mixed boundary conditions, 120
    stress boundary conditions, 115
brittle material, 268, 270
bulk modulus, 98

C
coefficient of thermal expansion, 104, 106
compatibility
    conditions, 76
    equations, 77
    of strains, 75
concentrated force, 24
constitutive equations, 98
contact, 188, 200, 319
    mechanics models, 201
continuity, 16
control volume, 285
curved beam, 230

D
deflection, 25, 180
deformation
    axisymmetric, 123
    elastic, 87
    plane-strain, 130
    plane-stress, 128
    plastic, 88, 238
    symmetry, 122
degree of freedom, 288
difference in plane-deformation states, 131
dilatation, 73
direct stress, 27
discretization error, 282
displacement, 66
displacement approach, 138
displacement boundary conditions, 115
displacement-strain relationship, 68
distortion energy, 240
ductile material, 268, 269

E
effective elastic modulus, 319
eigenvalue problem, 44
eigenvector, 45
elastic constants, 98
    relationships, 103
    physical indications, 98
elastic deformation, 87
elastic limit load, 246
elastic modulus, 89
elastic strain, 242
element connectivity, 289
energy
    strain energy, 240
    method, 232
    density, 240
equations of equilibrium, 56

equations of motion, 56
external force, 24

F
failure, 268
    curve, 272, 273
    mode, 268
    theories, 268
finite element method, 280
    approximation error, 282
    compatibility, 291
    completeness, 291
    considerations, 298
    control volume, 285
    degree of freedom, 288
    discretization error, 282
    element connectivity, 289
    global displacement, 289
    global load, 289
    isoparametric formulation, 291
    nodal displacement, 289
    nodal load, 289
    node, 282
    procedures, 296
    shape function, 290
    stiffness matrix, 292, 294
force, 23
    body force, 25, 54, 232
    concentrated force, 24
    external force, 24
    internal force, 23
    surface force, 25
    type of forces, 23
fracture, 268
fracture strain, 268

G
generalised Hooke's law, 89
geometric equations, 72
global displacement, 289
global load, 289

H
half space, 192, 1945, 197
hardening, 242
harmonic function, 190, 191
homogeneity, 17
Hooke's law, 89

I
initial stress, 18
initial yield, 241
internal force, 23
invariant of
    strain, 73
    stress, 50
isoparametric formulation, 291
isotropy, 17

L
Lamé's constants, 92
Laplace equation, 190
loading, 243

M
mixed approach, 145
mixed boundary conditions, 120
model(ling), 9, 114, 298
Mohr theory, 271
Mohr's circle, 42, 315
modulus of elasticity, 89

N
nodal displacement, 289
nodal load, 289
node, 282
nominal stress, 225
normal strain, 67
normal stress, 27

O
octahedral
    plane, 239
    shear stress, 239
onset of plasticity, 241

P
Papkovich-Neuber solution, 190
plane-strain deformation, 130
physical indications of elastic constants, 98
plane-stress deformation, 128
plastic deformation, 88, 238
plastic limit load, 246
plastic strain, 242
Poisson's ratio, 95, 100
potential energy, 287, 325
principal strain, 73
principal stress, 39, 43, 48

R
Ramberg-Osgood equation, 244
relationships among elastic constants, 103
residual stress, 247, 261, 263
rigid body displacement, 66

S
Saint-Venant's principle, 158
scalar method, 232
scale of analysis, 14
shape function, 290
shear modulus, 101
shear stress, 27
similarity in plane-deformation states, 131
small deformation, 18
solution, 114
    displacement approach, 138
    mixed approach, 145
    stress approach, 153
stiffness matrix, 292, 294
strain, 67
    compatibility of strains, 75
    direct strain, 67
    hardening, 244
    invariants, 73
    normal strain, 67
    principal strain, 73
    shear strain, 67
    strain-displacement relations, 68
    thermal strain, 106
    volume strain, 73
stress, 26
    invariants, 50
    mathematical definition, 26
    maximum shear stress, 39, 51, 269
    physical definition, 27
    sign of stress, 31
    transformation, 35
    vector, 27

stress approach, 153
stress boundary conditions, 115
stress concentration, 223
stress function, 209
stress-strain relationship, 86
superposition, 133

T
tensor
    strain, 71
    stress, 30
theorem of minimum potential
    energy, 287, 325
thermal deformation, 106, 181
thermal strain, 106
thermal stress, 107, 181
transformation matrix, 48
Tresca criterion, 249
type of forces, 23

U
unloading, 242

V
volume strain, 73
von Mises criterion, 250

W
work hardening, 242

Y
yield(ing), 88, 248
    condition, 248
    criterion, 248
    surface, 249
Young's modulus, 89, 99